NEURONAL
MECHANISMS
of HEARING

NEURONAL MECHANISMS
of HEARING

Edited by
Josef Syka

Czechoslovak Academy of Sciences
Prague, Czechoslovakia

and
Lindsay Aitkin

Monash University
Clayton, Victoria, Australia

PLENUM PRESS • NEW YORK AND LONDON

Library of Congress Cataloging in Publication Data

Main entry under title:

Neuronal mechanisms of hearing.

"A symposium on neuronal mechanisms of hearing, organized by the Czechoslovak Academy of Science as a satellite symposium to the 28th International Congress of Physiological Sciences."
Bibliography: p.
Includes index.
1. Hearing—Physiological aspects—Congresses. 2. Auditory pathways—Congresses. I. Syka, Josef. II. Aitkin, Lindsay. III. International Congress of Physiological Sciences (28th: 1980: Prague, Czechoslovakia) [DNLM: 1. Hearing—Physiology—Congresses. 2. Acoustic nerve—Physiology—Congresses.
WV 272 N494 1980]
QP461.N44   599.01'825   81-1626
                                                                        AACR2

ISBN-13: 978-1-4684-3910-6   e-ISBN-13: 978-1-4684-3908-3
DOI: 10.1007/978-1-4684-3908-3

Proceedings of a symposium on "Neuronal Mechanisms of Hearing,"
organized by the Czechoslovak Academy of Sciences as a
satellite symposium to the 28th International Congress
of Physiological Sciences, held July 20—23, 1980, in Prague,
Czechoslovakia

© 1981 Plenum Press, New York
Softcover reprint of the hardcover 1st edition 1981

A Division of Plenum Publishing Corporation
233 Spring Street, New York, N.Y. 10013

PREFACE

In contrast to the level of interest which is paid to the organization of meetings about the structure and function of the auditory periphery, the central auditory system has received little attention in the last several years. However, much recent data accumulated during this period has provided auditory physiologists with new ideas about the function of the central auditory system. The successful exploration of new anatomical tracing techniques (tritiated aminoacids, horseradish peroxidase, 2-deoxyglucose) together with the collection of electrophysiological data obtained with intracellular and extracellular recordings from the receptors and neurones in the auditory pathway have considerably deepened our understanding of central auditory function.

Particular interest was concentrated upon the development of the auditory system under normal conditions and in conditions of auditory deprivation. Although, from the methodological point of view, the conditions of reversible auditory deprivation are complicated, promising new data appeared in this field. Similarly the specific ability of the auditory system to encode communication signals and speech sounds has been examined in many laboratories all over the world.

A very fruitful method, based upon the results of electrical stimulation of cochlear nerve fibres in experimental animals, is the application of neuroprostheses in deaf patients. At the present time, the method still does not meet all requirements and many improvements will be necessary. Undoubtedly the exploration of the results of recent physiological experiments may help in the further improvement of neuroprostheses.

Therefore, besides discussion about the structure and function of the auditory system, the emphasis at the symposium on Neuronal Mechanisms of Hearing was placed on the development of the auditory system under normal conditions and conditions of auditory deprivation, the coding of communication and speech signals and the physiological background of neuroprostheses.

The lack of specialized meetings devoted to this topic in the past and the fact that the symposium was organized as a satellite symposium to the 28th International Physiological Congress in Budapest attracted to Prague a large audience of outstanding auditory physiologists as well as many young physiologists who represent the future of this scientific field. This book contains most of the papers presented at the symposium.

We wish to express our gratitude to some participants of the symposium, including Drs. Buchwald, Dallos, Hind, Merzenich and Woolsey, who did not contribute to the volume, but whose active participation on the meeting helped significantly to make the symposium a success. The preparation of this book was carried out by both authors, with the detailed editing and typing done in Prague.

                                    Josef Syka
                                    Lindsay Aitkin

                                    October 1980

CONTENTS

SESSION I. COCHLEAR MECHANISMS A
            Chairmen: Y. KATSUKI and P. DALLOS

The Responses of Hair Cells to Low Frequency
Tones and Their Relationship to the Extracel-
lular Receptor Potentials and Sound Pressure
Level in the Guinea Pig Cochlea   3
    I.J. RUSSELL and P.M. SELLICK

Some Comparative Aspects of a Cochlear Mechanism.   17
    Y. KATSUKI

The Parasympathetic Innervation of the Inner
Ear and the Problem of Cochlear Efferents:
Enzyme and Autoradiographic Studies   31
    M.D. ROSS and H. ROGER JONES

Adaptation and Dynamic Response Occurring at
Hair Cell — Afferent Fiber Synapse.   37
    T. FURUKAWA and S. MATSUURA

Phase Versus Frequency Plots from Caiman Primary Auditory Fibres: Is There a Travelling
Wave?   43
    J. SMOLDERS and R. KLINKE

SESSION II. COCHLEAR MECHANISMS B.
             Chairman: R. KLINKE

The Dynamics of pO_2-Changes in the Perilymphatic Perfusate of the Guinea Pig Cochlea
Depending on the Metabolism   51
    E.J. HABERLAND, K.O. KUHL and P. LOTZ

Analytical Studies on Biochemistry and Physiology of Perilymph (Guinea Pig)   57
    F. SCHEIBE

Distribution of Microphonic Potentials in the
Four Turns of the Guinea Pig Cochlea.   63
    J. SYKA, I. MELICHAR and L. ULEHLOVÁ

SESSION III.   CODING IN THE AUDITORY NERVE AND
             COCHLEAR NUCLEUS
               Chairmen: A.R. MØLLER and E.F.
                    EVANS

The Dynamic Range Problem: Place and Time
Coding at the Level of Cochlear Nerve and
Nucleus   69
    E.F. EVANS

Coding of Complex Sounds in the Auditory Nervous System   87
    A.R. MØLLER

On Predicting the Response of Auditory Nerve
Fibers to Complex Tones   105
    E. JAVEL

Effects of Masking Noise on the Representa-
tion of Vowel Spectra in the Auditory Nerve .   113
    H.F. VOIGT, M.B. SACHS and E.D. YOUNG

Neuronal Circuits in the Dorsal Cochlear Nucleus   119
    K.K. OSEN and E. MUGNAINI

The Internal Organization of the Dorsal Cochlear Nucleus.   127
    E.D. YOUNG and H.F. VOIGT

SESSION IV.   CENTRAL AUDITORY MECHANISMS A
              Chairmen: J.E. HIND and M. MOLNÁR

Functional Organization of the Inferior Colliculus   137
    J. SYKA, R. DRUGA, J. POPELÁŘ and
    B. KALINOVÁ

Integration and Segregation of Input to the   155
Cat Inferior Colliculus . . . .
    M.N. SEMPLE and L.M. AITKIN

Some Facets of the Organization of the Principal Division of the Cat Medial Geniculate
Body.   163
    L.M. AITKIN, M.B. CALFORD, C.E. KENYON
    and W.R. WEBSTER

Functional Organization of the Medial Geniculate Body Studied by Simultaneous Recordings
of Single Unit Pairs.   183
    P. HEIERLI, F. de RIBAUPIERRE, A. TOROS
    and Y. de RIBAUPIERRE

Possibilities of Recording Multiunit Activity
in the Auditory Pathway   187
    E. DAVID

Poststimulatory Effects in the Medial Genicu-
late Body of Guinea Pigs.   191
    CH. SCHREINER

How Biosonar Information Is Represented in
the Bat Cerebral Cortex   197
    N. SUGA, K. KUZIRAI and W.E. NEILL

Three-Dimensional Study of Evoked Field Po-
tentials in the Auditory Cortex of the Cat. .   221
    M. MOLNÁR, G. KARMOS and V. CSÉPE

Differential Diagnosis of Hearing Disorders-
Clinical Findings Contributing to Information
Processing in the Auditory Pathway.   225
    H. von SPECHT

SESSION V.   CENTRAL AUDITORY MECHANISMS B
           Chairman: F. de RIBAUPIERRE

Binaural Interaction in the Cat Inferior Colliculus: Comparison of the Physiological Data
with a Computer Simulated Model   233
    Y. SUJAKU, S. KUWADA and T.C.T. YIN

Coding Properties of the Different Nuclei of
the Cat's Medial Geniculate Body.   239
    A. TOROS-MOREL, F. de RIBAUPIERRE and
    E. ROUILLER

Interaural Delay Sensitive Units in the MGB
of the Cat.   245
    C. IVARSSON, Y. de RIBAUPIERRE,
    A. BAROFFIO and F. de RIBAUPIERRE

Temporal Information in the Medial Geniculate
Body.   251
    E. ROUILLER, Y. de RIBAUPIERRE, A. TOROS
    and F. de RIBAUPIERRE

Some Investigation of Acoustical Evoked Potentials from Peripheral and Central Struc-
tures of the Auditory Pathway in Rabbits. . .   257
    M. BIEDERMANN, E. EMMERICH and
    H. KASCHOWITZ

SESSION VI.   AUDITORY LOCALIZATION
           Chairmen: L. AITKIN and J. SYKA

Anatomical-Behavioral Analyses of Hindbrain
Sound Localization Mechanisms   263
    R.B. MASTERTON, K.K. GLENDENNING and
    R.J. NUDO

Effects of Unilateral Ablation of Anteroventral Cochlear Nucleus on Localization of
Sound in Space.   277
    J.H. CASSEDAY and H.A. SMOAK

Binaural Interaction Models and Mechanisms. .   283
    H.S. COLBURN and P.J. MOSS

Psychophysical and Neurophysiological Data on
the Sound Source Perception   289
    J.A. ALTMAN

SESSION VII.   NEURAL CODING OF SPEECH AND COMPLEX
           STIMULI
           Chairmen: J.S. BUCHWALD and K.
               SEDLÁČEK

Information Processing in Neuronal Populations of the Human Brain during Learning of
Verbal Signals.   303
    N.P. BECHTEREVA and Y.D. KROPOTOV

A Comparison of the Responses Evoked by Artificial Stimuli and Vocalizations in the Infe-
rior Colliculus of Squirrel Monkeys   307
    J.A. MANLEY and P. MÜLLER-PREUSS

Acoustic Properties of Central Auditory Pathway Neurons during Phonation in the Squirrel
Monkey.   311
    P. MÜLLER-PREUSS

Selectivity of Auditory Neurons for Vowels
and Consonants in the Forebrain of the Mynah
Bird.   317
    G. LANGNER, D. BONKE and H. SCHEICH

Some Aspects of Functional Organization of
the Auditory Neostriatum (Field L) in the
Guinea Fowl   323
    D. BONKE, B.A. BONKE, G. LANGNER and
    H. SCHEICH

14-C-Deoxyglucose Labeling of the Auditory
Neostriatum in Young and Adult Guinea Fowl. .   329
    H. SCHEICH and V. MAIER

Integration of Voco-Auditory Centers in Song
Birds   335
    N. SAITO and M. MAEKAWA

Response Properties and Spike Waveforms of
Single Units in the Torus Semicircularis of
the Grassfrog (Rana Temporaria) as Related to
Recording Site.   341
    J.J. EGGERMONT, D.J. HERMES, A.M.H.J.
    AERTSEN and P.I.M. JOHANNESMA

Coding of Amplitude-Modulated Sounds in the
Midbrain Auditory Region of the Frog.   347
    N. BIBIKOV and O. GORODETSCAYA

SESSION VIII.   DEPRIVATION AND DEVELOPMENTAL
              STUDIES
                Chairmen: C.N. WOOLSEY and J.
                      MYSLIVEČEK

Effects of Early Auditory Stimulation on Cortical Centers   355
    J. HASSMANNOVÁ, J. MYSLIVEČEK and
    V. NOVÁKOVÁ

Effects of Acoustic Deprivation on Morphological Parameters of Development of Auditory
Neurons in Rat   359
    J. COLEMAN

Behavioral and Anatomical Studies of Central
Auditory Development.   363
    J.B. KELLY

Input-Dependent 2-Deoxy-D-Glucose Uptake in
the Central Auditory System of Rana
Temporaria.   369
        H. FLOHR, R. AMMELBURG, H. KORTMANN and
        W. ELSEN

Plastic Changes in the Inferior Colliculus
Following Cochlear Destruction.   377
        I. TANIGUCHI

Developmental Changes of Auditory Evoked Re-
sponses in Normal and Kanamycin Treated Rats.   381
        S. MATSUURA and T. TOKIMOTO

SESSION IX.   AUDITORY PROSTHESES, PHYSIOLOGICAL
              BACKGROUND
                 Chairmen: W. D. KEIDEL and S. TICHÝ

Physiological Background of Hearing Prosthe-
ses   389
        W.D. KEIDEL

Electrical Stimulation of the Human Cochlea -
Psychophysical and Speech Studies   411
        Y.C. TONG AND G.M. CLARK

Preliminary Speech Perception Results through
a Cochlear Prosthesis   417
        I.J. HOCHMAIR-DESOYER, E.S. HOCHMAIR,
        R.E. FISCHER and K. BURIAN

Tactile Aid for the Deaf; Search of a Code
Allowing Somesthetic Processing of Acoustic
Messages.   423
        Y. de RIBAUPIERRE, P. HEIERLI, M. HOLDEN
        M. ROSSI, M. DEMOULIN and F. de
        RIBAUPIERRE

Participants   427

Index   435

SESSION I
COCHLEAR MECHANISMS A
CHAIRMEN: Y. KATSUKI AND P. DALLOS

THE RESPONSES OF HAIR CELLS TO LOW FREQUENCY TONES AND THEIR
RELATIONSHIP TO THE EXTRACELLULAR RECEPTOR POTENTIALS AND
SOUND PRESSURE LEVEL IN THE GUINEA PIG COCHLEA

I. J. Russell and P. M. Sellick*

Ethology and Neurophysiology Group
School of Biological Sciences
The University of Sussex
Falmer, Brighton
Sussex, BN1 9QG, U.K.

INTRODUCTION

This paper is concerned with a description of the receptor
potential recorded intracellularly from inner hair cells in the
basal turn of the guinea pig cochlea. It includes a discussion of
the special properties which permit the responses of hair cells to
high frequency auditory stimulation to be transmitted to auditory
nerve fibres. Finally, the intracellularly recorded receptor potentials from inner hair cells are compared with the extracellularly
recorded cochlear microphonic from the scala tympani and sound
pressure, measured at the tympanic membrane, in response to low
frequency tones. These measurements form a basis for comparing the
way in which the stereocilia of inner and outer hair cells are mechanically coupled to the overlying tectorial membrane.

THE NATURE OF INNER AND OUTER HAIR CELL RECEPTOR POTENTIALS

Extracellular potentials have been recorded from various regions
of the cochlea in response to tone bursts and they reflect the receptor currents flowing across the sensory membranes of the inner
and outer hair cells of the organ of Corti. The potentials consist
of the alternating waveform of the cochlear microphonic (CM) which

---

*Present address: Department of Physiology,
University of Western Australia,
Nedlands, W. A. 6009.

resembles the acoustic waveform and this is usually offset, either above or below the recording baseline, depending on where it is recorded, by the summating potential (SP). It is of considerable interest to discover what relative contribution the inner and outer hair cell populations make to these potentials, since extracellular potentials have formed the basis for comparison with neural responses and measurements of basilar membrane motion for many years. It is also evident from these extracellular measurements that cochlear hair cells respond to high frequencies of auditory stimulation. However afferent fibres in the acoustico lateralis system are excited by the presynaptic release of chemical transmitter (Furukawa and Ishii, 1967; Flock and Russell, 1973, 1976; Sand, Ozawa and Hagiwara, 1975), and since transmitter release is a voltage dependent phenomenon (Katz, 1969), it seems reasonable to expect that the high frequency potential changes associated with acoustic transduction would be filtered out by the capacitive impedance of the hair cell membranes. Thus, what special properties of the receptor potential permit inner hair cells to transmit their responses to high frequency auditory stimulation to the afferent nerve terminals?

Intracellular recordings have been made from inner hair cells in the basal turn of the guinea pig cochlea (Russell and Sellick, 1977, 1978; Sellick and Russell, 1980). The cells were identified by iontophoretic injection of Procion and Lucifer yellow, and the resting membrane potentials were found to be between -20 and -45mV. Potentials were recorded from a region of the organ of Corti where the characteristic frequencies of the hair cells were between 15-22 KHz, however it was possible to elicit receptor potentials from them in response to low frequency auditory stimulation (300Hz) providing the stimulus intensity was high (c. 100dB SPL).

Receptor potentials in responses to 300Hz and 3kHz tone bursts are shown in Fig. 1. In response to symmetrical, sinusoidal pressure changes measured at the tympanic membrane, the depolarizing phase of the receptor potential is two times greater than the hyperpolarizing phase. This rectification of the receptor potential in the direction of depolarization is a property common among hair cells of the acoustico lateralis system (Flock and Russell, 1973, 1976; Hudspeth and Corey, 1977; Fettiplace and Crawford, 1978), and is clearly seen in the transfer function illustrated for an inner hair cell in Fig. 2.

When the frequency of auditory stimulation is progressively increased, the phasic (AC) component of the receptor potential becomes attenuated about a DC level whose magnitude is determined by the stimulus intensity and the rectifying properties of the transduction process. Above 1kHz the DC component of the receptor potential dominates the voltage responses of the inner hair cells (Fig. 1).

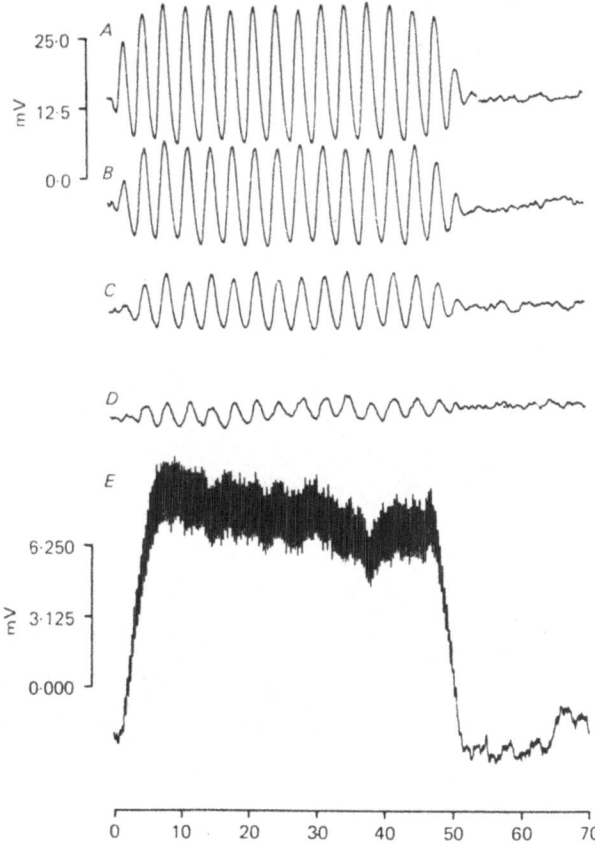

Fig. 1. Intracellular receptor potentials from a hair cell with
         a resting potential of -35mV recorded in the 16 kHz region
         of the basilar membrane. A-D: responses to 300Hz tone
         90-50 dB in 10 dB steps; E: response to 3kHz at 80 dB.
         Sound pressure levels in dB re 2 x 10^{-5} Nm^{-2}. (Reprinted
         from Russell, 1980)

It has been proposed that the attenuation of the AC component
of the receptor potential is due to the capacitative impedance of
the hair cell membrane (Russell and Sellick, 1978). In order to test
this hypothesis the time constants and AC/DC ratios of individual
hair cells were measured (Sellick and Russell, 1980). Typical
examples, illustrated in Fig. 3, show close correspondence between
the 3 dB cut off frequencies calculated from the time constants and
those determined from the AC/DC ratio. Below the cut-off frequency,
the ratio between the AC and DC components is 3, and above this frequency
the ratio decreases by about 6 dB/octave, which is to be expected from
a passive R-C network with time constants similar to those measured
in the hair cells (0.89-0.19 ms, mean 0.43 ms in 15 cells).

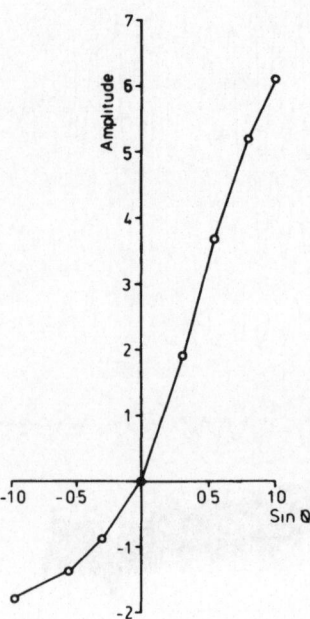

Fig. 2.   Transfer characteristics of a cochlear hair cell measured
          in response to a 200Hz tone at 100dB SPL. The ordinate
          represents the amplitude of the receptor potential in
          millivolts, and the abscissa is the sinusoidal sound pres-
          sure level measured at the tympanic membrane. (Reprinted
          from Russell, 1980)

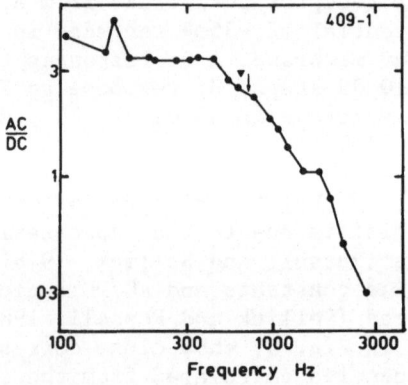

Fig. 3.   The relationship between the ratio of AC and DC components
          of the receptor potential versus frequency of a single
          hair cell. The arrow indicates the 3dB point and ▼ indicates
          the cut-off frequency determined from the electrical time
          constant. (Reprinted from Sellick and Russell, 1980)

These measurements support the earlier hypothesis that the AC component is reduced by the capacitative impedance of the hair cell, so that at high frequencies of auditory stimulation, the DC component predominates.

The frequency dependent form of the inner hair cell receptor potentials are closely related to the patterns of discharge activity which have been recorded from auditory nerve fibres, and presumably account for them. Below 1kHz, fibres in the auditory nerve are phase locked to the stimulus with a high degree of synchronization, above this, synchronization decreases with frequency and disappears above 4kHz (Rose, Brugge, Anderson and Hind, 1968). Stimulation by tones above 4kHz produce sustained discharges from auditory nerve fibres, but they have no time structure which can be related to the periodicity of the stimulus.

Thus the asymmetrical operating characteristics of the inner hair cells allows for the production of voltage responses from hair cells at high frequencies of auditory stimulation. It is significant that the point of maximum inflexions of this curve coincides with its operating point at the resting potential. Thus, a DC component will be generated at high frequencies of auditory stimulation even when the stimulus levels are small.

We have made intracellular recordings from two morphologically identified outer hair cells. The receptor potentials contrast with those recorded from inner hair cells in that they are symmetrical and almost an order of magnitude smaller (Fig. 4). Furthermore, large DC potentials were not recorded from them in response to high frequency tones, an observation which has been confirmed by Tanaka, Asanuma and Yanagusawa (1980) and Dallos (personal communication). Their small amplitude may reflect only a small receptor current flowing through them during excitation, or they may have low impedance. Support for the possibility that outer hair cells have low impedance is drawn from the observation that the tissues of the organ of Corti are dominated by extracellular potentials originating in the outer hair cells (the organ of Corti CM) (Fig. 4). The flow of current between the supporting cells may be facilitated by electrotonic coupling between adjacent supporting cells, based on the observation that fluorescent dyes flow readily between them. However, it remains to be seen if outer hair cells are electronically coupled to neighbouring cells.

On the basis of these observations it seems that the outer hair cells are predominantly responsible for generating the extracellular CM, and inner hair cells generate the extracellular summating potential. Thus we are in agreement with Dallos and Cheatham (1976) who first put forward this proposition based on extracellular receptor potential recordings from normal cochleae and in cochleae where the outer hair cells had been selectively destroyed by ototoxic poisoning.

   The symmetrical receptor potentials of outer hair cells have an
interesting implication for their functional limitations. At high
frequencies of auditory stimulation they generate little or no DC
component and will not excite afferent fibres at these frequencies
if it is assumed that neural transmission between hair cells and
their afferent innervation is a voltage dependent process. In view
of their ultrastructural peculiarities (Spoendlin, 1978), it is of
considerable interest to discover if the afferent innervation of
outer hair cells is functional at any frequency of auditory stimulation.

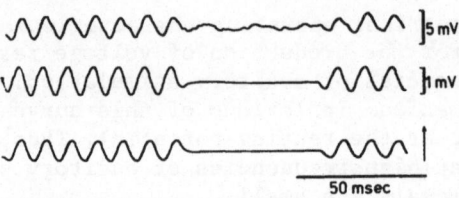

Fig. 4.   Intracellular recording of receptor potential from an outer
          hair cell in the basal turn of the guinea pig cochlea in
          response to a 125Hz tone at 100dB re 2 x 10^{-5}Nm^{-2}. Middle
          trace: organ of Corti microphonic potential. Lower trace:
          sound pressure recorded from auditory meatus. Arrows indi-
          cate direction of rarefaction. (Reprinted from Russell and
          Sellick, 1980)

INNER HAIR CELLS RESPONSE TO BASILAR MEMBRANE VELOCITY

   The different properties of the sensory epithelia of the acoustico lateralis system depend to a large extent on the properties of
the accessory structures and the way the hair cells are coupled to
them. In the cochlea, the way inner and outer hair cells are mechanically coupled to the tectorial membrane has been a central issue
in the study of its functional morphology for a number of years.
Electronmicroscopical observations by Kimura (1966) and Lim (1972)
reveal that the tips of the stereocilia of outer hair cells are
firmly embedded in the underside of the tectorial membrane, while
those of inner hair cells barely make contact with it. The implications of these morphological observations are that the stereocilia
of outer hair cells will be moved directly by the displacements of

the tectorial membrane in relation to the organ of Corti, but the stereocilia of inner hair cells will be moved by the viscous drag of fluid streaming around them. Thus the displacement of the free-standing stereocilia of inner hair cells will be proportional to basilar membrane velocity. The interpretation of these ultrastructural studies is plagued by the uncertainty that the relationships between inner and outer hair cells and the tectorial membrane may be a product of the histology (Engström and Engström, 1979). These uncertainties have, to some measure, been resolved by electrophysiology.

Dallos, Billone, Durrant, Wang and Raynor (1972) recorded CM differentially across the basilar membrane in different turns of the cochlea while producing trapezoidal movements of the basilar membrane by displacing the stapes with a low frequency triangular oscillation, since it has been established that basilar membrane motion in the guinea pig is proportional to stapes velocity (Dallos, 1970; Wilson and Johnstone, 1975). In those regions of the cochlea in which the outer hair cells remained undamaged by the kanamycin treatment, the CM reflected the basilar membrane displacement, and the potential waveforms were trapezoidal, while in other regions, usually the basal, high frequency regions, where only inner hair cells remained, the CM was reduced and its waveform reflected the differential or velocity of basilar membrane motion. The response of the inner hair cells to basilar membrane velocity was confirmed by comparing the amplitude and phase of CM recorded from regions of the basilar membrane devoid of outer hair cells and in normal cochleae in response to sinusoidal stimulation (Dallos, 1973). It was found that in kanamycin poisoned animals, the sensitivity of the CM increased at a rate which was 6dB/octave faster than in normal animals over the range 100-500Hz, and at 100Hz, the phase leads normal CM by 90°. This phase lead was not sustained however, and fell to about 45° for the frequency range 300-2000Hz. On the basis of these observations Dallos et al. (1972) proposed that inner hair cells responded to basilar membrane velocity while outer hair cells responded to basilar membrane displacement.

One objection to these experiments is that the responses of the inner hair cells to basilar membrane velocity, in those regions denuded of outer hair cells by kanamycin, is a product of experimental manipulation and not the normal response. An attempt was made to avoid this criticism by making intracellular recordings from inner hair cells and comparing their receptor potentials with the extracellularly recorded CM during sinusoidal and trapezoidal stimulation of the basilar membrane (Sellick and Russell, 1980; Rusell and Sellick, 1980).

The receptor potential recorded intracellularly from an inner hair cell, and the CM recorded from a gross electrode in the scala tympani, in response to triangular sound pressure variations are

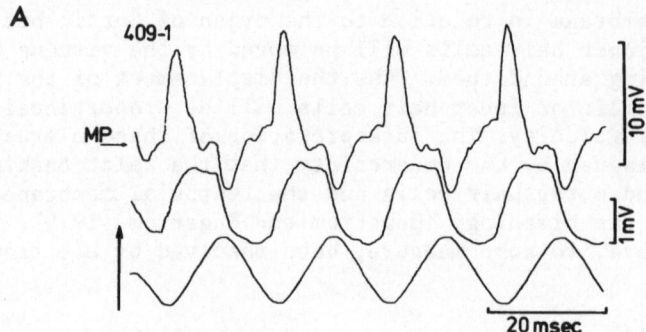

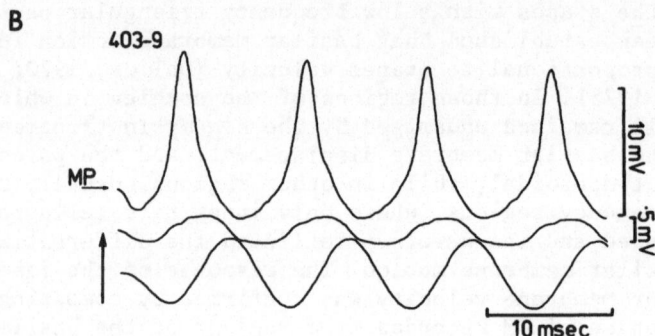

Fig. 5.   Receptor potentials recorded intracellularly from inner
          hair cells in the basal turn of the guinea pig cochlea in
          response to (A) a 52Hz triangular acoustic stimulus at
          100dB SPL and (B) a 102Hz sinusoidal tone at 80dB SPL. In
          both records upper trace: receptor potential; middle trace:
          CM recorded adjacent to the recording site; lower trace:
          sound pressure recorded at tympanic membrane. MP indicate
          membrane potential, (A) -40mV,(B) -30mV. Arrows in bottom
          trace indicate direction of rarefaction. Sound intensities
          in dB re 2 x 10^{-5} Nm^{-2}. (Reprinted from Sellick and Russell,
          1980).

illustrated in Fig. 5A. The trapezoidal shape of the CM indicates
a response to the first derivative of the sound pressure variation,
i. e. to the basilar membrane displacement, where as the inner hair
cell receptor potential is clearly the rectified second differential
of the sound pressure, which indicates that it is proportional to
basilar membrane velocity. Similar conclusions were drawn from
observations based on sinusoidal stimulation (Fig. 5B).

     The phase and amplitude of the inner hair cell receptor poten-
tials were measured to discover if the inner hair cells responded
to basilar membrane velocity over the frequency range 20-4000Hz. It

was anticipated that the intracellular potentials would be delayed and attenuated by the electrical time constants of the hair cell membranes and these were measured and taken into account in the calculation of the phase and amplitude relations of the receptor potentials.

The amplitude characteristics of the receptor potential for a cell with a measured time constant of 0.22 ms (3dB point 723Hz) are illustrated in Fig. 6A. Over the frequency range 28-300Hz, the receptor potential amplitude increases at 12dB/octave, while the CM increases at 6dB/octave. This is to be expected if the receptor potential is proportional to basilar membrane velocity, and the CM is proportional to displacement. Above 300Hz, the amplitude of the receptor potential plateaus and then declines at 12dB/octave. These relationships become more apparent if the ratio of the receptor potential amplitude and CM is plotted with respect to frequency, since the results are then standardized for constant stapes velocity.

This is illustrated in Fig. 6B where the relationship between the ratio of AC:CM versus frequency may be approximated as a plateau between 300-700Hz with rising and falling slopes of 6dB/octave. If the receptor potential/CM ratio is corrected for the decrement of the receptor potential produced by the hair cell time constant, the resulting curve has the characteristics of a high pass filter with a 2dB point close to 200Hz. One interpretation of these results is that the rising slope of the curve represents the velocity response of the hair cell, at 200Hz the hair cell begins to respond to basilar membrane displacement, and at 700Hz and above the phasic component of the receptor potential is attenuated by the capacitative impedance of the hair cell membrane. This interpretation is supported by analysis of the phase differences between the receptor potential and sound pressure measured at the tympanic membrane and the CM. These two relationships are illustrated for a single hair cell in Fig. 7. The two curves for receptor potential / sound pressure phase difference and receptor potential / CM phase difference are similar in shape but separated from each other by 90°. Data is not presented for phase with respect to sound pressure above 400Hz since phase changes occur in the middle ear above this frequency which were not taken into account in our experiments. Below 100Hz the receptor potential phase leads CM by 80°, which indicates that at these frequencies the hair cell behaves as a velocity detector, and the phase difference gradually declines to zero at 300Hz. Above 400Hz, the receptor potential phase-lags CM, presumably due to the membrane time constant of the hair cell. If this is taken into account, then the phase differences between the receptor potential and CM move from a lead of 80° to 0° at a rate which would be predicted for a high pass filter, with a 3dB point close to 200Hz. There is close agreement between the uncorrected data and a theoretical curve calculated from high and low pass filters with cut-off frequencies suggested by the results (Fig. 7).

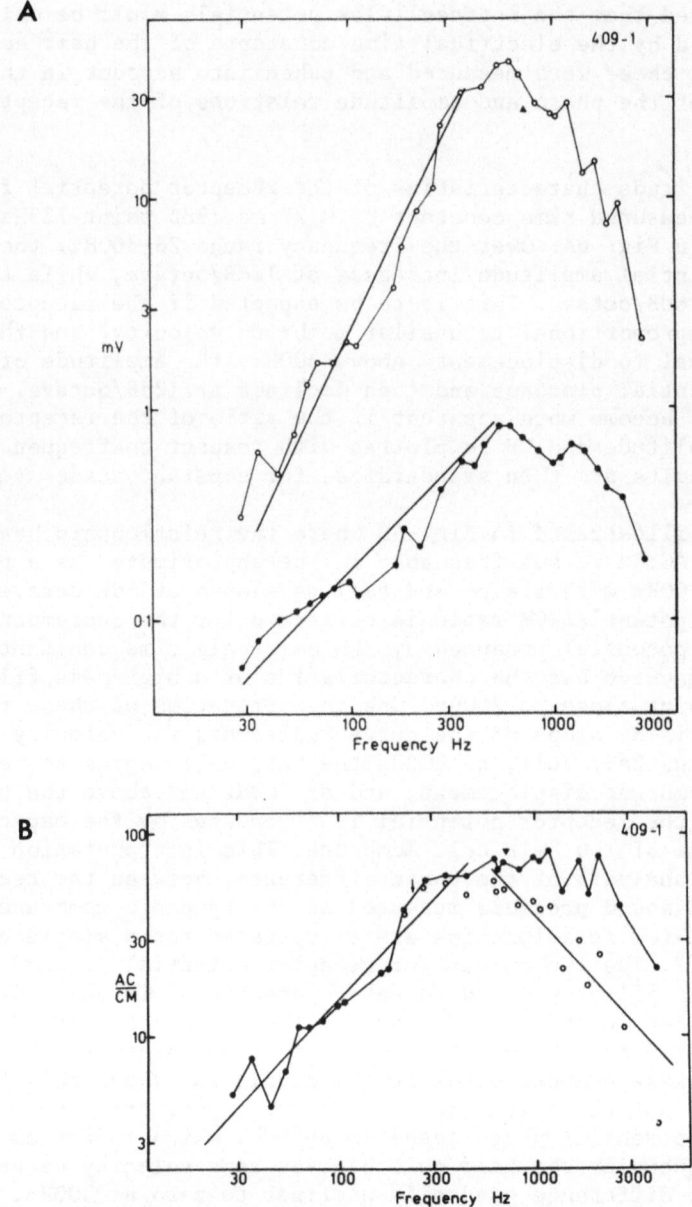

Fig. 6. (A) The relationships between the amplitude of the IHC
         intracellular receptor potential and CM versus frequency
         for constant SPL at 80dB SPL. O and ● indicate the amplitude
         of the receptor potential and CM respectively. The lines of
         the upper and lower curves indicate 12 and 6dB/octave slopes
         respectively. ▲ indicates the cut-off frequency determined
         from the electrical time constant, of the hair (cont'd)

(Fig. 6 cont'd) cell. (B) The relationship between the ratio
of IHC receptor potential and CM versus frequency for the
hair cell illustrated in Fig. 7. The arrow indicates the 3dB
point of the mechanical high pass filter and ▼ indicates the
cut-off frequency of the hair cell determined from the
electrical time constant. O: raw data; ●: data corrected for
membrane time constant. Rising and falling lines represent
slopes of 6dB/octave. (Reprinted from Sellick and Rusell,
1980)

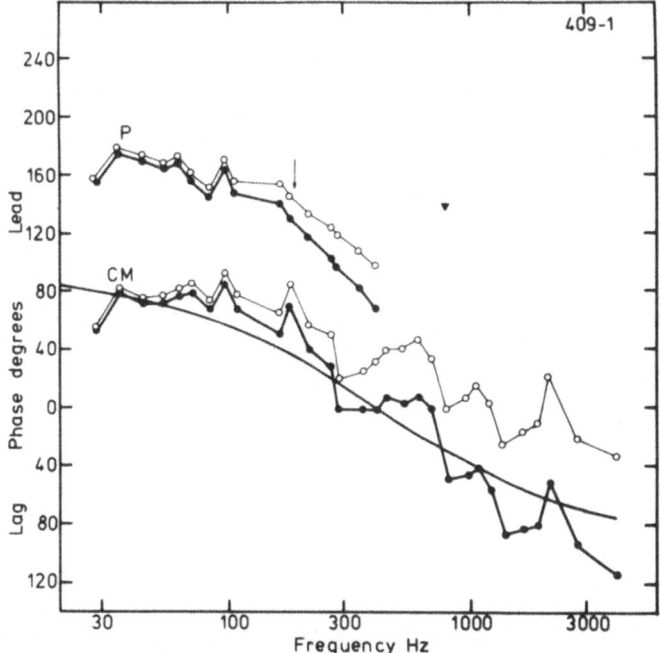

Fig. 7. The phase differences between the IHC receptor potential
         and the sound pressure for the cell illustrated in Fig. 6,
         measured in the external auditory meatus (P), and between
         the IHC receptor potential and the CM, measured adjacent
         to the recording site in an inner hair cell (CM), versus
         frequency in the first turn of a guinea pig cochlea. ●:
         raw data; O: data corrected for membrane time constant.
         The continuous curve represents phase behaviour of a high
         pass filter with 3dB cut-off of 200Hz and a series low pass
         filter with 3dB cut-off of 800Hz. (Reprinted from Sellick
         and Russell, 1980)

    Analysis of the phase and amplitude characteristics of the IHC
receptor potential indicate that for the hair cell illustrated, the

receptor potential is proportional to basilar membrane velocity below 200Hz, and above this, it behaves as a displacement detector. This process has the characteristics of a high pass filter, and in each cochlea the points at which the hair cells change their response from basilar membrane velocity to displacement are remarkably similar, although the measurements of membrane time constant may vary considerably in different cells.

The apparent viscous coupling of the stereocilia of the inner hair cells to the tectorial membrane may have a protective role at low frequencies of acoustic stimulation, preventing the stereocilia from being biased by any sustained, large amplitude displacements of the basilar membrane. Protection against biasing is important for the function of the IHCs at high frequencies when the AC component of the receptor potential is very small and transmitter release, from the afferent synapse, is governed by the DC component. This is a product of the asymmetrical transfer function of the hair cell and it is important that the operating point of the hair cell remains at a point of inflection on the transfer function so that a rectified receptor potential and, consequently, a DC component is produced.

ACKNOWLEDGMENTS

This work was supported by a grant from the M. R. C. and we thank J. Art, L. Cowley, E. Edelstein, A. Kroese and D. Lowe for their valuable criticism of the manuscript. Data presented in Figs. 1 and 2 were obtained at the Coleman Memorial Laboratory, University of California, San Francisco, in association with Dr. M. Merzenich.

REFERENCES

Dallos, P., 1970, Low frequency auditory characteristics: species dependence, J. Acoust. Soc. Am., 48: 489-499.
Dallos, P., 1973, Cochlear potentials and cochlear mechanics, in: "Basic Mechanisms in Hearing", A. Møller, ed. Academic Press, New York, London.
Dallos, P. and Cheatham, M. A., 1976, Production of cochlear potentials by inner and outer hair cells, J. Acoust. Soc. Am., 60: 510-512.
Dallos, P., Billone, M. C., Durrant, J. D., Wang, C-y. and Raynor, S., 1972, Cochlear inner and outer hair cells: functional differences, Science 177: 356-358.
Engström, H. and Engström, B., 1978, Structure of the hairs on cochlear sensory cells, Hearing Res., 1: 49-66.
Fettiplace, R. and Crawford, A. C., 1978, The coding of sound pressure and frequency in cochlear hair cells of the terrapin, Proc. R. Soc. Lond. B, 203: 209-218.

Flock, A. and Russell, I. J., 1973, Efferent nerve fibres: postsynaptic action on hair cells, Nature, 243: 89-91.

Flock, A. and Russell, I. J., 1976, Inhibition by efferent nerve fibres: action on hair cells and afferent synaptic transmission in the lateral line canal organ of the Burbot Lota lota, J. Physiol., 257: 45-62.

Furukawa, T. and Ishii, Y., 1967, Neurophysiological studies on hearing in goldfish, J. Neurophys., 30: 1377-1403.

Hudspeth, A. J. and Corey, D. P., 1977, Sensitivity, polarity and conductance changes in the response of vertebrate hair cells to controlled mechanical stimuli, Proc. Natl. Acad. Sci., 74: 2407-2411.

Katz, B., 1969,"The Release of Neural Transmitter Substances", Liverpool University Press, Liverpool.

Kimura, R. S., 1966, Hairs of the cochlear sensory cells and their attachment to the tectorial membrane, Acta Oto-laryngol., 61: 55-63.

Lim, D. J., 1972, Fine morphology of the tectorial membrane, Arch. Otolaryngol. 96: 199-215.

Rose, J. E., Brugge, J. F., Anderson, D. J. and Hind, J. E., 1968, Patterns of activity in single auditory nerve fibres of the squirrel monkey, in: "Hearing Mechanisms in Vertebrates", A. V. S. de Reuck and J. Knight, eds., Churchill, London.

Russell, I. J., 1980, The responses of vertebrate hair cells to mechanical stimulation, in: "Neurones without Impulses", A. Roberts and B. Bush, eds. Cambridge University Press, in press.

Russell, I. J. and Sellick, P. M., 1977, Tuning properties of cochlear hair cells, Nature, 267: 858-860.

Russell, I. J. and Sellick, P. M., 1978, Intracellular studies of hair cells in the mammalian cochlea, J. Physiol., 284: 261-290.

Sand, O., Ozawa, S. and Hagiwara, S., 1975, Electrical and mechanical stimulation of hair cells in the mudpuppy, J. Comp. Physiol., 102: 13-26.

Sellick, P. M. and Russell, I. J., 1980, The responses of inner hair cells to basilar membrane velocity during low frequency auditory stimulation in the guinea pig cochlea, Hearing Res., in press.

Spoendlin, H., 1978, The afferent innervation of the cochlea, in: "Evoked Electrical Activity in the Auditory Nervous System", R. F. Naunton and C. Fernandez, eds., Academic Press, New York, San Francisco, London.

Tanaka, Y., Asanuma, A. and Yanagisawa, K., 1980, Potentials of outer hair cells and their membrane properties in cationic environments, Hearing Res., in press.

SOME COMPARATIVE ASPECTS OF A COCHLEAR MECHANISM

Yasuji Katsuki

National Center for Biological Sciences

Okazaki, Aichi 444, Japan

Today I am very honored to be given this opportunity to talk about our recent studies of some comparative aspects of cochlear mechanism at this symposium.

When I graduated from medical school, Dr. Georg von Békésy was actively investigating Corti's organ and published several articles. I was very much impressed by them and started to study auditory mechanisms.

Dr. von Békésy was born in Budapest in 1899. He received his basic education in Munich and Switzerland. After graduation from the University of Hungary, he worked at the post office research laboratory in Budapest. His famous experiments in his early stage were done there. Thereafter he moved to Stockholm and then to the U.S.A., where he stayed at Harvard for some time. In 1952 I met him for the first time at Harvard on the last day of that year. From that time carried on a continuously increasing correspondence, and finally, we worked together in Hawaii for two years in 1967-68. During that period we could talk every day about our common interests.

Dr. von Békésy died on June 13, 1972.

When I was with him in Hawaii, our first work was on the lateral-line organ of the shark to investigate the receptive mechanism of mechanical stimulation of that organ, because it is known that the lateral-line organs of aquatic animals are homologous to the mammalian inner ear.

The strict regulations against importation of foreign animals
to Hawaii prevented our working on the inner ear of higher animals.

Since then I and my colleagues have been continuing our studies
on the lateral-line organ of various aquatic animals.

I. RESPONSES OF THE LATERAL-LINE ORGAN TO IONS

1) When we worked on the lateral-line organ of shark in Hawaii,
we happened to find that various metalic monovalent ions could
stimulate that organ. The order of effectiveness of these ions
was $K^+ = Rb^+ > Na^+ > Cs^+ = Li^+$
On the other hand divalent cations suppressed the effect of monovalent ones and the order of the suppressive effect was
$$Ca^{++} \gg Mg^{++} = Sr^{++}$$
Furthermore, the suppressive effect of Ca^{++} on the monovalent ion
stimulating effect was rather easily removed by washing the end
organ with distilled water. (Katsuki et al., 1970).

2) In the case of teleosts the order of stimulating effectiveness
of the monovalent cations was $K^+ = Rb^+ = NH_4^+ > Cs^+ > Na^+$ and the
order of the suppression of divalent cations on the monovalent stimulating effect was
$$Ca^{++} > Sr^{++} = Mg^{++}$$ (Katsuki et al., 1971)

3) In the lateral-line organ of the tadpole of _Rana catesbeiana_,
the order of effectiveness of monovalent cations was
$$Ag^+ \gg Tl^+ > Na^+ > K^+ > Li^+$$
and the order of divalent cations was
$$Sr^{++} > Mg^{++} \geq Ba^{++} > Ca^{++} > Cd^{++} \geq Co^{++} > Mn^{++} > Zn^{++}.$$
The order of the suppressive effects of divalent cations on the
stimulating effect of K^+ was $Cd^{++} > Mg^{++} > Ca^{++} > Mn^{++} > Zn^{++}$.
Certain divalent cations suppressed the stimulating effects of
Ag^+ in the order of $Cd^{++} > Zn^{++}$, while neither, Ca^{++}, nor Mg^{++}
produced any effect on the Ag^+ stimulating effect. The stimulating
effects of Ag^+, Cd^{++} and Zn^{++} were abolished by Dithio threitol
(DTT) but not by a simple washout with water. Based on the experimental results obtained, the ions which were investigated were
classified into two groups. (Nakagawa et al., 1974).

A group: Na^+, K^+, Li^+, Ca^{++}, Mg^{++}, Sr^{++}, Ba^{++}

B group: Ag^+, Tl^+, Cd^{++}, Zn^{++}

The physiological characteristics of the A group ions were:

1) The effects were relatively short-lived, but repeated stimulation was possible.

2) The responses obtained by these ions could be removed by water.

3)   The stimulating effects of the monovalent cations which belong to this group were suppressed by all divalent cations.

The physiological characteristics of the B group ions were:

1)   The stimulating effects, if they occurred, continued for many hours.

2)   The effects were removed by DTT, but not by simple water wash-out.

3)   The stimulating effects of monovalent cations belonging to B group could be suppressed only by divalent cations belonging to the same group.

4)   The effects of the B group cations were generally stronger than those of the A group ions.

Guided by these experimental results, the chemical characteristics of the ions belonging to these two groups were examined.

II. THE HARD AND SOFT ACID AND BASE (HSAB) PRINCIPLE.

Among those ions belonging to A group Na^+ and Ca^{++} were examined, while from B group Ag^+ and Cd^{++} were selected as examples. A simple classification can be arrived at by the chemical reaction between sulfur and these cations. According to chemical bond theory, Na^+ and Ca^{++} do not readily bind with sulfur while both Ag^+ and Cd^{++} form stable, insoluble compounds with this element. Such distinct chemical differences are thought to be a result of the respective atom's electron configuration. Similar differences can also be seen between Ca^{++} and Cd^{++}. In 1968, Pearson proposed a new hypothesis which differentiates between hard acids and bases, and soft acids and bases. When metallic ions react chemically with the ligand of a counter substance, the metallic ions behave as acid while the ligand behaves as a base. There are many metallic ions and many ligands. However, they can all be divided into two groups. One of these groups reacts strongly with nitrogen, oxygen, fluorine and so on, while the other reacts with sulfur, phosphorus, iodine and similar elements. Based on such facts Pearson (1968, 1973) proposed that acids and bases each be classified into two groups, a hard one and a soft one.

According to his classification the hard base does not polarize easily while the soft base polarizes very easily. In additions the hard bases react with the hard acids to form stable compounds while soft bases react similarly with soft acids. The degree of reaction is correlated with a softness parameter, o, with acids and bases which have equal values of o tending to react most strongly. Soft-

ness parameter values are given by Pearson and Ahrland. The hard
group has smaller values, while the s oft group has larger values.
The classiffication of metallic ions in their effects on membrane,
which I mentioned before, corresponds exactly to Pearson˙s hard and
soft acids.

III. BASES ON THE RECEPTOR MEMBRANE

It is well known that biomembranes are composed of proteins
and lipids in a ratio which is very close to 1:1. Those molecules on
the surface of biomembranes which have negative charges can be
considered to be base.

The protein amino acids which are exposed at the membrane
surface are mostly polar amino acids. In physiological pH those
amino acids which are negatively charged are glutamic acid and
asparagic acid. Both of these are hard acids. Two amino acids,
cystein and methionin, contain sulfur atoms in the form of SH. These
are soft bases. On the other hand, the main components of lipids in
the biomembrane are phospholipids (phosphatidylcholin, phosphatidyl-
ethaı plamin, phosphatidylserin, phophatidylinositol). All of these
react aɛ hard bases. To summarize:
   The hard bases are asparagic acid, glutamic acid and the
   phospholipids.
   The soft bases are cysteine and methionine.

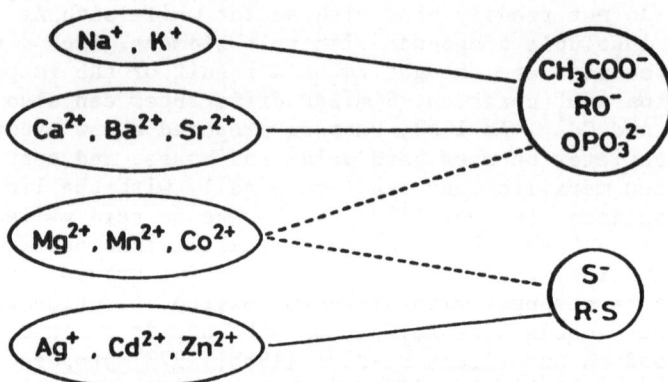

Fig. 1. Hard (upper) and soft (lower) acids (left) and bases
         (right) affinity —— high, --- low.

When metallic ions, either hard or soft acids, encounter receptors
on the membrane, they may chemically bind to these so-called sites
at points where negative charges are located. Such relations are
schematically represented in Fig. 1.

IV. CHEMORECEPTION IN THE LATERAL-LINE ORGAN.

1)   Fig. 2 shows the relationship between concentration and neuronal responses for various monovalent ions in tadpoles.

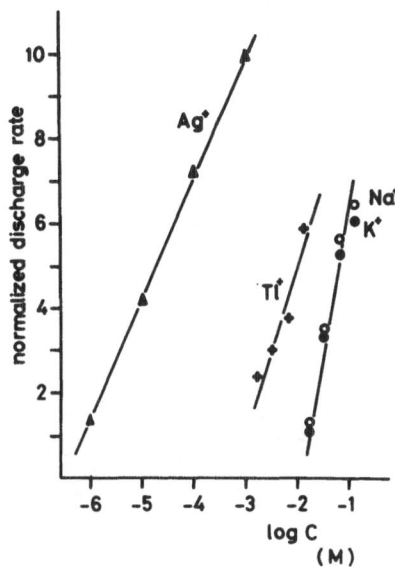

Fig. 2. Normalized afferent discharge rates are shown against con-
         centrations of various monovalent metallic salt solutions,
         $AgNO_3$, $TINO_3$, $NaNO_3$, and KNO_3. The molar concentrations
         are shown in logarithmic scale on abscissa. Note linear
         relationships between discharge rate and log C in all ions.

The abscissa shows molar concentration of metallic ions in a
logarithmic scale and the ordinate shows normalized discharge rate
of a single lateral-line nerve fiber. The effect of the NH_4^+ solution is almost the same as that of the K^+ solution while Li^+ is less
effective than Na^+ solution. Ag^+ and Tl^+ which are soft acids show
very high effectiveness. Moreover stimulation of the receptor organ
by Ag^+ cannot be stopped by water wash, but only by DTT. There are
linear relationships between discharge rate and log C for all ion
types.

2)   Neuronal responses to divalent cations

    When similar curves are plotted between discharge rate and
log C for divalent cations and compared to the previous monovalent

cation curves, it is quite noticeable that the increase of discharge rate against logarithm of concentration has a very small slope and for Mn^{++} and Zn^{++} the slope is almost 0 or sometime negative. This means that a decrease of discharge rate occurr as concentration increases. (Yoshioka et al., 1976) (Fig. 3).

According to the classification of the periodic table Mg, Ca, Sr and Ba are known as group II-a and Zn and Cd are elements of group II-b. Mn and Co belong to group VII-a and VIII respectively. The stimulating effects of Sr^{++} and Mg^{++} were remarkable, but their

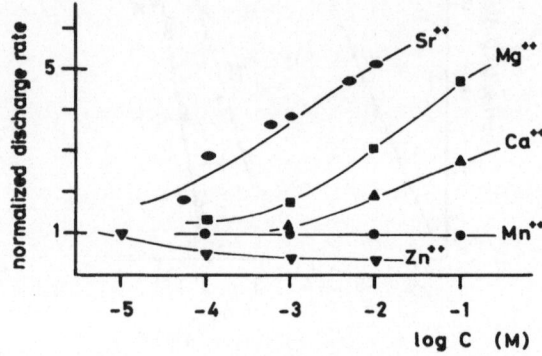

Fig. 3.   Relation between logarithms of molar concentrations of divalent cations, Sr^{++}, Mg^{++}, Ca^{++}, Mn^{++}, and Zn^{++} (abscissa) and normalized discharge rates of a single lateral-line nerve fiber (ordinate).

effects disappeared promptly upon rinsing with water. The effects of Ca^{++} and Ba^{++} could not be so easily removed with water. The effect of Ca^{++} were eliminated easily with a $10^{-5}M$ EGTA solution. Mn^{++} itself produced neither stimulation nor suppression of the K^+ or Na^+ effects as other divalent ions did. Fig. 4 shows the suppressive effects of divalent cations on the stimulating effects of monovalent cations.

The ordinate shows the degree of suppression or the effective suppressive ratio ε where

$$\varepsilon = \frac{N-N^{\cdot}}{N_o} \quad , \text{ and}$$

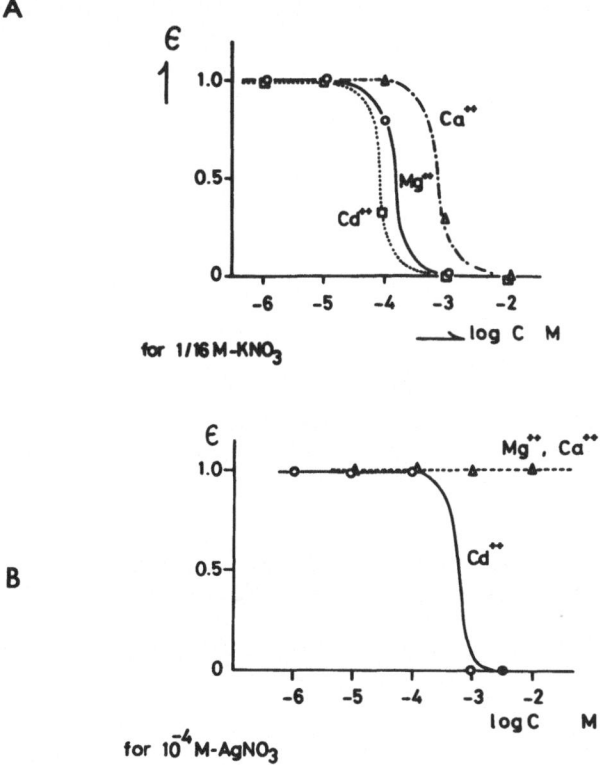

Fig. 4. Effective suppressive ratios, ε, of various divalent ions plotted against logarithms of concentration. Measurements were made against the stimulatory effect of $1/16$ M KNO_3. Δ, Ca^{++}; O, Mg^{++}; $\square$, Cd^{++}, 10^{-4} M $AgNO_3$. Δ, Ca^{++} and Mg^{++}; O, Cd^{++}.

N, N', N_o are increments of impulse frequency above the level of spontaneous discharge which are caused by the mixture of $Mg(NO_3)_2$ and $1/16$ M KNO_3, by $Mg(NO_3)_2$ alone and by $1/16$ M KNO_3, respectively. When $\varepsilon = 1$ there is no suppression; when $\varepsilon = 0$ there is complete suppression. In the Fig. 4 A responses to $1/16$ M KNO_3 solution are suppressed by Ca^{++}, Mg^{++} and Cd^{++} solutions. In the Fig. 4 B responses to 10^{-4} M $AgNO_3$ solution are suppressed by Cd^{++} but not by Mg^{++} or Ca^{++}.

In 1974 Yanagisawa et al. measured the membrane potential and the resistance of hair cells of lateral-line organ of Necturus maculosus. They found a linear relationship between the ionic concentration and the membrane potential following Beidler's taste equation (Beidler, 1954). This relationship, however, satisfies only a necessary condition for the adsorption model, but not necessary and sufficient condition (Yoshioka et al., 1975). The effect of Ag^+ is very high, although it is known experimentally that there is no movement of Ag^+ across the membrane. This seems to give powerful support to the adsorption model (Asanuma, 1975).

Analysis of experimental results by use of the adsorption model. Beidler's taste equation is:

$$\frac{C}{R} = \frac{C}{Rm} + \frac{1}{KRm} \quad \ldots\ldots\ldots \quad (1)$$

where C is the concentration of metallic ions;
R is the magnitude of response;
Rm is maximum response to a given substance; and
K is an association constant in the reaction.
This equation is used by many authors to express the binding of small ions to proteins.

In the present case, that of a site binding model, the maximum number of binding sites per mole of protein, n, was used in place of the Rm used in the taste equation (Friedberg, 1974).

The equation for the Scatchard plot of binding behavior then becomes

$$\frac{R}{C} = Kn - KR \quad \ldots\ldots\ldots \quad (2)$$

If we plot R/C vs. R the result is a straight line. A plot of R/C vs. R yields Kn as the intercept on the R/C axis, and n as the intercept on the R axis. If the receptor membranes have the sort of functional proteins that can react with some cations, we can use this equation to analyse the chemical response of the lateral-line organ.

The normalized discharge rate and concentrations of various monovalent and divalent metallic salts used in our previous article were plotted using the form of Eq(2). The replotted data are shown in Fig. 5 using a semilog scale in order to present all of the data together. The values of n and K obtained from Equation (2) for various ions are shown in Table 1.

A large value for the binding constant K expresses a large affinity of ions to the receptor membrane. The chemical affinity can be expressed by the softness parameter, σ. The relation between σ and K is shown in Fig. 6.

The abscissa shows σ and the ordinate the logarithm of K. The A, B groups could both be plotted on a straight line. Our original

TABLE I

	n	K
Na, K	8.2	$5.0 \times 10_{2}$
Tl	7.9	1.9×10^{5}
Ag	7.3	2.1×10^{3}
Ca	1.8	8.4×10^{3}
Mg	2.5	1.0×10^{4}
Cd	2.2	1.0×10^{4}

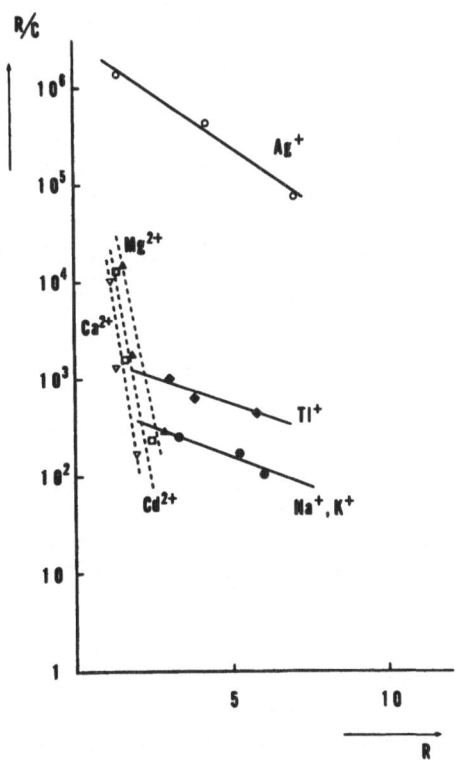

Fig. 5. Logarithmic plot of binding behavior. Normalized afferent
discharge rates against the concentration of various mono-
and divalent salt solution was plotted. Open marks represent soft cation group and filled marks mean hard cation
group.

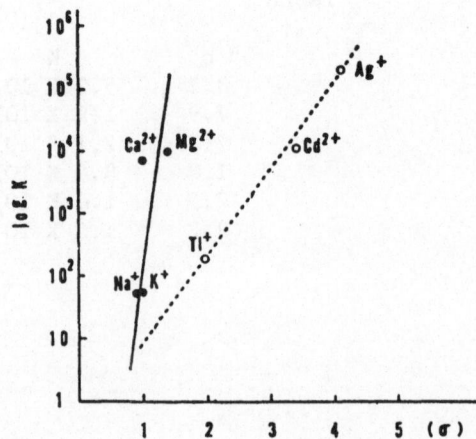

Fig. 6.   Correlation between softness parameter(σ) and the associa-
          tion constant. Data can be divided into two groups. One
          group which is shown by solid line represents hard cation
          group (filled circle). Cations plotted on the dotted line
          belong to soft group (open circle).

idea that metallic ions would be classified into a hard and a soft
group was thus confirmed.

     The suppressive effects of divalent cations on the stimulatory
effects of monovalent cations must be shown by the same adsorption
model. In this case a competitive adsorption model was assumed. The
results are shown in Fig. 7.

Areas shown with oblique lines are the experimentally effective
concentrations and the arrow in each figure shows the theoretical
value calculated for competitive adsorption. Divalent cations
belonging to the A group show nice agreement with theoretical
values.

V. MODEL OF RECEPTOR SITES

     In light of the above discussions, the following conclusions
might be derived; the initial events of chemical reception could
be initiated by the adsorption of metallic cation (acid) on the
various basic receptor sites. Let us consider a molecular mechanism
of chemical reception.

Soft sites. The soft sites may consist, in part, of sulfur containing proteins which include Cystein and Methionin as amino acids.
In such cases, PCMB (parachloromercuribenzoate) and DTT will affect
the receptivity of the sites. In Fig. 8, the effectiveness of DTT

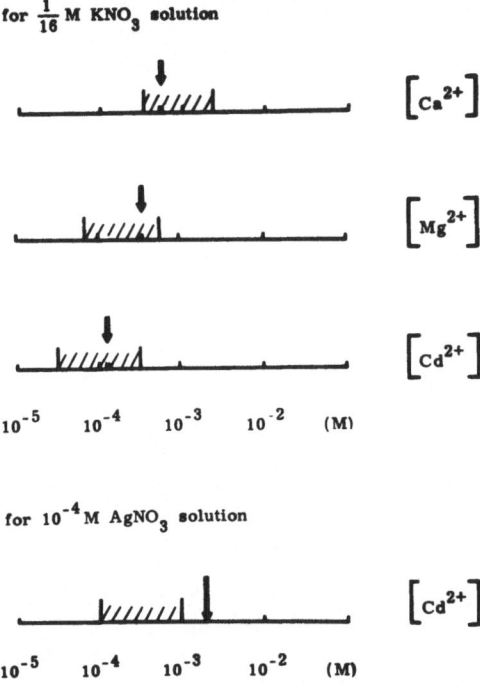

for $\frac{1}{16}$ M KNO$_3$ solution

$\left[Ca^{2+}\right]$

$\left[Mg^{2+}\right]$

$\left[Cd^{2+}\right]$

10^{-5} 10^{-4} 10^{-3} 10^{-2} (M)

for 10^{-4} M AgNO$_3$ solution

$\left[Cd^{2+}\right]$

10^{-5} 10^{-4} 10^{-3} 10^{-2} (M)

Fig. 7. Comparison between experimental values and the calculated
         ones for the suppressive effect of divalent cations on the
         stimulating effect of 1/16 M KNO$_3$ solution and that of
         10^{-4} M AgNO$_3$ solution. Shaded area in the figure means
         experimentally obtained "half suppression concentration."
         Big arrows show the calculated value.

in suppressing Ag^+ stimulation is demonstrated.

Attention should be payed to the fact that the DTT treated receptor
membrane showed no response to Ag^+ stimulation, whereas it did
react with K^+. This kind of experiment leads to the conclusion that
K^+ receptive membrane has no sulfur atom in the K^+ receptor site
protein. A powerful new technique, X-ray microanalysis, gave further
support to this model. The lateral-line organ of Xenopus laevis,
after exposure to 2 x 10^{-3} M AgNO$_3$ solution, showed an intense
silver L line in X-ray analysis of the surface of the hair cell. In
conclusion, a soft receptor site of the membrane must include some
sulfur containing protein, which soft cations can selectively
attack.

Hard sites. There are two candidates for the role of the hard site
on the receptor membrane. One of these is the carboxylic group polar
amino acid and the other is a phospholipid. The importance of the

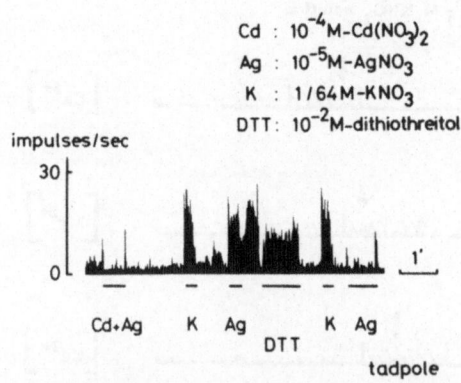

Fig. 8. The effect of DTT on that of KNO_3 and $AgNO_3$ at the lateral-
line receptor. Ordinate, discharge rate per second. Each
horizontal bar at the bottom indicates the duration of
stimulation. After application of distilled water the effect
of Ag^+ remained the same, but soon after 10^{-2} M DTT application the discharge rate of the nerve fiber returned to the
original spontaneous level. After application of DTT, the
response to K^+ was the same as before, but that to Ag^+ was
not. The last small response was probably the mechanical
response to rinsing water.

carboxylic group was also suggested by a study of excitable membrane.
The sodium channel might have a carboxylic group at the mouth of the
channel. An excitable sodium channel, however, has severe ion selectivity so that the carboxylic group is necessary, but not sufficient
to explain a complete model of the chemoreceptive mechanism.

Inositol phospholipids have been known to be present in a wide
variety of mammalian tissues and their various physiological activities have been noticed. Triphosphorinositide (TPI) in particular
has been regarded to be an important constituent of the nervous
system because of its high turn over rate. TPI has unique physico-
chemical characteristics. Since it has five negative charges at
physiological pH, it is easily dissolved in water. It shows an extra-
ordinarily high affinity for calcium ion which in turn results in
its becoming highly hydrophobic. On the other hand, in the presence
of K^+ it becomes hydrophilic. Thus TPI shows drastic changes in
hydrophilicity, when univalent-divalent cation exchange reactions
occur.

A hypothesis is proposed to explain the molecular mechanism of the chemical stimulation by mono- and divalent cations, and the suppressive effect of divalent cations on monovalent cation stimulation. (Hayashi, 1977).

In the natural state, TPI in the receptor membrane may exist as a TPI-Ca complex. When the K^+ concentration in the surrounding medium is increased, Ca in the TPI is replaced by K and membrane conductance begins to increase. When the K concentration exceeds a critical value, TPI is abruptly changed from its hydrophobic state to hydrophilic state. This results in rapid changes in membrane conductance or potential.

If a receptor membrane which contains TPI is treated by concentrated Ca^{++}, the receptor membrane conductance will drop gradually from the value it has in a high monovalent ion concentration. This is the reverse process of Ca-K exchange. Suppression of the monovalent chemical effect by a divalent cation may occur in this way.

The stimulating effect which divalent cations have on chemical receptors may be more complex. In order to change the membrane conductance, a stimulating metallic cation should replace Ca^{++} on the monoester phosphorus of inositol ring. In this case, conformational changes, such as hydrophobia to hydrophilia cannot be expected. Consequently, gradual changes in conductance are possible, since conductance will change when different cations replace Ca^{++} at the site. Moreover, the diester part of TPI takes part in the reaction between a metallic ion and TPI. All of the phenomena observed in the lateral-line organ experiments can be explained by using the TPI conformational change model. The electrical response of hair cells to mechanical stimulation still remains unclear, even after the TPI model is introduced. This kinds of hypothesis could be tested with artificial membrane systems.

Recently Corey and Hudspeth (1979) reported on mechanosensitivity relations to membrane potential and ion content. The use of voltage clamp technique on the single hair cell is quite valuable. Our TPI model of the hair cell membrane is well valid for their conclusion.

In my abstract for this symposium it is told that the origin of cochlear microphonics might not be electric resistance change in the top surface of the hair cell, but rather be ion exchange with complex proteins or lipids in the membrane. At present it will be better to say that resistance change of cell membrane may not be the whole mechanism but a part of its mechanism.

Electron microscopists in Japan revealed recently the microstructure of hair cells of the inner ear. I will show you some of their fine works. The main part of my present idea came principally

from their results. Hama uses the freeze fracture method and Naga-
saka uses one million volt ultra high voltage electron microscopy.
We can now discuss the molecular structure of those tissues.

For those of you who would like more details about any part of
my today's discussion, I refer you to my book "Mechanisms of Sound
Reception" which is now in preparation.

REFERENCES

Beidler, L. M., 1954, A theory of taste stimulation, J. Gen.
     Physiol., 38: 133-139.
Corey, D. P. and Hudspeth, A. J., 1979, Ionic basis of the receptor
     potential in a vertebrate hair cell, Nature, 281: 675-677.
Friedberg, F., 1974, Effect of metal binding on protein structure,
     Q. Rev. Biophys., 7: 1-33.
Hayashi, K., 1977, Role of the water-soluble complex lipids in the
     biological membrane, Membrane, 2: 86-96 (in Japanese)
Katsuki, Y., Hashimoto, T. and Yanagisawa, K., 1970, The lateral-
     line organ of shark as a chemoreceptor, Adv. Biophys., 1:
     1-51, Univ. Tokyo Press.
Katsuki, Y., Hashimoto, T. and Kendall, J. I., 1971, The chemo-
     reception in the lateral-line organs of teleosts, Jap. J.
     Physiol., 21: 99-118.
Pearson, R. G., 1968, Hard and soft acids and bases, HSAB part 1,
     J. Chem. Education, 45: 581-587.
Pearson, R. G., 1973, "Hard and soft acid and base". R. G. Pearson
     ed., Dowden, Hutchinson and Ross. Inc., Stroudsburg, Pa.
Yanagisawa, K., Taglietti, V. and Katsuki, Y., 1974, Responses to
     chemical stimuli in the hair cells of the lateral-line organ
     of mudpuppy, Proc. Jap. Acad., 50: 526-531.
Yoshioka, T., Kawai, K. and Katsuki, Y., 1975, The chemisorption
     of the cation to the chemoreceptor membrane, Seitai no Kagaku,
     26: 338-346. (in Japanese)
Yoshioka, T., Kawai, K. and Katsuki, Y., 1976, The receptive
     mechanism of various metallic ions in the lateral-line organ
     of the tadpoles of Rana catesbeiana, Jap. J. Physiol., 26:
     441-453.
Yoshioka, T., Asanuma, H., Yanagisawa, K. and Katsuki, Y., 1978,
     The chemical receptive mechanism in the lateral-line organ,
     Jap. J. Physiol., 28: 557-567.

THE PARASYMPATHETIC INNERVATION OF THE INNER EAR AND THE PROBLEM

OF COCHLEAR EFFERENTS: ENZYME AND AUTORADIOGRAPHIC STUDIES

Muriel D. Ross and H. Roger Jones

Dept. of Anatomy
University of Michigan
Ann Arbor, Mi., U.S.A.

It has been assumed for many years that the matter of the existence of olivocochlear and vestibular efferents terminating on hair cells in the end organs of the inner ear is non-controversial. Opposing concepts that the centrally originating efferent fibers of the cochlear and vestibular nerves are parasympathetic (Ross 1969; Eyries and Chouard, 1970), and that the efferent-type terminals on hair cells might be collateral branches of spiral and vestibular ganglion nerve fibers (Ross, 1973) attracted little notice. However, two recent anatomical observations have served to refocus attention on these ideas. These are 1) the finding of multipolar neurons with synapses on their dendrites in human spiral ganglia (Kimura et al., 1979; Ota and Kimura, 1980), which supports the concept of a parasympathetic innervation of the inner ear; and 2) the observation that reciprocal synapses occur on outer hair cells in the human organ of Corti (Nadol, 1980), which substanties the notion that both afferent and efferent-type terminals can originate from the same parent nerve fiber. Additionally, a combined histochemical-physiological study (Goldberg and Fernandez, 1980) has shown that vestibular efferents originate bilaterally in the brain stem, lateral to the facial genu, in the monkey. This site corresponds to the general location of the lacrimal nucleus in primates (Chouard, 1962; Crosby and DeJonge, 1963). The elegant study of Goldberg and Fernandez further showed that stimulation of the vestibular efferents resulted in excitatory rather than in inhibitory influences on afferent vestibular output. Relatively long latent periods were required, even when the responses were characterized as "fast". These results taken together suggest that classical interpretations of the anatomical organization and functional significance of the inner ear efferents require reassessment.

In this paper we discuss the central origin and pathway of the
cochlear and vestibular efferents in the rat based upon histochemical observations, and reveal for the first time some autoradiographic evidence in support of the concept that the efferent-type
terminals in the organ of Corti are of peripheral origin. For part
of this work, the brain stems of rats 21-30 days of age were prepared by a modification of the Koelle technique for acetylcholin-
esterase detailed elsewhere (Ross, 1969). For the autoradiographic
studies, guinea pigs (150-300 g) were used. Animals were injected
with 1 quantities of ^{3}H-proline into the spiral ganglion of the
basal cochlear turn. Two to five hours later the ears were fixed
in 4% paraformaldehyde in Millonig buffer (pH 7.3) to minimize
fixation of free amino acids. After decalcification (Baird et al.,
1967) and embedment, tissues were prepared for light and electron
microscopic autoradiography, using up to one year exposure in the
cold (4°C) and dark for the ultrathin sections used in the ultrastructural portion of the study.

Histochemical results showed that the general visceral efferent
cell column located at the level of the facial nucleus in the rat
contained the cells of origin of both the inferior and the superior
salivatory components. That part of the complex giving origin to
fibers that later traveled with the ninth cranial nerve were located
in a compact mass at the caudal end of the cell column. Their axons
exited from the brain stem caudoventrally to join the ninth cranial
nerve. The neurons that projected their axons with the seventh
cranial nerve (the facial) were located more rostralward in the cell
column and were more scattered in the tegmentum (Fig. 1). This nucleus corresponds to what is called the "superior salivatory" nucleus in the rat, but it must be remembered that this is an all
inclusive term. Subgroups providing preganglionics for lacrimal and
nasal glands cannot be distinguished even though they must be
present in the nucleus. Neither can the precise locus of the preganglionic neurons supplying the inner ear be determined.

Some of the axons of these preganglionic neurons or else their
collaterals were sent across the midline in a decussation between
the facial genus. Crossed and uncrossed components gathered immediately ventral to the ascending portion of the facial genu on each
side. They formed several fascicles and then coursed across the
dorsolateral tegmentum. Some fibers turned ventralward out of these
fascicles at several points along the way, and provided a major
contribution to the motor intermediate nerve. The remaining fibers
continued lateralward to reach the upper part of the spinal tract
of the trigeminal nerve where they turned ventralward medial to
the sensory facial and the vestibular nerve roots. A small component
joined the motor intermediate nerve but the bulk of these fibers
anastomosed with the vestibular and then with the cochlear nerves.
These anastomoses correspond to the acousticofacial anastomoses
described first by Arnold (1851). The further, peripheral courses

of these preganglionic fibers in the vestibular and cochlear nerves were similar to those already described for the mouse (Ross, 1969).

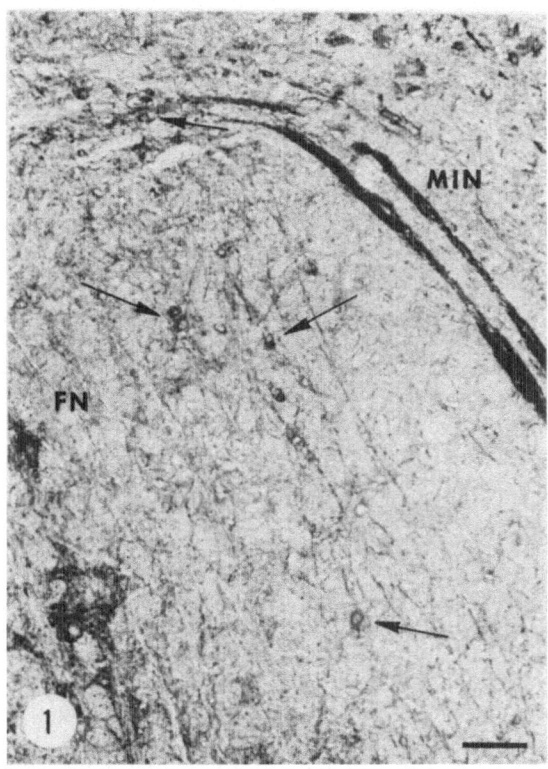

Fig. 1.   Preganglionic fibers destined for the inner ear arise in the superior salivatory nucleus in the rat. The rostral end of this nucleus has its small, acetylcholinesterase-positive, multipolar neurons (arrows) scattered in the dorsolateral tegmentum. MIN, motor intermediate nerve root fascicles; FN, facial nerve fibers coursing dorsorostrally into the ascending limb of the facial genu. The bar equals 0.1 mm.

Ross (1973) proposed in a prior report that the acetylcholinesterase-positive fibers reaching the inner ear from the brain stem were not continuous with the reactive terminals in the organ of Corti, and suggested that collaterals of spiral ganglion nerve fibers supplied both the afferent and the efferent-type terminals. Our autoradiographic study was designed to test this concept.

It is well known that only neuronal cell bodies and not fibers of passage are able to take up amino acids for synthesis into pro-

teins which are then transported into the cell's processes and
terminals. After our experimental procedures in which radioactive
proline was placed in the spiral ganglion of the basal cochlear
turn, we observed bidirectional transport of labeled material away
from the spiral ganglion cells. Peripherally, the labeling extended
into the organ of Corti of the basal turn only: the apical cochlear
turn did not show radioactivity. Because light microscopic autoradiography could not resolve the issue of the precise localization
of the silver grains in the organ of Corti, transmission electron
microscopy was next employed. Background labeling of the inter-
cellular spaces was very low at the ultrastructural level, indicating
that unincorporated amino acids had been washed away and that label
observed could be interpreted to be associated with proteins. Silver
grains were present over many efferent-type terminals ending on the
outer hair cells and over some afferent terminals (Fig. 2). However, outer hair cells also showed labeling, prominently over their
nuclei, but also over stereocilia and in the immediate vicinity of
synaptic clefts. Deiters'cells showed a few silver grains. The
number of silver grains present was small for all labeled structures
in our ultrathin sections; this precluded meaningful quantitative
analysis.

One interpretation of our findings that would be in agreement
with general autoradiographic observations is that the efferent-
type nerve endings were labeled by the transport of protein into
them from the cell bodies of neurons in the spiral ganglion. Because
silver grains were also commonly observed at the synaptic cleft
regions between efferent-type terminals and hair cell membranes,
and because resolution of the silver grains to the precise locations
of the radioactive source is not perfect, the question of possible
transfer of material or of breakdown products from the nerve endings
to the hair cells arises (Grafstein, 1971). On the other hand, it
would appear that diffusion of ^{3}H-proline to the organ of Corti of
the basal turn occurred and, indeed, may be impossible to prevent.
Thus, the question of a retrograde transfer of labeled material
from the hair cells to the nerve endings must also be considered,
even though there is little present evidence to support such a notion.

The sum of many recent anatomical and physiological observations as well as those reported upon here suggest that the innervation of the inner ear is enormously complex rather than a simple
matter of an afferent system under direct central, "efferent"
regulation. Autonomic influences on afferent nerve activity are
possible through presently little-understood mechanisms of fluid
and ion regulation in the inner ear. In light of our findings,
Warr and Guinan's (1975) light microscopic results cannot be considered to be conclusive evidence of labeling of efferent fibers after
central administration of radioactive material, especially in view
of the fact that, in their case, ultrastructural proof was lacking.

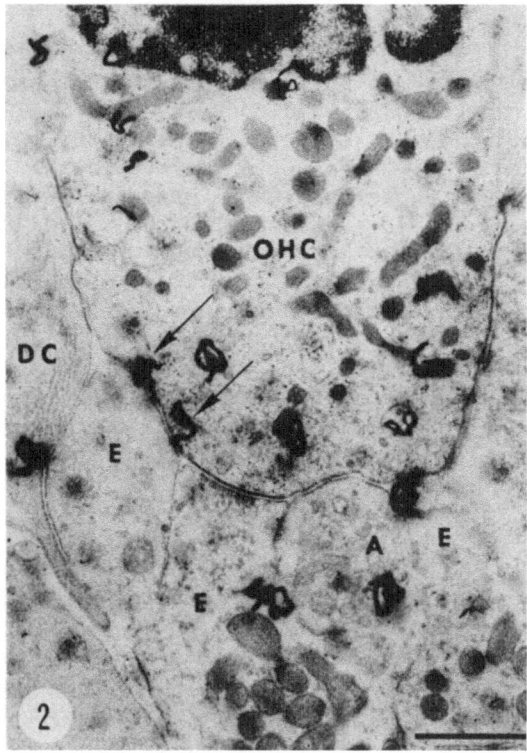

Fig. 2.   After injection of ^{3}H-proline into the spiral ganglion,
labeling occurred in some of the efferent terminals (E) in
the organ of Corti, and also in some nerve fiber endings
that appear to be of afferent-type (A). Silver grains were
often found near or over synaptic clefts arrows, to the
left , and in the outer hair cells (OHC). DC, Deiters cells.
The bar equals 1 μm.

On the other hand, Goldberg and Fernandez's (1980) careful
investigation may indicate that, in the primate, the centrally located vestibular efferent neurons, which we consider to be autonomic,
have their cell bodies among the "lacrimal" group of parasympathetic
preganglionic neurons of the seventh cranial nerve.

ACKNOWLEDGMENTS

Supported by Public Health Service grant NS 13428.

REFERENCES

Arnold, F., 1851, "Handbuch der Anatomie des Menschen", vol. 2, Herdersche Verlagshandlung, Freiburg im Breisgau.

Baird, I. L., Windborn, W. B. and Bockman, D. E., 1967, A technique of decalcification suited to electron microscopy of tissue closely associated with bone, Anat. Rec., 159: 281-289.

Chouard, C.-H., 1962, Recherches sur l'organisation intra-axiale des formations motrices et parasympathiques du nerf facial, These Med., Paris.

Crosby, E. C. and DeJonge, B. R., 1963, Experimental and clinical studies of the central connections and central relations of the facial nerve, Ann. Otol. Rhin. Laryng., 72: 735-755.

Eyries, C. and Chouard, C.-H., 1970, Les anastomoses acoustico-faciales, Ann. Otolaryngol. (Paris), 87: 321-326.

Goldberg, J. M. and Fernandez, C., 1980, Efferent vestibular system in the squirrel monkey: anatomical location and influence on afferent activity, J. Neurophysiol., 43: 986-1025.

Grafstein, B., 1971, Transneuronal transfer of radioactivity in the central nervous system, Science, 172: 177-179.

Kimura, R. S., Ota, C. Y. and Takahashi, T., 1979, Nerve fiber synapses on spiral ganglion cells in the human cochlea, Ann. Otol., 88: Supplement 62: 1-17.

Nadol, J. B., Jr., 1980, Reciprocal synapses at the base of outer hair cells in the organ of Corti of man, Ann. Otol., (in press)

Ota, C. Y. and Kimura, R. S., 1980, Ultrastructural study of the human spiral ganglion, Acta Otolaryngol., 89: 53-62.

Ross, M. D., 1969, The general visceral efferent component of the eighth cranial nerve, J. Comp. Neurol., 135: 453-477.

Ross, M. D., 1973, Autonomic components of the VIIth nerve, Adv. Oto-Rhino-Laryng., 20: 316-336.

Warr, W. B. and Guinan, J. J., Jr., 1979, Efferent innervation of the organ of Corti: two separate systems, Brain Res., 173: 152-155.

ADAPTATION AND DYNAMIC RESPONSE OCCURRING AT HAIR CELL-AFFERENT

FIBER SYNAPSE

T. Furukawa and S. Matsuura

Department of Physiology, Tokyo Medical
and Dental University, Tokyo
Department of Physiology, Osaka City University
Osaka, Japan

The rate of afferent discharges in the cochlear nerve increases at the start of sound, but gradually declines thereafter until it reaches a steady level. Upon cessation of the sound, the rate goes down for a while below the level of spontaneous firing (Harris and Dallos, 1979). Also when a small increment or decrement is added to the intensity of the sound, afferent discharges show characteristic dynamic changes (Smith and Zwislocki, 1975). Since no corresponding changes were found in the microphonic potentials, i.e., in the activity of sensory hair cells, it is generally assumed that these adaptive phenomena may occur at the synapse between hair cells and afferent fibers. We report here results on the goldfish saccule, in which adaptive processes with the above mentioned properties were demonstrated as changes in the amplitude of the excitatory postsynaptic potentials (e.p.s.p.s) produced in the primary afferent fiber terminals. We analyzed the phenomena in some detail and reached the conclusion that the observed adaptive and dynamic changes are produced in relation to the mechanism of the transmitter release at the afferent synapse (Furukawa and Matsuura, 1978; Furukawa et al., 1978).

Experiments were carried out in a sound-proof room on anesthetized goldfish about 12 cm long. E.p.s.p.s were recorded intracellularly with glass pipette microelectrodes from large afferent fibers which distribute over the rostral part of the saccular macula. Tetrodotoxin (10^{-5} g/ml) was applied locally to block spike potentials. The sound stimulus (frequency, 400-500 Hz; duration, 50-200 ms) was delivered once every one or two seconds from a loudspeaker placed in front of the fish. To observe the dynamic changes

in the amplitude of the e.p.s.p.s, an increment or decrement was
given to the sound intensity 50-100 ms after its start.

Fig. 1 shows the relationship between the amplitude of the
e.p.s.p.s and sound intensity under different conditions. The left-
most curve (CON) shows the intensity-response relation at the start
of sound. It rises as a sigmoid curve with increase of the sound
intensity until it saturates at about 96 dB SPL. As shown in inset

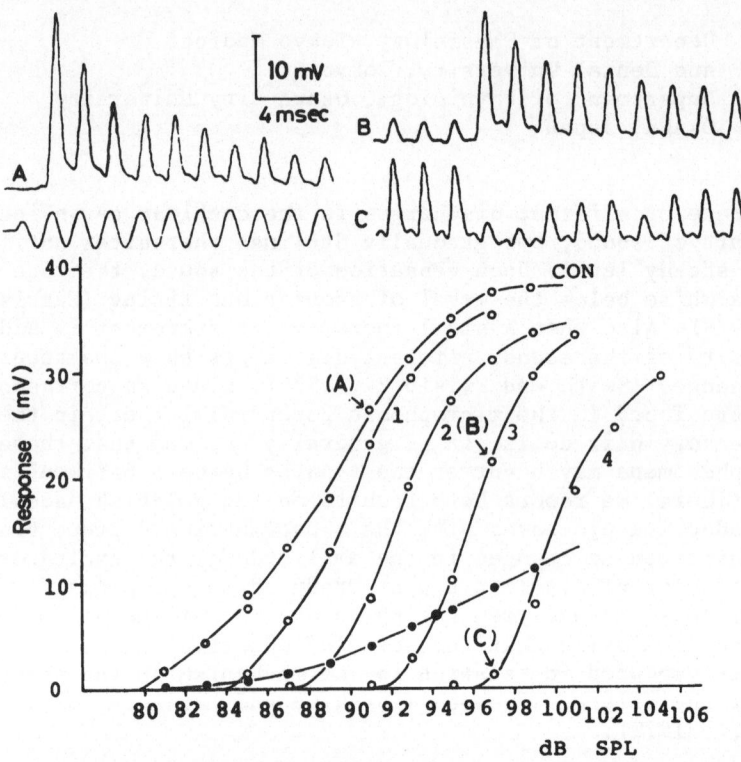

Fig. 1.   Amplitude of the e.p.s.p.s plotted against the sound inten-
           sity under different conditions. CON, input-output relation
           for e.p.s.p.s at the start of sound; filled circles, steady
           state level of e.p.s.p.s; curves 1-4, dynamic responses
           obtained for different intensities of adapting sound.
           Intensity of adapting sound, 85, 89, 94 and 99 dB SPL for
           curves 1 to 4 respectively. Inset records show (cont´d)

(Fig. 1 cont'd) sample records for the start of sound (A),
incremental response (B) and decremental response (C).
Sound intensity: A, 91 dB; B, increased from 94 to 97 dB;
C, decreased from 99 to 97 dB SPL. Frequency, 500 Hz.

A, the e.p.s.p.s run down from the amplitude at the start of sound
to much smaller ones. The steady state level of the e.p.s.p.s for
different sound intensities are shown with filled circles in Fig. 1.

The dynamic responses of this fiber to a step increment or
decrement of the sound intensity are shown by open circles in
curves 1 to 4. Intensity of the adapting sound was 85, 89, 94 and
99 dB SPL for points plotted in curves 1, 2, 3 and 4 respectively.
Amplitudes of the e.p.s.p.s were measured from the base-line prior
to the sound application. In response to an increment of the sound
intensity, e.p.s.p.s increased temporarily to a size much larger
than the sustained level for the new sound level (inset B). The
decremental responses are just as marked. The e.p.s.p.s decrease
in size to a level much lower than that of the sustained e.p.s.p.s.
After dropping in size, the e.p.s.p.s started to increase in size
roughly exponentially toward the new sustained level (inset C).
But the recovery time course was not the same for different magnitude of decremental steps, being much slower for a greater decrement.

It must be noted that four intensity-response curves (curves
1-4) in an adapted state run parallel each other. This was regular-
ly observed in all the cases so far studied (12 cases in a total).
But, the curve for the start of sound (CON) was less steep in most
cases. This finding seems to indicate that the input-output relation of hair cells rises more gradually in a weak stimulus intensity
range. The parallel shift of response curves indicates that adaptation brings about a real change in the sensitivity. The shift of
the response curve produced by adaptation approximately amounted
to 13 dB in the case of Fig. 1. The same range of values was encountered in most other cases studied, although in a single case a shift
of 23 dB was observed.

It must be noted that the results such as shown in Fig. 1 are
qualitatively very similar to adaptive behaviors observed in other
sensory systems. For example, our results resemble those observed
in photoreceptor cells, although the range of adaptation is more
limited than in the latter (Normann and Werblin, 1974). Despite the
similarity in over-all appearance, the mechanisms underlying these
adaptive effects are quite different in these two organs. Namely,
adaptation in photoreceptor cells mostly occurs in the process
leading to the generation of receptor potentials, while the adaptation in hair cell organs takes place at the afferent synapse, i.e.,
at a stage later than the generation of receptor potentials.

As is well known, the transmitter release from sensory hair cells is graded to the intensity of the stimulus. It is the mechanism for this graded release that is most intimately related to the adaptive phenomena. To begin with, let us consider the problem of how the input-output curve for the start of the sound appears as a sigmoid curve. In our multiple release sites model, we assume the presence within single hair cells of many release sites whose sensitivities for release are different among different release sites (Furukawa and Matsuura, 1978). That is to say, upon a weak depolarization of the hair cell membrane, the transmitter may be released only from the most sensitive sites. As the sound intensity is increased, however, more release sites may be brought into activity. The rising phase of the input-output curve can be explained in this way. On the other hand, the saturation of the response curve seems to occur at the postsynaptic membrane by the effect of non-linear summation of the e.p.s.p.s. It is shown in Fig. 1 that the amplitude of the e.p.s.p.s at the saturation level is close to 40 mV. But there are grounds to believe that the amplitude of the e.p.s.p.s at the site of their generation may be a little larger than this. We found that the input-output curve could be greatly extended toward the stronger intensity range by hyperpolarizing the postsynaptic membrane with anodal current flowing through the recording microelectrode. Therefore, it seems quite likely that the amount of transmitter released from hair cells may increase with the sound intensity over a very wide range.

Now, the rundown of the e.p.s.p.s may be accounted for from the depletion of the available transmitter quanta at the presynaptic release sites. In our multiple release sites model, depletion should occur only in those release sites whose threshold are below the intensity of the stimulus used. The incremental response (inset B, Fig. 1) can readily be explained with the present model, for it is expected that the release sites whose threshold is above the intensity of the adapting sound should remain intact and be available for release in response to an increase in the sound intensity.

To substantiate these reasonings, we resorted to statistical analysis based on the quantal release hypothesis. We thus determined the value of release parameters $\underline{m}$, number of transmitter quanta released in an e.p.s.p.; $\underline{n}$, number of available release sites at the presynaptic membrane; and $\underline{p}$, release fraction. These parameters are not independent, since the relation, $\underline{m} = \underline{n} \times \underline{p}$, holds. Results of these analyses are summarized as follows: 1) Intensity of the sound influenced the amplitude of the e.p.s.p.s. mostly by changing the value of parameter $\underline{n}$; 2) adaptive decline in the amplitude of the e.p.s.p.s was attributed to a decline in the value of $\underline{n}$; 3) an increment (or decrement) in the sound intensity brought about an increase (or decrease) in the amplitude of the e.p.s.p.s by increasing or decreasing the value of $\underline{n}$. Thus, throughout these different occasions, the parameter that changed was $\underline{n}$, while parameter $\underline{p}$

stayed generally unchanged. These results generally conform to our multiple release sites model (Furukawa et al. 1978; Furukawa and Matsuura, in preparation).

In the above we briefly described how adaptation and dynamic responses in fish ear are produced at the afferent synapse with mechanisms that are closely related to the phenomena of graded transmitter release. It is not clear yet, however, to what extent our results will be applicable to the mammalian cochlea. Detailed mechanisms about synaptic transmission and impulse generation in the cochlea must be elucidated.

REFERENCES

Furukawa, T. and Matsuura, S., 1978, Adaptive rundown of excitatory post-synaptic potentials at synapses between hair cells and eighth nerve fibres in the goldfish, J. Physiol., 276: 193-209.

Furukawa, T., Hayashida, Y. and Matsuura, S., 1978, Quantal analysis of the size of excitatory post-synaptic potentials at synapses between hair cells and afferent nerve fibres in goldfish, J. Physiol., 276: 211-226.

Harris, D. M. and Dallos, P., 1979, Forward masking of auditory nerve fiber responses, J. Neurophysiol., 42: 1083-1107.

Normann, R. A. and Werblin, F. S., 1974, Control of retinal sensitivity. I. Light and dark adaptation of vertebrate rods and cones, J. Gen. Physiol., 63: 37-61.

Smith, R. L. and Zwislocki, J. J., 1975, Short-term adaptation and incremental responses of single auditory-nerve fibers, Biol. Cybernetics, 17: 169-182.

PHASE VERSUS FREQUENCY IN CAIMAN PRIMARY AUDITORY FIBRES: IS THERE A TRAVELLING WAVE?

J. Smolders and R. Klinke

Zentrum Physiologie
J. W. Goethe-Universität
D 6000 Frankfurt, BRD

INTRODUCTION

Anatomically the inner ear of the caiman is similar to that of mammals (v. Düring et al., 1974). Its primary afferent auditory fibres also show a number of properties similar to those of mammals (Klinke and Pause, 1977, 1980). A major difference is the shift of the characteristic frequency of caiman auditory fibres towards lower frequencies with decreasing body temperature (Klinke and Smolders, 1977). This phenomenon was also shown in other lower vertebrates (toad, Moffat and Capranica, 1976 and gecko, Eatock and Manley, 1976), but does not occur in cat primary fibres (Smolders and Klinke, 1977). Likewise, no, effect of temperature shift could be found in humans possessing genuine absolute pitch (Emde and Klinke, 1977).

Therefore we wondered if the basic mechanical properties of the inner ear of submammalian and mammalian vertebrates are the same. This hypothesis is supported by Peake and Ling (1979) who doubt the existence of mechanical tuning in the basilar papilla of the alligator lizard on the basis of their Mössbauer measurements.

We used the method of Pfeiffer and Kim (1975) and Kim and Molnar (1979) to check whether the vibration patterns in the caiman inner ear are similar to those shown in the cat. Their method has the disadvantage that it is an indirect one but offers the advantage that the inner ear need not be disturbed and that information about a large portion of a single basilar papilla can be obtained.

43

METHODS

As many single primary fibres with different CFs as possible have to be collected from one ear in order to cover a large enough range of the basilar papilla. The test stimulus, a pure tone of fixed frequency and intensity is identical for each fibre. So in all cases identical vibration patterns of the basilar membrane are achieved as any phase and amplitude distortions introduced through the acoustic system are identical for all fibres.

Three caiman (Caiman crocodilus) were used. Technical details are described in Klinke and Pause (1980). Ten different test stimuli were used: 88, 177, 354, 707 or 1414 Hz at 25 dB attenuation (0 dB attenuation corresponds to 110 dB SPL). Best frequency was determined using iso-intensity-frequency contours recorded at about 20 dB above threshold. (BF$\doteq$CF). Additionally spontaneous rate and click responses were recorded. Period histograms were computed in 256 bins /period and Fourier-transformed. Four response measures were used: Spontaneous discharge rate (SR), average response rate (R$\emptyset$), amplitude and phase of the first harmonic of the Fourier transform (R1 and P1), SR, R$\emptyset$ and R1 are measured in spikes/s; P1 in radians. Phase was calculated relative to the positive zero crossing of the electrical signal. Details are found in Kim and Molnar (1979).

From these four response measures average rate (R$\emptyset$/SR) and phase locking response measures (R1/SR or R1/R$\emptyset$) were derived to allow comparison between the different fibres. These response measures were plotted as a function of BF since the spatial distribution of CF on the caiman basilar papilla is unknown. We assume, however, that the basilar papilla is tonotopically organised. Data point selection depended upon estimates for the standard deviation of amplitude RLIMIT = $\sqrt{2}$/N and phase PLIMIT = 1/A . RLIMIT where A is the normalized phase locking measure (R1/SR or R1/R$\emptyset$, see Littlefield, 1973, Pfeiffer and Kim, 1975). Values of RLIMIT accepted were from 0.100 - 0.058 and of PLIMIT between 3.142 - 0.220 rad.

Click post stimulus time histograms were computed and the click latencies to condensation and rarefaction clicks measured.

RESULTS

The data shown are from 409 single neurones from one auditory nerve. Fig. 1 shows the results for two stimulus frequencies, 177 and 354 Hz both at 75 dB SPL. The value for the amplitude of the response can be taken as an indirect measure for the vibration amplitude of the papilla. The phaselag between stimulus and response varies systematically in the region of best frequencies close to the stimulus frequency. The phaselag decreases towards higher best frequencies. Fig. 1A, stimulus frequency 177 Hz, clearly shows that

in regions of best frequencies far above the stimulus frequency
there is hardly any phase change with best frequency. The slopes in
the region of BFs close to the stimulus frequency are 1.2 πrad./oct
(88 Hz), 1.7 πrad./oct (177 Hz), 1.8 πrad./oct (354 Hz), 2.2 πrad./
oct (707 Hz). A slope at 1414 Hz could not be calculated.

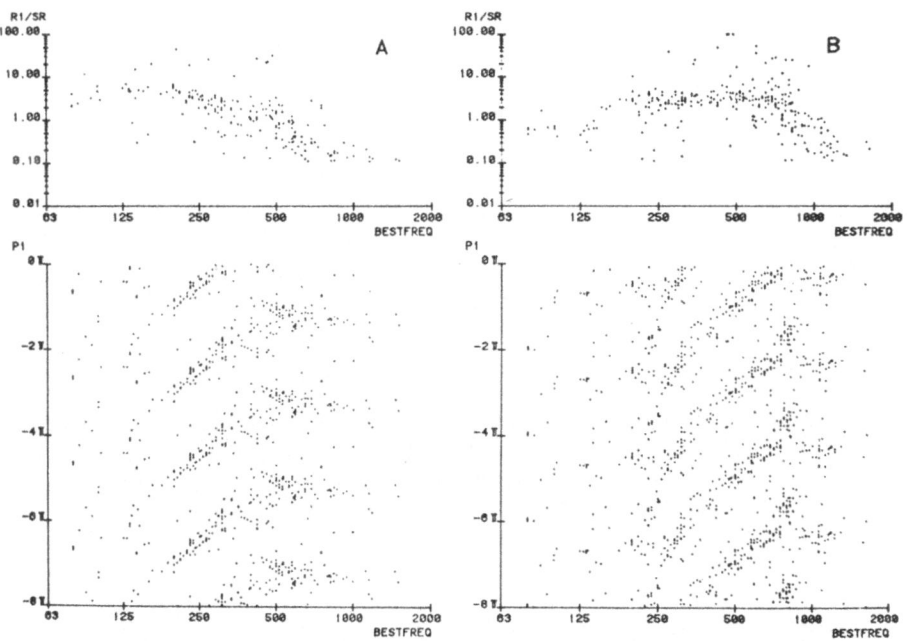

Fig. 1.   Amplitude and phase of phaselocked response of primary
          auditory fibres of one papilla of Caiman crocodilus to
          steady pure tone stimuli as a function of the best fre-
          quency of the fibres (≙location along the basilar papilla).
          R1/SR: Amplitude of the fundamental component of the
          Fourier transform of the period histogram of the response.
          Data from the different fibres are normalised by division
          by their spontaneous discharge rates. P1: Phaselag between
          the pure tone stimulus and the fundamental component of the
          neural response. Negative values represent phaselag. The
          data are plotted modulo·2π (each data point occurs 4 times,
          each time shifted by 2 π rad. A: Stimulus 177 Hz, 75 dB
          SPL. B: Stimulus 354 Hz, 75 dB SPL. Selection criteria:
          RLIMIT 0.058, PLIMIT 0.44.

     Out of 409 fibres 348 were tested with both condensation and
rarefaction clicks at 100 dB SPL. Compound PST histograms were
calculated. In 181 out of these 348 fibres a first peak was clearly

visible in the PSTH. Out of these fibres 60 had the shortest latency
with rarefaction clicks, 121 with condensation clicks. Fig. 2 shows
the amplitude and phase for a stimulus frequency of 177 Hz, of those
neurones, presented in Fig. 1A for which click data were available
and a decision could be made whether response latency was shortest
for condensation or for rarefaction clicks. "Rarefaction units" and
"condensation units" do not show a separation.

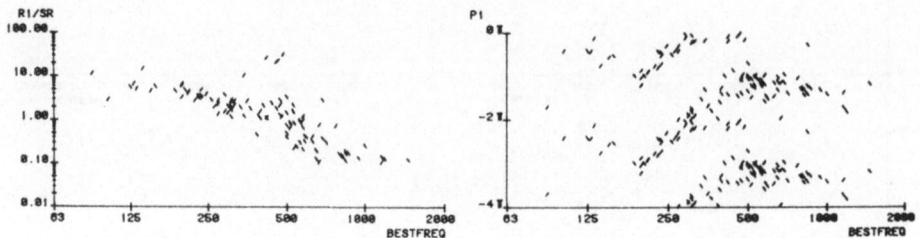

Fig. 2.   Amplitude (left) and phaselag (right) of phaselocked re-
          sponses of primary auditory fibres of one papilla of Caiman
          crocodilus as a function of the fibres best frequency
          ($\doteq$location along the basilar papilla). The fibres are
          labelled according to their responses to click stimuli at
          100 dB SPL: / = Fibres responding with shortest latency to
          rarefaction clicks. \ = Fibres responding with shortest
          latency to condensation clicks. Stimulus 177 Hz, 75 dB SPL
          pure tone. Selection criteria: RLIMIT 0.058, PLIMIT 0.44.

CONCLUSIONS

     The phase data measured in the caiman show properties compar-
able with those obtained in cat (Pfeiffer and Kim, 1975, Kim and
Molnar, 1979) and thus fulfil a necessary, although not a sufficient,
criterion for the existence of a travelling wave in caiman.

     The scatter in the data is larger than in cat although the same
selection criteria as in cat were used (Pfeiffer and Kim, 1975).

     The difference in innervation pattern of the caiman hair cells
may explain for this since afferent fibres from IHC contact about 4
receptor cells and those from OHC about 20 (v. Düring, pers.
commun.).

Klinke and Pause (1980) could separate their fibres according to the click latency: 34% responding to condensation clicks with shortest latency, the remainder to rarefaction clicks. In our phase data separation of the neurones according to type of click response (at 100 dB SPL) does not lead to two distinct groups in the phase plots as would be expected if two populations of neurones reacting to opposite phases of the stimulus would exist. Moreover in our data about 33% of the cells where a clear peak was discernable in the PSTH respond first to condensation clicks. The method used seems inadequate to solve the problem whether two fibre types with respect to click responses exist. At higher intensities new earlier peaks in the PSTH´s may occur or non detectable ones may grow large enough. Still it is striking that this phenomenon is not described for mammalian primary fibres where at higher intensities all neurones react with shorter latency to rarefaction clicks.

ACKNOWLEDGMENT

This work was supported by the Deutsche Forschungsgemeinschaft (SFB 45)

REFERENCES

During v., M., Karduck, A. and Richter, H.-G., 1974, The fine structure of the inner ear in Caiman crocodilus, Z. Anat. Entwickl.-Gesch. 145: 41-65.

Eatock, R. A. and Manley, G. A., 1976, Temperature effects on single auditory nerve fiber responses, J. Acoust. Soc. Am., 60: S 80.

Emde, C. and Klinke, R., 1977, Does absolute pitch depend on an internal clock? in: "Inner Ear Biology", M. Portmann and J. M. Aran, eds., INSERM, Vol. 68, Paris.

Kim, D. O. and Molnar, C. E., 1979, A population study of cochlear nerve fibers: Comparison of spatial distributions of average-rate and phase-locking measures of responses to single tones, J. Neurophysiol.,42: 16-30.

Klinke, R. and Pause, M., 1977, The performance of a primitive hearing organ of the cochlea type: Primary fibre studies in the caiman, in: "Psychophysics and Physiology of Hearing", E. F. Evans and J. P. Wilson, eds., Academic Press, London.

Klinke, R. and Pause, M., 1980, Discharge properties of primary auditory fibres in caiman crocodilus: Comparisons and contrasts to the mammalian auditory nerve, Exp. Brain Res.,38: 137-150.

Klinke, R. and Smolders, J., 1977, Effect of temperature shift on tuning properties, Addendum to Klinke and Pause.

Littlefield, W. M., 1973, Investigations of the linear range of the peripheral auditory system, D. Sc. Thesis, Washington University, St. Louis.

Moffat, A. J. M. and Capranica, R. R., 1976, Effects of temperature
    on the response properties of auditory nerve fibers in the
    American toad (Bufo americanus), J. Acoust. Soc. Am., 60:
    S 80.
Peake, W. T. and Ling, A., 1979, Basilar membrane motion in the
    alligator lizard: Its relation to tonotopic organization and
    frequency selectivity, Submitted to J. Acoust. Soc. Am.
Pfeiffer, R. R. and Kim, D. O., 1975, Cochlear nerve fiber responses:
    Distribution along the cochlear partition, J. Acoust. Soc. Am.,
    58: 867-869.
Smolders, J. and Klinke, R., 1977, Effect of temperature changes on
    tuning properties of primary auditory fibres in caiman and cat,
    in: "Inner Ear Biology", M. Portmann and J. M. Aran, eds.,
    INSERM, Vol. 68, Paris.

SESSION II
COCHLEAR MECHANISMS B
CHAIRMAN: R. KLINKE

THE DYNAMICS OF pO_2-CHANGES IN THE PERILYMPHATIC PERFUSATE OF THE GUINEA-PIG COCHLEA DEPENDING ON THE METABOLISM

E.-J. Haberland, K.-D. Kuhl and P. Lotz

Department of Otolaryngology
Medical School, Martin-Luther University
Halle, GDR

We perfused the perilymphatic space of the guinea-pig cochlea with an artificial perilymph of a balanced ionic composition. The indicators for the metabolic situation of the cochlea are the cochlear microphonics (CM), the oxygen partial pressure and the concentration of different metabolites of the perfused artificial perilymph. CM and oxygen tension are recorded simultaneously. The fractioned perfusate drops are examined later biochemically. The animal is narcotized by urethane, relaxed and machine-made respired. Changes of the metabolic status are induced by respiring with chosen gas compositions or by admixtures to the artificial perilymph before perfusion.

The connection between the cochlea and the electrochemical oxygen measuring chamber is a one millimeter polyethylene tube of high flexibility. The oxygen diffusion through the tube wall into the perfusate before the pO_2 is measured must be considered. Absolute values of oxygen tension are obtained by using the following formula (Haberland et al., 1979):

$$C_E = C_A - (C_A - C_L) \exp (-k_s/V)$$

C_E and C_A are the oxygen tensions in the outgoing capillary and in the surrounding air, C_L is the measured pO_2, k_s the tube constant, calculated by a known value of C_E and V is the perfusion rate. The exponential dependence is shown in Fig. 1. Reducing the perfusion rate of an oxygen poor solution stepwise causes the same stepwise increase in the measured pO_2. Best agreement of the transient amplitudes shows the logarithm of V. A further source of errors is the peculiarity of the electrochemical oxygen measuring chamber, containing only 7 microliters (VEB Metra Radebeul, GDR). A curve

shift of some percent occurs following a variation of a practical
value of V (Fig. 2). In addition to this pure physical event we
should consider the physiological effect. Fig. 3 shows a similar

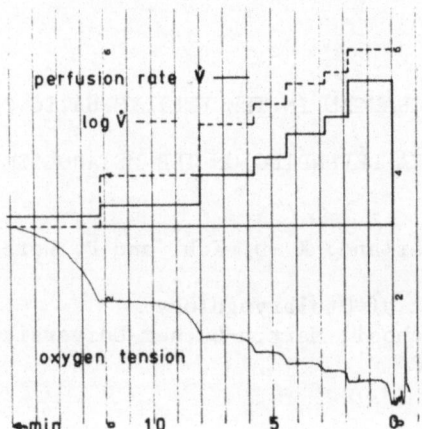

Fig. 1.   Relationship between the difusion rate and oxygen tension.

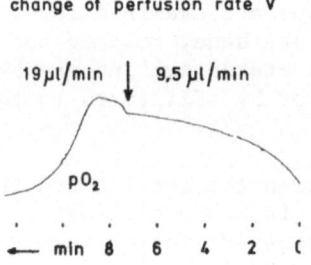

Fig. 2.   Shift in oxygen tension curve following changes in the
          perfusion rate.

behaviour as in Fig. 1 when the perfusion rate is increased. The
air-saturated artificial perilymph is flowing through the perilymphatic space of the cochlea and its pO_2 decreases by difusion to a
level below the arterial oxygen partial pressure, also dependent upon
the perfusion rate. The pO_2 increases again along the tube wall to
the oxygen detector. One may see in this curve the superposition of
both described effects. The small peaks are caused by the rate-
dependent detector oxygen consumption as shown before in Fig. 2.
The intracochlear difusion does not seem to be proportional to V,

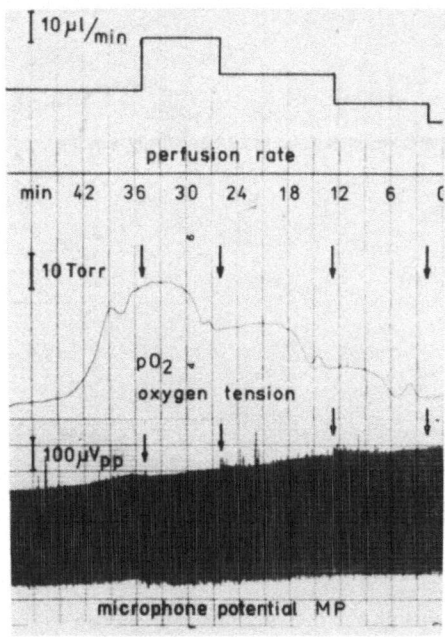

Fig. 3.   Influence of increased perfusion rate on oxygen tension and
          microphonic potentials.

as the last step shows. Therefore interpretable time-dependencies
of pO_2 registrations require constant perfusion rates.

The next figures show the application of this method in experiments with metabolic inhibitors. Cyanide ions block the oxygen
dependent metabolism. As consequence of such perfusion (Fig. 4)
the oxygen consuption of the cochlea decreases and the pO_2 level
approaches the input value of 100%. Iodine acetic acid blocks
anaerobic metabolism. Its perfusion in low concentrations stops
the decrease of CM, but ten millimols of this agent causes a further
CM-depression. Nitrogen breathing causes a typical decrease in the
pO_2 registration. The recurrent level during perfusion of 10 mM
iodine acetic acid is higher than the initial level.

Most rapid changes of pO_2 were found at the beginning and the
end of an asphyxia. Drop time and rise time were 1.5 minutes for
the 1/e-value. Respiration of pure nitrogen (Fig. 5) doubles this
time. Even a shock such as high KCN-concentration perfusion shows
a time constant in this region. The time for 95% intimation of the
oxygen detector is about 30 s.

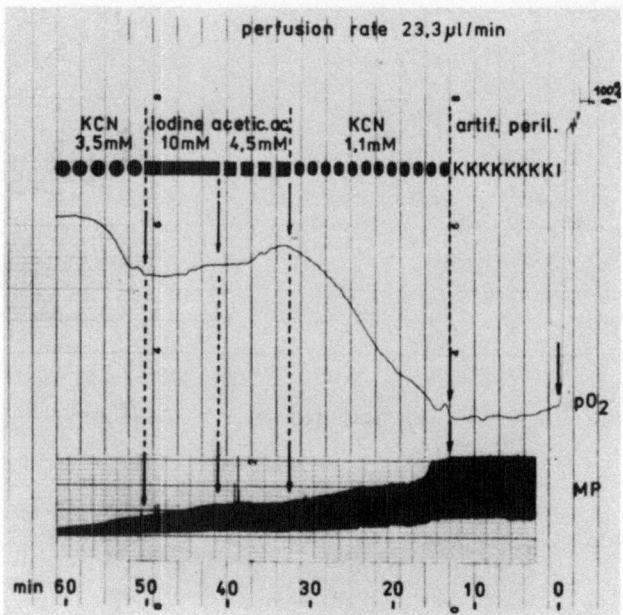

Fig. 4. Effect of metabolic inhibitors on oxygen tension and CM.

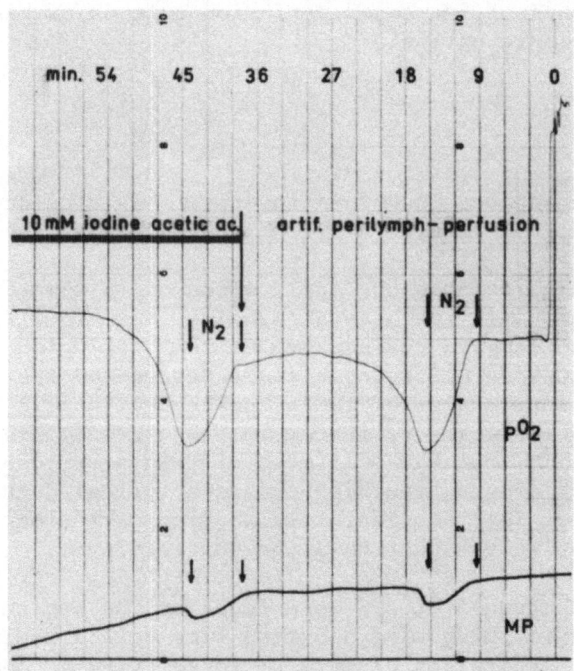

Fig. 5. Effect of respiration of nitrogen on oxygen tension and CM.

Results:

- oxygen tension measuring in this manner requires a correction of the registration curve to obtain absolute values
- the reason for short oscillations are often purely physical
- pO$_2$-changes in the minute region are influenced strongly by the time constants of the measuring system
- this method permits a continuous comparison between cochlear microphonics, oxygen consumption and, after biochemical analysis, the synchronous behaviour of some metabolites

REFERENCE

Haberland, E.-J., Kuhl, K.-D. and Lotz, P., 1977, Zu Problemen der perilymphatischen Cochleaperfusion, Wiss. Beitr. Martin-Luther-Univ. Halle, 43: 123-128.

ANALYTICAL STUDIES ON BIOCHEMISTRY AND PHYSIOLOGY OF PERILYMPH

(GUINEA PIG)

F. Scheibe

ENT Clinic, Humboldt University of Berlin

Schumannstr. 20-21, 1040 Berlin, GDR

INTRODUCTION

The genesis and the biochemical nature of perilymph (PL) –
ultrafiltrate from the cochlear blood vessels or/and open influx of
cerebrospinal fluid (CSF) –, and even more its metabolic function
for the organ of Corti, are basic questions of the biochemistry and
the physiology of PL which have not been solved completely yet
(Schindler, 1965; Gershbein et al., 1974; Lawrence, 1974; Schnieder,
1974; Jung, 1975, 1979; Lawrence et al., 1975; Kellerhals, 1976;
Scheibe et al., 1977; Cabezudo, 1978; Kommos and Giebel, 1978).
A contribution to solve this problem is a comprehensive biochemical
comparison between PL, CSF and blood plasma. The analytical data
available in the literature are insufficient to compare the chemical
compositions of these fluids exactly.

The present paper deals with comparative analytical studies of
seven selected chemical components in PL, CSF and blood plasma in
anaesthetized guinea pigs. The analytical data presented in a con-
densed review are discussed with regard to their meaning for the
biochemical nature and the genesis of PL.

MATERIALS AND METHODS

Total protein, total activity of lactate dehydrogenase (LDH),
sodium, potassium, glucose, pyruvate and lactate were studied
comparatively in PL of scala tympani and scala vestibuli, CSF and
arterial blood plasma (serum) of guinea pigs weighing 150-300g. The
animals used in the metabolic studies (glucose, pyruvate, lactate)
had fasted for 15-20 h before the experiments were started to

stabilize the blood glucose levels. The samples were obtained from living animals anaesthetized with ethyl urethane. Special care was taken to minimize possible artefacts during the sampling, mainly in the case of PL (Scheibe et al., 1976, 1977; Bergmann et al., 1979).

Lactate, pyruvate, glucose and total LDH were determined enzymatically using the fluorometric technique described previously (Haupt et al., 1979; Scheibe et al., 1979). Total protein was analyzed according to Lowry's micromethod (Scheibe et al., 1975). Sodium and potassium were measured simultaneously by micro flame photometry (Giebel and Scheibe, 1980).

In each case the mean value and the standard deviation for the number of samples indicated (Table 1) were calculated from the analytical data. The pyruvate and the lactate plasma figures were calculated from data of native blood samples. The mean values of the individual components in PL, CSF and plasma (serum) were tested statistically (t-test). The mean PL/CSF and PL/plasma concentration ratios (Table 2) were calculated from the mean concentrations of the components using PL values averaged from both scalae.

RESULTS AND DISCUSSION

The concentrations of the chemical components in PL of scala vestibuli and scala tympani, CSF and arterial blood plasma (serum) are compiled in Table 1. There are significant differences ($p < 0.01$) in total protein, total LDH, pyruvate and lactate between PL and both CSF and plasma (serum). The glucose difference between PL and plasma is also significant.

Table 1. Concentrations of the chemical components in perilymph of scala vestibuli (PL_v) and scala tympani (PL_t), cerebrospinal fluid (CSF) and arterial plasma (serum) in guinea pig

components	PL_v	PL_t	CSF	plasma[1]
total protein (g/l)	2.08 ±0.66(33)	1.78 ±0.76(20)	0.32 ±0.18(40)	43.06 ±8.94(71)
total LDH (U/l)	58 ±41(30)	74 ±47(24)	19 ±13(42)	115 ±52(31)
sodium[2] (mmol/l)	133 ±6(6)	143 ±4(5)	137 ±9(25)	130 ±5(21)
potassium[2] (mmol/l)	9.0 ±3.1(4)	4.9 ±1.8(4)	3.4 ±0.8(25)	5.0 ±0.8(21)

Table I (continued)

components		PL_v	PL_t	CSF	plasma[1]
glucose	(mmol/1)	3.4 $\pm$1.3(20)	3.4 $\pm$1.0(35)	3.3 $\pm$0.6(15)	7.3 $\pm$0.9(18)
pyruvate	(mmol/1)	0.35 $\pm$0.10(20)	0.34 $\pm$0.11(35)	0.20 $\pm$0.05(14)	0.12 $\pm$0.02(31)
lactate	(mmol/1)	3.9 $\pm$0.9(17)	4.0 $\pm$1.2(32)	1.3 $\pm$0.4(14)	1.4 $\pm$0.4(23)

[1] total protein, sodium and potassium were determined in serum

[2] PL data are only tentative

In order to compare the chemical composition of PL with both those of CSF and plasma (serum) the/PL/CSF and the PL/plasma concentration ratios of the individual components are presented in Table 2. According to the original CSF hypothesis of PL (open influx of CSF) the PL/CSF ratios of the investigated components should amount to about one. However, most of the ratios are markedly higher than one, and moreover, there are clear differences between the individual ratios. The concentrations of most of the components are in PL twice to six times higher than in CSF. That means the chemical compositions of both fluids differ clearly.

Table 2. Mean PL/CSF and PL/plasma concentration ratios of the chemical components in guinea pig

components	PL/CSF	PL/plasma
total protein	6.0	0.045
total LDH	3.4	0.56
sodium	1.0	1.1
potassium[1]	1.4	1.0
glucose	1.0	0.54
pyruvate	1.8	2.9
lactate	3.0	2.8

[1] scala tympani PL

On the other hand, according to the ultrafiltration hypothesis of PL (plasma ultrafiltration within the cochlea) the concentrations of the low molecular nonprotein-bound components (electrolytes and metabolites) should be in PL approximately the same as in blood plasma, e.g., the PL/plasma ratios of these components should also amount to about one. But as shown in Table 2 the metabolic ratios deviate markedly from one. The concentrations both of pyruvate and lactate are in PL about three times as high as in plasma, while the glucose is only a half of the plasma glucose. This means the metabolite concentrations of PL do not correspond to that of an ultrafiltrate of blood plasma. These differences suggest an intracochlear origin of the perilymphatic pyruvate and lactate and a blood-PL barrier for glucose and probably for pyruvate and lactate too (Scheibe et al., 1979; Haupt et al., 1980).

In the case of the high molecular components it is remarkable that the differences in the PL/plasma ratios of total protein (0.045) and total LDH (0.56) amounts to one order of magnitude. This means the percentage of LDH in the perilymphatic proteins is about ten times higher than in the plasma proteins. We suppose that this difference is due to a direct LDH formation within the cochlea and not to a special permeability of the cochlear vessels for this relatively high molecular enzyme.

CONCLUSIONS

The differences in the concentrations of the chemical components found between PL and both CSF and blood plasma (serum) reveal that the chemical composition of PL in the anaesthetized and operated guinea pigs corresponds neither to that of CSF (PL/CSF ratios ≠ 1) nor to that of a simple ultrafiltrate of blood plasma (PL/plasma ratios ≠ 1 in the case of the metabolites). These differences contradicts the both original opinions on the genesis and the biochemical nature of PL. The current analytical results suggest PL has an own specific chemical composition apparently determined by intracochlear metabolic processes and by a so-called blood-PL barrier also for low molecular components not bound to protein. However, it is to take into account that the effect of the anaesthesia and the surgery on the concentration ratios under discussion cannot be estimated hitherto.

REFERENCES

Bergmann, K., Haupt, H., Scheibe, F. and Rogge, I., 1979, Der Verschluss des Aquaeductus cochleae für Perilymphuntersuchungen am Meershweinchen, Arch. Otorhinolaryngol., 224: 257-266.
Cabezudo, L. M., 1978, The ultrastructure of the basilar membrane in the cat, Acta Otolaryngol., 86: 160-175.

Giebel, W. and Scheibe, F., Zur mikroflammenphotometrischen
    Elektrolytbestimmung in Innenohrflüssigkeiten, in preparation.
Gershbein, L. L., Manshio, D. T. and Shurrager, P. S., 1976, Bio-
    chemical parameters of guinea pig perilymph sampled according
    to scala and following sound presentation, Enviroment. Health
    Perspect., Vol., 157-164.
Haupt, H., Bergmann, K. and Scheibe, F., 1979, Zur Gesamtaktivität
    der Laktatdehydrogenase von Perilymphe, Plasma und Liquor
    cerebrospinalis normaler und schallbelasteter Meerschweinchen,
    in: "Cochlea-Forschung'77", H. Jakobi and K.-D. Kuhl, eds.,
    Martin-Luther-Universität, Halle (Saale).
Jung, W. K., 1975, Zur Resorptionskinetik der cochleären Perilymphe,
    gemessen mit Radionucliden nach minimaler Störung, Arch. Oto-
    rhinolaryngol., 211: 113-127.
Jung, W. K., 1979, Results in evaluating cochlear kinetics of carbon
    -14 labelled metabolites, Revue de Laryngologie, 100: 207-214.
Kellerhals, B., 1976, Quantitative assessment of perilymph sources,
    ORL 38: 193-197.
Kommoss, J. and Giebel, W., 1978, Die Ausbreitung löslicher
    Substanzen nach Applikation in die Cisterna cerebellomedularis,
    Arch. Otorhinolaryngol., 221: 67-76.
Lawrence, M., 1976, Direct visualization of living organ of Corti
    and studies of its extracellular fluids, Laryngoscope, 84:
    1767-1776.
Lawrence, M., Nuttal, A. L. and Burgio, P. A., 1975, Cochlear
    potentials and oxygen associated with hypoxia, Ann. Otol.,
    84: 499-508.
Scheibe, F., Haupt, H. and Hache, U., 1976, Vergleichende Unter-
    suchungen der Laktatkonzentration von Perilymphe, Blut und
    Liquor cerebrospinalis normaler und schallbelasteter Meer-
    schweinchen, Arch. Otorhinolaryngol., 214: 19-25.
Scheibe, F., Haupt, H., Hache, U. and Bergmann, K., 1977, Analy-
    tischer Vergleich einiger chemischer Bestandteile in Perilymphe,
    Liquor cerebrospinalis und Blut des Meerschweinchens, in:
    "Inner Ear Biology", M. Portmann and J.-M. Aran, eds., INSERM,
    Paris.
Scheibe, F., Hache, U. and Haupt, H., 1979, Zur Laktat- und Pyruvat-
    konzentration von Perilymphe, Blut und Liquor cerebrospinalis
    des Meerschweinchens, in: "Cochlea-Forschung'77", H. Jakobi
    and K.-D. Kuhl, eds., Martin-Luther-Universität, Halle (Saale).
Schindler, K., 1965, Perilymphe als Ultrafiltrat des Serums, Arch.
    Ohr.-, Nas.- u. Kehlk.-Heilk., 185: 586-592.
Schnieder, E.-A., 1974, A contribution to the physiology of the
    perilymph. Part I: The origins of perilymph., Ann. Otol.,
    83: 76-83.

DISTRIBUTION OF MICROPHONIC POTENTIALS IN THE FOUR TURNS OF

THE GUINEA PIG COCHLEA

J. Syka, I. Melichar and L. Úlehlová

Institute of Experimental Medicine
Czechoslovak Academy of Sciences
128 08 Prague 2, U nemocnice 2, Czechoslovakia

Frequency selectivity of the recording of cochlear microphonics (CM) with the electrode near the round window is low; the output is dominated by the CM generated in the basal turn. The differential technique of the CM recording, with microelectrodes inserted into the scala tympani and scala vestibuli of individual cochlear turns (Tasaki and Fernández, 1952; Dallos, 1969) provides better frequency selectivity. Similar results as those with the differential technique may be obtained with the microelectrode introduced into the scala media (Honrubia and Ward, 1968). Although the effect of the CM generators located in other turns is not fully excluded with this type of recording, the results are representative for the estimation of the functional state of the individual cochlear turn and may demonstrate some special properties of the CM, e.g. the shift of the maximum CM voltage towards the base with the increased sound intensity. We attempted in our experiments to repeat some of the Honrubia and Ward measurements and to explore the method for the estimation of the effects of the narrow band noise exposure on the CM in individual turns of the guinea pig cochlea. In addition to CM data values of the endocochlear potential, the anoxic endocochlear potential and the distribution of hair cells in four turns of the cochlea were measured.

In guinea pigs anaesthetized with urethane (20% solution, 0.8 ml/100 g body weight) and immobilized with Tricurane (gallamine iodide) the bulla was opened and small fenestrae at four cochlear turns above the scala media (approximately 4.5, 11.0, 14.5 and 18.0 mm from the base of the cochlea) were made. The animals were artificially ventilated. Glass microelectrodes filled with 3M KCl were introduced through the fenestrae for the measurement of the EP and CM. Acoustical stimulation was performed with a sealed sound system

with a miniature piezoelectric transducer. The acoustical pressure
before the eardrum was controlled. Tone pips with a frequency band
of 100 Hz to 5 kHz served as stimuli. One group of guinea pigs was
exposed to 142 dB SPL, 1/3-octave-band noise centered at 1 kHz, for
1 hour. The CM and EP in the exposed animals were measured 20 days
after exposure. The distribution of hair cells in all animals was
assessed on the basis of the surface specimen technique.

In control non-exposed animals the distribution of CM isointensity curves was similar to that described by Honrubia and Ward
(1968). In Fig. 1 in the upper part the isointensity curves in a
typical experiment for 50, 60, 70 and 80 dB SPL and frequencies
2400 Hz and 600 Hz are shown. The data are plotted on semilogarithmic
coordinates. The locus of the maximum CM voltage shifts towards the
base of the cochlea with increased sound intensity.

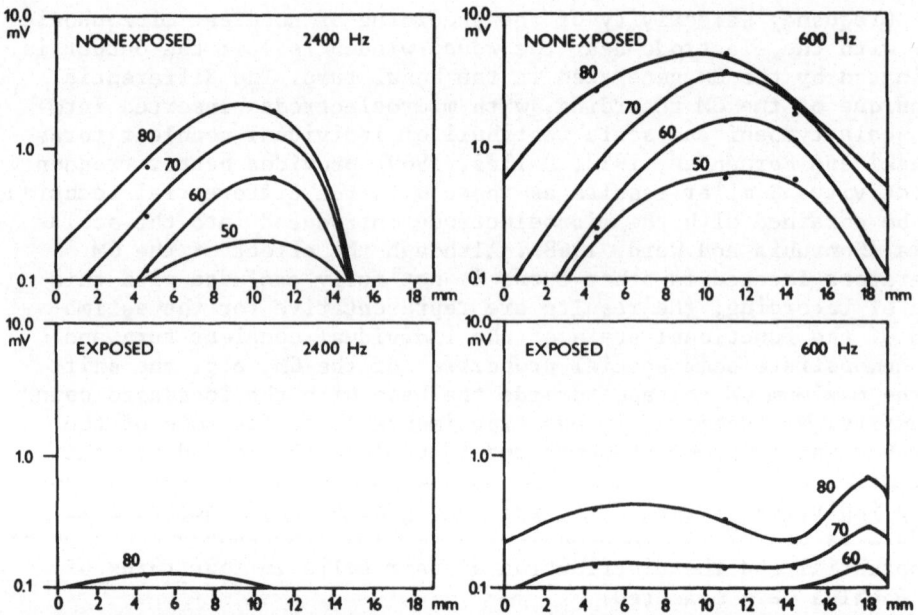

Fig. 1.   Longitudinal distribution of the CM isointensity curves
          to 2400 Hz tone (on the left) and 600 Hz tone (on the
          right) in a control non-exposed guinea pig (upper part)
          and in an exposed guinea pig (lower part). Abscissa-
          distance from the base of the cochlea in mm; ordinate – CM
          amplitude in mV.

In animals 20 days after the noise exposure a wide variety of reductions of the CM amplitudes, ranging from total unresponsiveness to 90 dB SPL sounds to a relatively mild decrease in the CM values was observed. The values of the positive EP in noise exposed animals measured 20 days after exposure in four cochlear turns were similar to those in controls. Immediately after exposure the EP, however, decreased evenly in all four cochlear turns and returned to normal physiological values in about 5 days (Melichar et al., 1980).

In guinea pigs with the relatively mild decrease in the CM values, the CM amplitudes in individual turns were related to the loss of outer hair cells. In guinea pig 79-10-11, where the outer hair cell loss in the second, third and fourth turn was not severe (50%, 60% resp. 90% persisting outer hair cells in comparison with average values found in normal non-exposed animals) it was possible to evoke CM at 80 dB SPL sound intensity, however, reduced in amplitude in comparison with those in normal animals. Fig. 1 in the lower part shows that at 600 Hz the CM voltage in the fourth turn is practically normal, whereas in the first turn (with less than 10% persisting outer hair cells) the amplitude is significantly reduced. The decrease in the CM amplitude is more marked at a frequency of 2400 Hz, in accordance with the generation of high frequency CM in upper cochlear turns (detailed discussion in Dallos, 1973). In contrast to the mild loss in the CM amplitude observed in animal 79-10-11, in some animals the cochleas were unresponsive to 90 dB SPL tones even though the loss of outer hair cells was not severe.

In addition to the positive EP in animals 20 days after the exposure values of the negative anoxic EP 15-60 minutes after interruption of artificial ventilation were measured. Values of the anoxic negative EP were less negative than values found in normal animals (Melichar et al., 1980; Konishi et al., 1979) and the EP usually attained less negative values in the second turn, i.e. in the turn where the narrow band noise exerted the major destructive effect. The differences in the anoxic EP between the second turn and other turns were not significant. The values of the anoxic negative EP correlated, however, with the distribution of persistent hair cells in individual cochlear turns.

In conclusion: the data confirm the longitudinal distribution of CM within the cochlea, when recorded in the scala media in four cochlear turns (Honrubia and Ward, 1969) and demonstrate that this type of measurement is a useful method for the estimation of changes along the cochlear duct. The anoxic EP and the CM voltage depend on the functional state of the organ of Corti in individual turns, however, a great variability in the CM output among animals was observed.

REFERENCES

Dallos, P., 1969, Comments on the differential electrode technique,
    J. Acoust. Soc. Amer., 45: 999-1007.
Dallos, P., 1973, "The auditory periphery", Academic Press, New
    York, London.
Honrubia, V. and Ward, P. H., 1968, Longitudinal distribution of the
    cochlear microphonics inside the cochlear duct (guinea pig),
    J. Acoust. Soc. Amer., 44: 951-958.
Konishi, T., Salt, A. N. and Hamrick, P. E., 1979, Effects of
    exposure to noise on ion movement in guinea pig cochlea,
    Hear. Research, 1: 325-342.
Melichar, I., Syka, J. and Ůlehlová, L., 1980, Recovery of the
    endocochlear potential and the K^+ concentration in the cochlear
    fluids after acoustic trauma, Hear. Research, 2: 55-63.
Tasaki, I. and Fernández, C., 1952, Modifications of cochlear micro-
    phonics and action potentials by KCl solution and by direct
    currents, J. Neurophysiol., 15: 497-512.

SESSION III
CODING IN THE AUDITORY NERVE AND
COCHLEAR NUCLEUS
Chairmen: A. R. Møller and E. F. Evans

THE DYNAMIC RANGE PROBLEM: PLACE AND TIME CODING AT THE LEVEL OF

COCHLEAR NERVE AND NUCLEUS

E. F. Evans

Department of Communication and Neuroscience
University of Keele
Staffs. ST5 5BG, U.K.

The greatest obstacle to a straightforward acceptance of the
˜place˜ theory of the coding of frequency in the auditory system is
the ˜dynamic range problem˜ (Evans, 1977ab, 1978ab, 1980a). It is
an everyday experience that our ears can operate over an amazingly
large dynamic range approaching 100dB, yet the dynamic range of
the great majority of cochlear nerve fibres is remarkably limited.
The solution to this problem has implications both for our understanding of the fundamental mechanisms of the neural coding of
auditory stimuli (Evans, 1978a) and for the design of multi-channel
electrical prostheses intended to impart speech information to the
profoundly hearing impaired (Evans, 1978b). This paper briefly
reviews the problem and discusses various possible solutions in the
light of recent physiological data.

Measurements of the psychophysical dynamic range for the auditory system of animals is unfortunately lacking, except for our own.
In man, the ear´s acuity for intensity discrimination is maintained
without deterioration for signal levels 100dB above threshold
(e.g. Riesz, 1928; Miller, 1947). Our ears´ ability to analyse
complex sounds like speech is maintained relatively unchanged over
this range, although small but significant deterioration in frequency selectivity does occur at sound levels above about 70dB SL,
as determined psychophysically (Scharf and Meiselmann, 1977; Pick,
1977) and physiologically (Evans, 1977b).

This remarkable perceptual dynamic range becomes a problem in
the face of the relatively restricted dynamic range demonstrated by
physiological measurements at the cochlear nerve level. As Kiang
(1968) first pointed out, the dynamic ranges of individual cochlear
nerve fibres (that is to say the range of stimulus levels - at the

characteristic frequency of the fibre - over which the fibre can
indicate the signal level by its mean discharge rate is restricted
to some 30-50dB, in the great majority of cases. Furthermore, as
Kiang (1968) also pointed out, the range of minimum thresholds of
most cochlear nerve fibres at any one frequency is also restricted
to about 20dB or less in any one individual*. Cochlear nerve fibres
therefore appear to be incapable of specifying the level of a
stimulus in terms of their mean discharge rate over a sufficiently
wide range to account for the psychophysical dynamic range.

This therefore poses a severe problem for classical, and
particularly pattern recognition, ˉplaceˉ models of the coding of
frequency - how does the auditory system ˉkeep the lines separateˉ
corresponding to the spectral components of a complex stimulus at
moderate and high stimulus levels when most, if not all, of the
cochlear fibres should be saturated, and therefore apparently not
able to signal fine variations in energy distribution across the
frequency spectrum?

For single tone discrimination, it has been assumed that the
level of the signal could be coded by the degree of spread of activity to unsaturated fibres of adjacent characteristic frequency
(e.g. Allanson and Whitfield, 1956). This undoubtedly occurs, as in
Fig. 1. However, for frequencies of about 1kHz and less, the Frequency Threshold (ˉTuningˉ)Curves (FTCs) of cochlear fibres in cat
and guinea pig are sufficiently broad that the spread of activity
becomes so large that at moderate to high sound levels any specificity of activation corresponding to the frequency ˉplaceˉ (1kHz
in Fig. 1B) is lost (at 45 and 70dB SPL, Fig. 1B). Admittedly, at
higher frequencies, where the cat is most sensitive, the FTCs are
much sharper and the situation is less serious (e.g. 8kHz, Fig. 1A):
the increasingly steep high frequency cut-off of the FTCs with level
means that the sharp border between active and inactive fibres is
maintained even at the highest stimulus levels. However, the spread
of activity towards fibres with higher CFs means that, in principle,
stimuli having multiple frequency components cannot be resolved in
terms of discrete places of neural activity at the cochlear nerve
level.

Even for single component stimuli, the coding of intensity need
not necessarily require the spread of activity to fibres of adjacent
characteristic frequency. It has been shown that the dynamic range
for intensity discrimination is maintained in the presence of
simultaneous low- and high-pass masking noise (i.e.: band-stop noise)
designed to mask out spread of activity to fibres of lower and higher
characteristic frequency (Viemeister, 1974; Hellman, 1974; Moore
and Raab, 1975). We have investigated the behaviour of single fibres
in the cochlear nerve, and cells in the cochlear nucleus, under these

---

* More recent measurements appear to contradict this. See later.

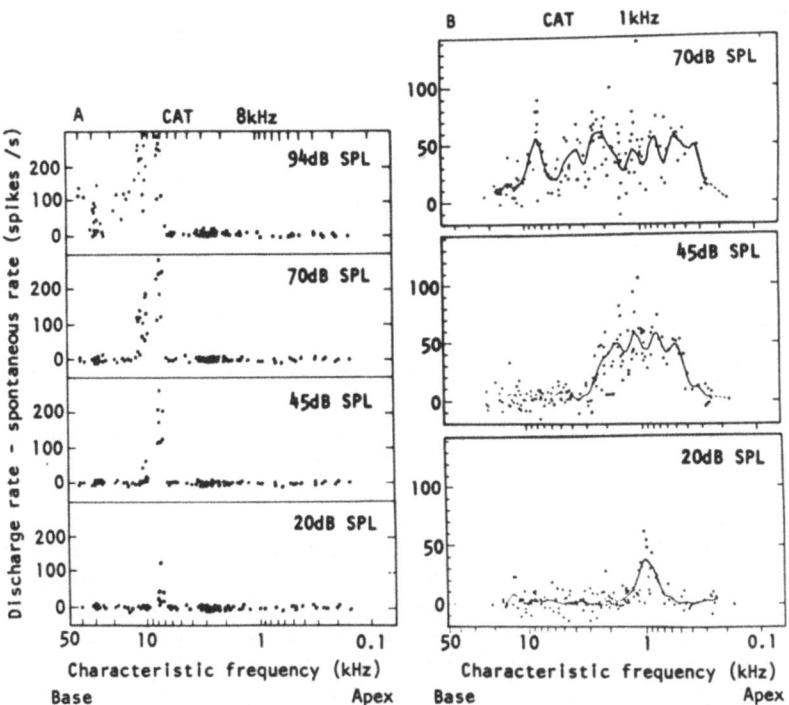

Fig. 1.   Distribution of evoked activity across cat cochlear nerve
          fibre array in response to single frequencies at different
          sound levels. Each point represents the increase in dis-
          charge rate above spontaneous activity for a given single
          cochlear nerve fibre plotted at its characteristic fre-
          quency (and therefore inferred position along the cochlear
          partition). Left half: responses to 8kHz tone at four sound
          levels (Evans and Palmer, unpublished data).
          Right half: responses to 1kHz tone at three sound levels
          (from Kim and Molnar, 1979).

conditions (Evans and Palmer, 1975). Apart from relatively small
effects due to lateral (¯two-tone¯) suppression, the saturation and
dynamic range of the cochlear fibres is unchanged by the bandstop
noise. In contrast, approximately two-thirds of the cells in the
dorsal division of the cochlear nucleus signalled, by their discharge
rate, differences in level of a tone or noise signal in the presence
of bandstop noise over a very wide dynamic range (over 100dB in some
cases). These cells (in contrast to cochlear nerve fibres) have

strong inhibitory ˜side-bands˜. Bandstop noise energy falling in
these side-bands inhibit the cells˜ discharge rates to below saturation so that they are ˜biassed˜ within their working range and can
respond to different levels of the signal. Analogous behaviour has
been found in these cells for ˜comb-filtered˜ noise stimuli, i.e.:
noise having multiple spectral peaks and valleys (Evans, 1977b).

It thus appears that the dynamic range problem, at least for
multi-component stimuli (where it is most acute) is circumvented at
the level of the dorsal cochlear nucleus. But we are still left with
the problem at the cochlear nerve level - what is the nature of the
input to these cochlear nucleus cells?

We have explored a number of possible answers to this question.

(A) NOT ALL COCHLEAR FIBRES HAVE RESTRICTED DYNAMIC RANGES

While the majority of cochlear fibres have dynamic ranges
restricted to about 40dB, the existence of fibres with ranges in
excess of 60dB has been reported by Nomoto et al. (1964) and Sachs
and Abbas (1974). The proportion and properties of these fibres
have been recently investigated in the cat (Palmer and Evans, 1979).
Fig. 2 illustrates the findings in a single cochlear nerve. For each
of the 121 fibres the dynamic range is plotted as a vertical bar at
the characteristic frequency: each bar extends from the just sub-
threshold point in spontaneous activity up to the first level at
which the discharge rate had saturated. The great majority of fibres
had saturated, as expected, by levels of 60-80dB SPL, and had a mean
dynamic range of 41dB. However, 9% of the fibres had dynamic ranges
in excess of 60dB, and 5% in excess of 70dB; several were not
completely saturated at the highest stimulus levels employed (arrows
in Fig. 2).

The characteristics of these fibres are of interest. They have
higher thresholds than average for fibres of their CF, (but the
thresholds are still within the 20dB overall variation mentioned
above). The slopes of their rate-level functions above 30dB above
threshold are substantially less steep than near threshold (the
˜sloping saturation˜ of Sachs and Abbas, 1974), by a factor of about
4. Thirdly, these fibres had lower spontaneous activity than the
average (Evans and Palmer, 1980a), almost all falling within a sub-
population of fibres having spontaneous discharge rates below 15/s.

In principle, therefore, these fibres could be responsible for
carrying information about stimulus component energies at high
levels. However, because their numbers are small, and the slope of
their rate-level functions is reduced at the high levels, a deterioration in performance might be expected at high stimulus levels
whereas, at least for intensity discrimination (although other
factors might here be operating), the reverse is the case.

(B) BECAUSE OF LATERAL SUPPRESSION, NOT ALL COCHLEAR FIBRES ARE SATURATED.

The possibility exists that lateral suppression effects might act for multicomponent stimuli in an analogous manner to the lateral inhibition in dorsal cochlear nucleus, thus ˜biassing˜ the discharges of cochlear fibres below saturation. In the great majority of fibres studied in the bandstop noise paradigm referred to above (Evans and Palmer, 1975), the lateral suppression effects are sufficiently weak to be negligible. In the fibres in the lower spontaneous rate population, however, the lateral suppression effects appear to be stronger

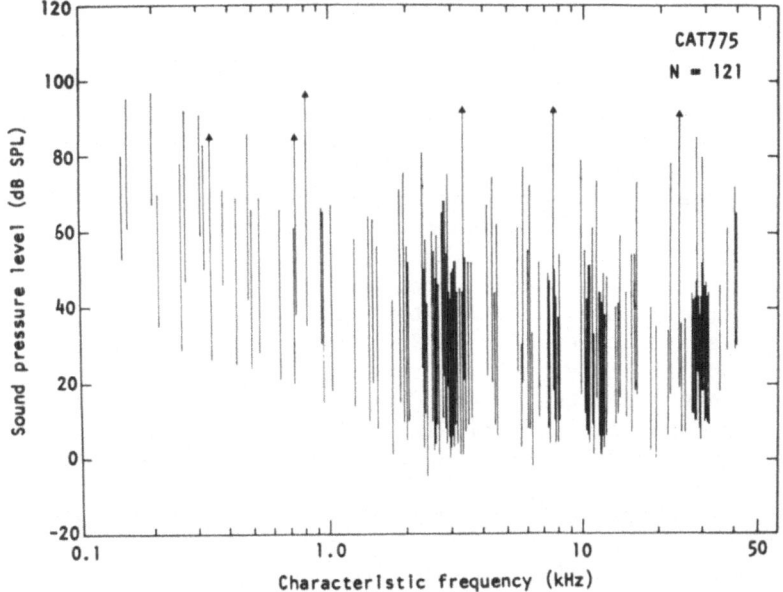

Fig. 2.   Dynamic ranges of mean discharge rate for 121 cochlear fibres in one cat cochlear nerve. Each vertical line represents the dynamic range of a single cochlear fibre plotted at its characteristic frequency. The extent of the line indicates the range of levels of 100 ms tones at the characteristic frequency between threshold (lower limit of line) and saturation of the mean discharge rate (upper limit). Arrows indicate fibres not saturated at higher sound levels employed. (From Palmer and Evans, 1979).

(Sachs and Young, 1979). These fibres do indeed show mild ˜biassing˜
effects, and it is conceivable that the cochlear nucleus cells
receive specific input from this sub-population of fibres and
˜amplify˜ their responses.

(C) CODES OTHER THAN MEAN DISCHARGE RATE ARE EMPLOYED: TEMPORAL
    PATTERN OF DISCHARGE.

It does not seem parsimonious that such a small proportion of
cochlear fibres should be responsible for carrying information on
the level of spectral components at high stimulus levels. The question has been asked therefore whether the information could be
carried by the majority of cochlear fibres in the temporal patterns
of discharge instead of or in addition to their mean rate (Moore,
1977; Goldstein and Srulowicz, 1977; Goldstein, 1978; Evans, 1977a,
b, 1978a; Young and Sachs, 1979).

It is well known that, for low frequencies at least (up to
5kHz), the dynamic range over which ˜phase-locking˜ of discharges
of cochlear fibres occurs to cycles of the stimulus waveform extends
well beyond that of the mean discharge rate (Rose et al., 1971). The
˜threshold˜ for the appearance of phase-locking of discharges can
be 5-20dB below the mean discharge rate threshold (Rose et al.,
1967; Evans, 1980a). Furthermore, the phase-locking is unaffected
by saturation of the mean discharge rate, and it has been suggested
that the degree of phase-locking could signal the level of stimuli
(Moore, 1977). For single component stimuli, however, the degree of
phase-locking is no more useful as an index of the absolute stimulus
level than the mean discharge rate. In fact it saturates at even
lower levels, e.g. 20-30dB below the level at which mean discharge
rate saturates (McGee et al., 1979; Evans, 1980a). Some form of
combined index of activity i.e.: rate and degree of phase-locking
would therefore yield the widest dynamic range for the purposes
of signalling the level of stimuli, but this in itself would not
solve the problem at high levels, where both measures are saturated.

The above considerations apply to the coding of the level of
single stimulus components. It is implicit in the data of Young
and Sachs (1979) that the level of components in a multicomponent
stimulus relative to one another could be signalled by the relative
degree of phase-locking to those components even above the saturation of the mean discharge rate. In their experiments, it was found
that the relative levels of the formants in steady-state (synthesized) vowels could be adequately signalled, at all sound levels, by
the degree of phase-locking of cochlear fibres to the frequency corresponding to their ˜place˜. What is not clear, however, from these
experiments, is the degree to which a given cochlear fibre can
demonstrate in its phase-locking behaviour, its frequency selective
properties. In other words, whether the relatively sharp filtering
properties of cochlear fibres (which in mean discharge rate terms

are more than adequate to account for psychophysical frequency
selectivity, up to saturation levels, e.g. Evans and Wilson, 1973;
Evans, 1977a, 1978a), continue to be manifest above saturation
levels in the relative degree of phase-locking to stimulus components of different frequencies. The present experiments were designed to answer this question.

The methods and preliminary results have been given elsewhere
(Evans, 1979, 1980a). Briefly, the discharges of single cochlear
nerve fibres in the anaesthetized cat were examined in response to
3 x 10s periods of stimulation, at different levels, of harmonic
tone complexes consisting of the second to eleventh harmonics of
a missing fundamental. The harmonics were mixed in cosine or random
phase, and were present at equal levels. The fundamental was chosen
for each fibre from a range of 75-600 Hz so that the fibre's
characteristic frequency was centred approximately at the arithmetic
mean of the harmonics. The period histograms of the discharges were
computed, synchronized in turn to the (missing) fundamental and to
each of the harmonics. The relative degree of phase-locking to each
harmonic component was computed either from the relative vector
strengths of the period histograms locked to each harmonic (Evans,
1980a) or more conveniently, from Fourier transform of the period
histogram locked to the fundamental. Both methods yielded the same
results.

Full data were obtained on 15 fibres, over sound levels from
threshold to 90dB above. The fibres had characteristic frequencies
ranging from 0.5 to 3.8kHz. The following generalizations can be
made with reference to Figs. 3-5.
(a) The relative levels of phase locking to the individual harmonics almost exactly correspond to the Frequency Threshold Curve of
the fibre (determined by an automatic threshold-following procedure
with pure tone bursts), for sound levels up to between 20-60dB above
threshold. That is to say, the amplitude of the phase-locking of
components is progressively attenuated the further they are from
the characteristic frequency, and this attenuation quantitatively
matches the FTS (Fig. 4, curves -90 to -60dB; Fig. 5, curves
-80 to -60dB).
(b) For fibres of characteristic frequency above 1kHz (1.5kHz to
3.8kHz), there is a progressively steeper attenuation of phase-
locking to stimulus components above the characteristic frequency,
and reduction in steepness below (Figs. 3, 4; curves -40 to -10dB).
At the highest levels (40-80dB above threshold; -10dB in Figs. 3, 4)
there is an actual shift of the ˊcharacteristic frequencyˊ towards
lower frequencies, amounting to between about 0.2-0.5 octaves. These
effects appear to be due to the lower frequency components dominat-
ing the phase-locking of the fibre, and hence reducing the effectiveness of the higher frequency components. This appears to be analogous to
the ˊsynchrony suppressionˊdescribed by Rose et al. (1974) for two-tone
complexes, and by Young and Sachs (1979) for steady state vowels.

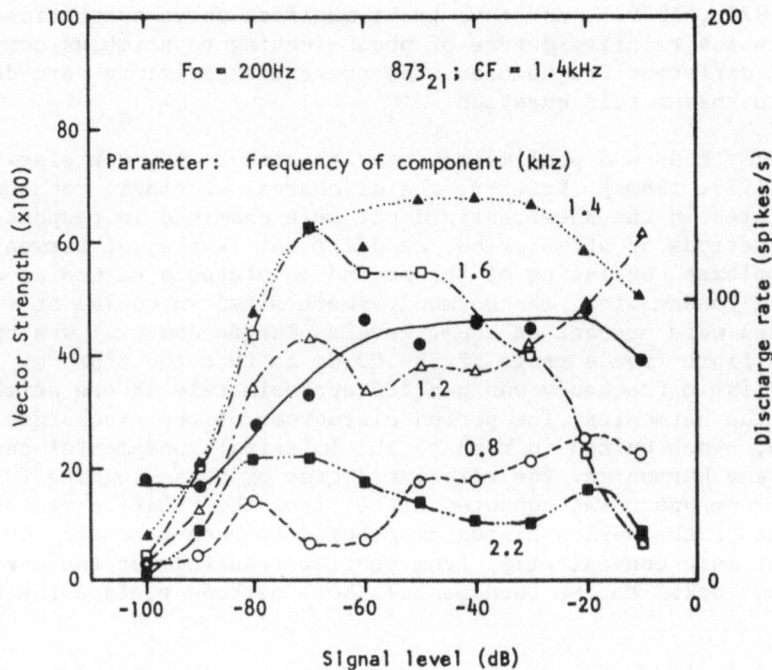

Fig. 3.   Mean discharge rate and phase-locking indices of activity
          for a single cat cochlear fibre of CF = 1.4kHz, against the
          level of a ten-component harmonic complex. Continuous line:
          mean discharge rate in response to 3 presentations each of
          10s duration of 10-component harmonic complex. Interrupted
          lines: vector strengths of phase-locked responses to the
          given individual harmonic components.
          0dB = 106dB SPL.

(c)   For fibres of characteristic frequency below about 1kHz (0.4-
0.75kHz), the converse effect to (b) occurs. Components with frequencies above the characteristic frequency tend to dominate the
phase-locking, so that at the higher stimulus levels the low frequency cut-off becomes steeper, and the ˉcharacteristic frequencyˉ
shifts towards higher frequencies (Fig. 5). The shift in characteristic frequency amounts to 0.2 to 0.5 octaves at levels 50-70dB
above threshold.

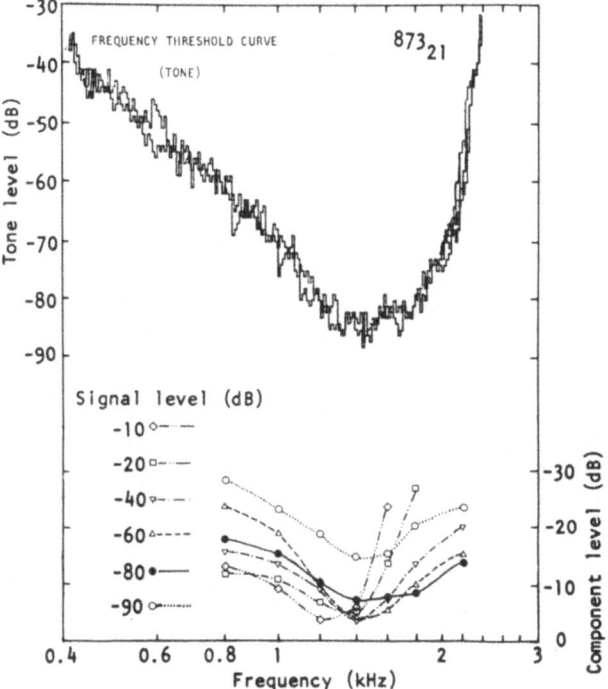

Fig. 4. Mean discharge rate threshold (Frequency Threshold Curve) of the single cochlear fibre of Fig. 3, compared with the vector strengths of its phase-locked response to individual harmonics of a ten harmonic tone complex at different stimulus levels. Vector strengths plotted on a log power scale for comparison. OdB = 106dB SPL. (Data from Evans, 1980a).

These effects are accompanied by little or no changes in the half-power bandwidth of the attenuation functions. In some cases, the bandwidth actually decreases (Fig. 5).

The changes in the cut-offs and ˉcharacteristic frequency of
the derived attenuation functions with stimulus level and their
dependence upon the characteristic frequency of the fibre, are
virtually identical to the changes observed in the filter functions
derived from cochlear fibres using the reverse correlation technique
of de Boer (1969), (Evans, 1977b). This technique allows one to
derive the impulse response of the cochlear filter responsible for
the tuning of a cochlear fibre by analysis of the time-locked

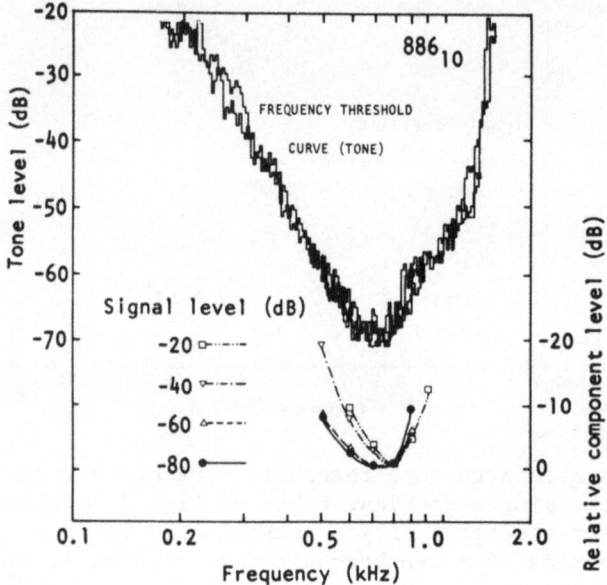

Fig. 5. As Fig. 4 for a single cat cochlear fibre of CF = 0.75kHz.
          0dB = 104dB SPL.

activity of the fibre in response to broad-band noise stimuli. For
fibres with characteristic frequencies above 1kHz, the filter
functions have steeper high frequency cut-offs and incur a shift
towards lower frequencies, whereas for fibres with lower characteristic frequencies the converse is the case (Fig. 6).

There is corroborating evidence from these shifts in the most effective frequency at high stimulus levels, in the mean discharge rate profiles of fibres with frequency in response to single tones (Rose et al., 1971; Evans, 1974; 1978a). For fibres of characteristic frequency above about 1kHz, the most effective frequency shifts towards lower values at higher stimulus levels, and vice-versa. These shifts are in the appropriate direction to account for the well known shifts in the pitch of pure and complex tones with intensity, depending upon the frequency (Evans, 1978a).

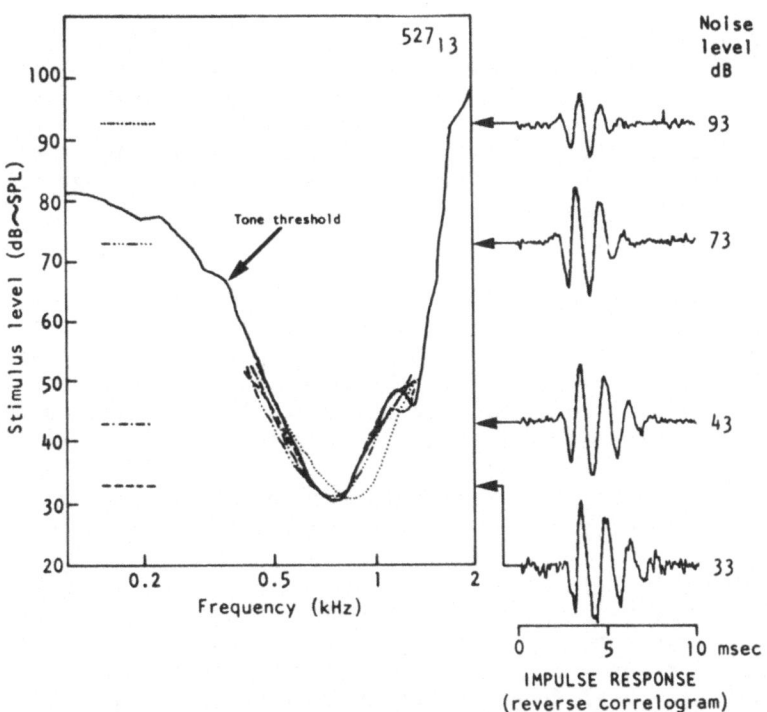

Fig. 6. Mean discharge rate threshold (FTC)of single cat cochlear fibre of CF = 0.75kHz compared with the power filter functions derived from its impulse response as determined by reverse correlation at different noise levels.

It is clear therefore from these results and from those of Young and Sachs (1979) that, in principle, adequate information is contained in the temporal discharge patterns of cochlear fibres on the relative levels of stimulus components in a broadband noise or multicomponent tone stimulus over a dynamic range approaching 100dB. The first-order neurone system behaves to a first approximation as if the saturation of mean discharge rate takes the form of an automatic gain control (AGC) before the probabilistic spike generator, so that the relative level of frequency components are preserved over a wide dynamic range in spite of the saturation of the mean rate. The present results demonstrate, in addition, that the threshold filtering properties of cochlear fibres are preserved, again relatively unchanged to a first approximation, over the same very large dynamic range. In fact, the period histograms obtained in these experiments can be very closely modelled by the output of a cochlear nerve fibre analogue consisting of a linear bandpass filter (to simulate the FTC attenuation function), and an AGC stage (providing saturation of mean discharge rate) feeding an ideal half wave rectifier and probabilistic spike generator (Evans, 1980b).

The question remains, of course, whether the higher levels of the auditory system can make use of this potential information. In any case, it is not obvious how information on stimulus frequencies much above 4kHz could be carried in this way, so that the problem of coding mechanism still remains for fibres providing input to dorsal cochlear nucleus cells of high CF where the dynamic range in the bandstop noise masking situation is still very large (e.g. at 20kHz, Evans and Palmer, 1975).

An alternative possibility is that the cochlear nucleus or other higher order cells may be able to cross-correlate the discharge patterns arriving at a number of input cochlear nerve fibres converging on the cell. Perhaps the coherence of the temporal discharge patterns could provide the requisite information on level. This will be extremely difficult to test.

In the face of these uncertainties concerning the problem of the dynamic range, a number of other relevant factors need to be mentioned.

(i) Firstly, physiological measurements of dynamic range are affected by the duration of the stimulus. Signals having durations in excess of seconds often produce smaller dynamic ranges and non-monotonic rate-level functions compared with those having durations of the order of 10-100ms (Evans, 1980a). In the latter case, order effects and adaptation artificially ‾narrow‾ the dynamic range for mean discharge rate (without, of course, affecting that for the preservation of the information in the temporal discharge patterns).

This effect does not mean, however, that the dynamic range for rapid amplitude fluctuations is thereby extended: the dynamic range of cochlear fibres for signalling amplitude modulation (100-400c/s) of a tone signal at CF corresponds very closely with that derived from the rate-level function for pure-tones of 50-100ms duration (Evans and Palmer, 1980b).

(ii) A subpopulation of cochlear fibres may exist having high thresholds. This has been suggested by Liberman (1978). Some cochlear fibres are encountered in apparently normal cochleas, having thresholds substantially higher than the majority whose thresholds fall within the generally restricted distribution referred to above (Evans, 1972; Liberman, 1978). The fibres tend to have low spontaneous discharge rates and are therefore easily missed. In the former study at least, however, these fibres showed, in their broad tuning, evidence of cochlear pathology. Such fibres, even if they exist in quantity in normal cochleas, could not easily serve to convey information on the level of closely spaced stimulus components.

(iii) The presence of background wideband noise of long duration (>10s) can ˉbiasˉ the dynamic range of the rate-level function for characteristic frequency tones towards higher levels, as shown in Fig. 7 (Evans, 1974). The higher the level of the continuous noise, the greater extent to which the rate-level function is shifted. This effect is not obtained with noise signals simultaneously gated with the tone bursts, and presumably reflects some sort of ˉadaptiveˉ AGC type mechanism (of much longer time constant than that referred to above), rather than lateral suppression, the effects of which are apparent immediately. Thus, while the cochlear fibre dynamic range itself is not extended at any given level of background noise (actually the converse occurs), it is ˉbiassedˉ in a potentially useful way by the background noise.

(iv) Finally, caution may be required in comparing the dynamic ranges determined psychophysically in awake human subjects with intact middle ear and efferent neural systems, with those derived from experiments on anaesthetised animals where the effects of the descending control systems are thereby eliminated. There is some evidence, in fact, that the dynamic range for multicomponent stimuli such as speech, may be more restricted than at first sight appears. In patients lacking middle-ear protective mechanisms, deterioration in speech intelligibility has been reported to occur at levels above about 80dB SPL (e.g. Borg and Zakrisson, 1973).

ACKNOWLEDGMENTS

I am grateful for the collaboration of A. R. Palmer in the

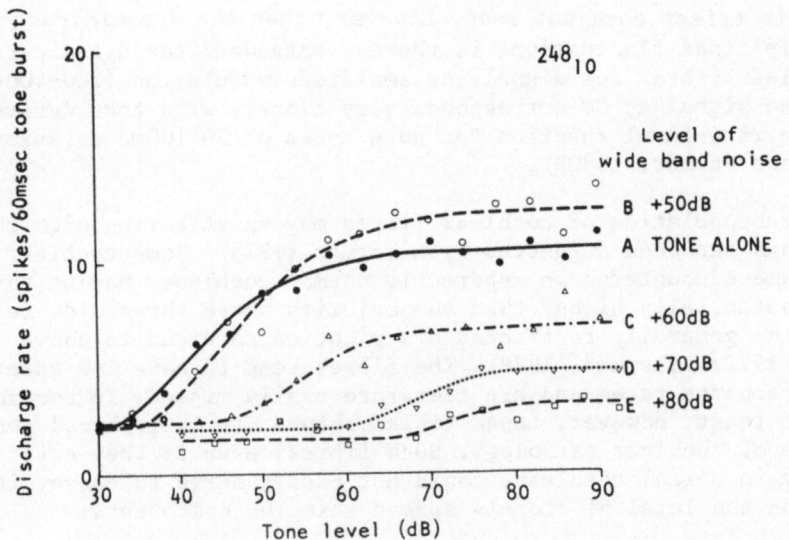

Fig. 7. Discharge rate-level functions for a single cat cochlear
         fibre at characteristic frequency (18kHz) in the presence
         of different levels of continuous background wideband noise.
         Levels in dB SPL, noise measured in 40kHz bandwidth. (From
         Evans, 1974).

cited studies. The work was supported by grants from the Medical
Research Council.

REFERENCES

Allanson, J. T. and Whitfield, I. C., 1956, The cochlear nucleus
    and its relation to theories of hearing, in: "3rd London
    Symposium on Information Theory", C. Cherry, ed., pp. 269-284,
    Butterworths, London.
Borg, E. and Zackrisson, J.-E., 1973, Stapedius reflex and speech
    features, J. Acoust. Soc. Am., 54: 525-527.
de Boer, E., 1969, Reverse correlation. II. Initiation of nerve
    impulses in the inner ear, Proc. k. ned. Akad. Wet., 72: 129-
    151.
Evans, E. F., 1972, The frequency response and other properties of
    single fibres in the guinea pig cochlear nerve, J. Physiol.,
    Lond., 226: 263-287.
Evans, E. F., 1974, Auditory frequency selectivity and the cochlear
    nerve, in: "Facts and models in hearing", E. Zwicker and
    E. Terhardt, eds., pp. 118-129, Springer, Berlin.

Evans, E. F., 1977a, Some interactions between physiology and psychophysics in acoustics, in: Proc. 9th Int. Congr. on Acoustics; volume of invited review lectures, pp. 55-65, Spanish Acoustical Society, Madrid.

Evans, E. F., 1977b, Frequency selectivity at high signal levels of single units in cochlear nerve and nucleus, in: "Psychophysics and Physiology of Hearing", E. F. Evans and J. P. Wilson, eds., pp. 185-192, Academic Press, London.

Evans, E. F., 1978a, Place and time coding of frequency in the peripheral auditory system: some physiological pros and cons. Audiol., 17: 369-420.

Evans, E. F., 1978b, Peripheral auditory processing in normal and abnormal ears: physiological considerations for attempts to compensate for auditory deficits by acoustic and electrical prostheses, in: "Sensorineural hearing impairment and hearing aids", C. Ludvigsen and J. Barfod, eds., pp. 9-47, Scand. Audiol., 6, suppl.

Evans, E. F., 1979, Single unit studies of the mammalian auditory nerve, in: "Auditory Investigations: The Scientific and Technological Basis", H. A. Beagley, ed., pp. 324-367, Oxford University Press, Oxford.

Evans, E. F., 1980a, "Phase-locking" of cochlear fibres and the problem of dynamic range, in: "International Symposium on Psychophysical, Physiological and Behavioural Studies in Hearing", G.v.d. Brink and F. Bilsen, eds., pp. 300-309, Delft University Press, Delft.

Evans, E. F., 1980b, An electronic analogue of single unit recording from the cochlear nerve for teaching and research, J. Physiol., 298: 33-34P.

Evans, E. F. and Palmer, A. R., 1975, Responses of units in the cochlear nerve and nucleus of the cat to signals in the presence of bandstop noise, J. Physiol., Lond., 252: 60-62P.

Evans, E. F. and Palmer, A. R., 1980a, Relationship between the dynamic range of cochlear nerve fibres and their spontaneous activity, Exp. Brain Res., 40: 115-118.

Evans, E. F. and Palmer, A. R., 1980b, Dynamic range of cochlear nerve fibres to amplitude modulated tones. J. Physiol., 298: 33-34P.

Evans, E. F. and Wilson, J. P., 1973, Frequency selectivity of the cochlea, in: "Basic mechanisms of hearing", A. R. Møller, ed., pp. 519-551, Academic Press, New York.

Goldstein, J. L., 1978, Mechanisms of signal analysis and pattern perception in periodicity pitch, Audiol., 17: 421-445.

Goldstein, J. L. and Srulovicz, P., 1977, Auditory-nerve spike intervals as an adequate basis for aural frequency measurement, in: "Psychophysics and Physiology of Hearing", E. F. Evans and J. P. Wilson, eds., pp. 337-346, Academic Press, London.

Hellman, R. P., 1974, Effect of spread of excitation on the loudness function at 250Hz, in: "Sensation and measurement", Moscowitz, et al., eds., pp. 241-249, Reidel, Dordrecht.

Kiang, N. Y.-s., 1968, A survey of recent developments in the study
    of auditory physiology, Ann. Otol. Rhinol. Lar., 77: 656–676.
Kim, D. O. and Molnar, C. E., 1979, A population study of cochlear
    nerve fibres: Comparison of spatial distributions of average
    rate and phase-locking measures of responses to single tones.
    J. Neurophysiol., 42: 16–30.
Liberman, M. C., 1978, Auditory nerve responses from cats raised in
    a low-noise chamber, J. Acoust. Soc. Am., 63: 442–455.
McGee, J.-A., Walsh, E.J. and Javel, E., 1979, Discharge synchroni-
    zation in auditory nerve and AVCN neurons, J. Acoust. Soc. Am.,
    65: Suppl. 1, S83.
Miller, G. A., 1947, Sensitivity to changes in the intensity of
    white noise and its relation to masking and loudness,
    J. Acoust. Soc. Am., 19: 609–619.
Moore, B.C.J., 1977, "An Introduction to the Psychology of Hearing",
    Macmillan, London.
Moore, B.C.J. and Raab, D. H., 1975, Intensity discrimination for
    noise bursts in the presence of a continuous bandstop back-
    ground: effects of level, width of the bandstop, and duration,
    J. Acoust. Soc. Am., 57: 400–405.
Namoto, M., Suga, N. and Katsuki, Y., 1964, Discharge pattern and
    inhibition of primary auditory nerve fibres in the monkey,
    J. Neurophysiol., 27: 768–787.
Palmer, A. R. and Evans, E. F., 1979, On the peripheral coding of
    the level of individual frequency components of complex sounds
    at high sound levels. Exp. Brain Res., Suppl. II, 19–26.
Pick, G. F., 1977, Comment on paper by Scharf and Meiselman, in:
    "Psychophysics and Physiology of Hearing", E. F. Evans and
    J. P. Wilson, eds., pp. 233–234, Academic Press, London.
Riesz, R. R., 1928, Differential intensity sensitivity of the ear
    for pure tones, Phys. Rev., 31: 867–875.
Rose, J. E., Brugge, J. F., Anderson, D. J. and Hind, J. E., 1967,
    Phase-locked response to low frequency tones in single audi-
    tory nerve fibres of the squirrel monkey, J. Neurophysiol.,
    30: 769–793.
Rose, J. E., Hind, J. E., Anderson, D. J. and Brugge, J. F., 1971,
    Some effects of stimulus intensity on response of auditory
    nerve fibers in the squirrel monkey, J. Neurophysiol., 34:
    685–699.
Rose, J. E., Kitzes, L. M., Gibson, M. M. and Hind, J. E., 1974,
    Observations on phase-sensitive neurons of anteroventral
    cochlear nucleus of the cat: nonlinearity of cochlear output,
    J. Neurophysiol., 37: 218–253.
Sachs, M. B. and Abbas, P. J., 1974, Rate versus level functions
    for auditory-nerve fibers in cats: tone-burst stimuli,
    J. Acoust. Soc. Am., 56: 1835–1847.
Sachs, M. B. and Young, E. D., 1979, Encoding of steady-state vowels
    in the auditory nerve: representation in terms of discharge
    rate, J. Acoust. Soc. Am., 66: 470–479.

Scharf, B. and Meiselman, C. H., 1977, Critical bandwidth at high
     intensities, in: "Psychophysics and Physiology of Hearing",
     E. F. Evans and J. P. Wilson, eds., pp. 221-232, Academic
     Press, London.
Viemeister, N. F., 1974, Intensity discrimination of noise in the
     presence of band reject noise, J. Acoust. Soc. Am., 56: 1594-
     1600.
Young, E. D. and Sachs, M. B., 1979, Representation of steady-state
     vowels in the temporal aspects of the discharge patterns of
     populations of auditory-nerve fibers, J. Acoust. Soc. Am., 66:
     1381-1403.

CODING OF COMPLEX SOUNDS IN THE AUDITORY NERVOUS SYSTEM

Aage R. Møller

Division of Physiological Acoustics
Department of Otolaryngology
University of Pittsburgh School of Medicine
Pittsburgh, Pennsylvania 15213, U.S.A.

Auditory research has for many years been dominated by the assumption that the ear is a frequency (or spectrum) analyzer. The origin of this assumption can be traced back to the formulation of the Place Theory of Hearing by von Helmholtz more than one hundred years ago. This hypothesis developed into one of the most successful theories in history and has influenced auditory research in a fundamental way. It has guided the design of innumerable experiments and has influenced the interpretation of results. Thus, studies of the ear using physiological methods have been aimed mainly at the determination of the ear's frequency selectivity. In the auditory nervous system, recordings of the electrical events in single auditory nerve cells and fibers have been used in studies of frequency selectivity. These studies have generally confirmed that the ear possesses a frequency selectivity which is maintained throughout the ascending auditory pathways. In addition, the tonotopic arrangement of cells in the various nuclei of the ascending auditory pathway including the auditory cortex has been taken as a further indication of the importance of the ear's frequency selectivity. Pure tones have been the dominant stimulus for such experiments. The threshold of firing of single nerve cells or nerve fibers or threshold of a just detectable increase in the discharge rate is the response parameter most commonly measured. Studies of the frequency selectivity in the nervous system have been concerned not only with exact determination of the value of this frequency selectivity, but also with an explanation of the different mechanisms which contribute to the degree of frequency selectivity in the inner ear and the various levels of the ascending auditory pathway.

    The Place theory of frequency discrimination relies on the fact
that different nerve cells are able to transmit information about
the intensity in a certain frequency band and that they can do so
over a large range of intensities. Using pure tones, it is generally
found that the intensity range where the discharge rate of auditory
nerve fibers is a function of the sound intensity is limited to a
range from threshold to 20-40 dB above threshold (see Kiang et al.,
1965). Above that intensity value the discharge rate reaches a
plateau and does not change even when there is an increase in the
sound intensity. This means that the nerve cells in the auditory
periphery cannot transmit information about the intensity of a sound
over a range larger than 20-40 dB above threshold. Since the physiological range of hearing spans a range of at least 80 dB, these
results no doubt serve to question the validity of the Place theory
of hearing.

    Although there is no question that the ear possesses a fre-
quency selectivity, responses to sounds such as pure tones at threshold do not indicate that the ear's frequency selectivity in response
to complex sounds at physiological intensity levels is of such a
nature that it can explain the ear's frequency discrimination power
as determined in psychoacoustic experiments. On the contrary, a
rather different response pattern is obtained from single auditory
nerve fibers and nerve cells when complex sounds at physiologic
intensities are used as stimuli. Specifically, stimulation with
broadband sounds such as noise at physiological levels shows a frequency selectivity of single auditory nerve fibers that decreases
as the sound level increases (Møller, 1977, 1978 a, b). At the same
time, the center frequency shifts downwards with increasing sound
intensity.

    Recordings from single nerve cells in the cochlear nucleus
reveal that the responses to tones, whose frequency is varying
rapidly, cannot be predicted on the basis of the responses to steady
state tones at threshold (Møller, 1969, 1971). Even more pronounced
is the fact that these cells respond to bands of noise in a substantially different way when the center frequency of the noise band
is changed rapidly compared to when the center frequency is kept
constant or changed slowly (Møller, 1974a, 1978c).

    In the following, some examples will be given of responses
obtained from single auditory nerve fibers to broadband noise, as
well as responses from cells in the cochlear nucleus to tones and
noise bands with changing frequency. Also, the responses to intensity modulated tones and noise will be considered.

    The frequency selectivity of the peripheral auditory analyzer
was assessed in experiments in anesthetized rats by stimulating the
ear with pseudorandom noise and recording the discharges from single
auditory nerve fibers (Møller, 1977, 1978a, b). Period histograms

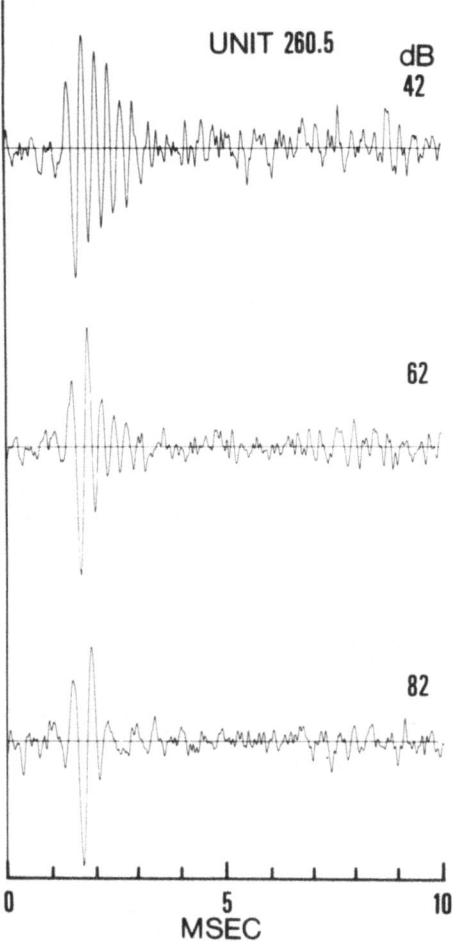

Fig. 1. Typical cross correlograms of the responses of an auditory
         nerve fiber to stimulation with pseudorandom noise and
         intensities between 42 and 82 dB SPL. The amplitude of the
         individual correlograms is normalized to the same peak
         value. (from Møller, 1977)

that were locked to the periodicity of the noise were compiled and
cross-correlated with one period of the noise. These histograms
represent the probability of firing during different instances along
one period of the pseudorandom noise. The modulation of these
histograms represents the degree to which the discharges are phase-

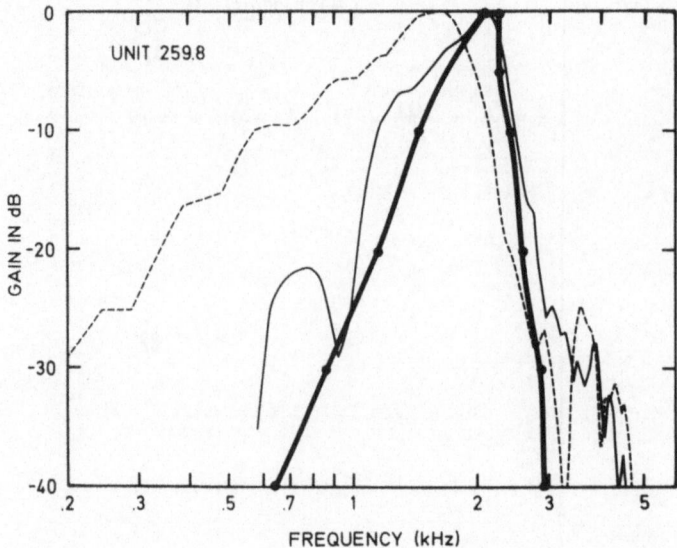

Fig. 2.   Frequency selectivity of a single auditory nerve fiber
          obtained using noise stimuli 15 dB above threshold (thin
          continuous line) compared with the frequency threshold
          curve obtained using tonebursts (heavy solid lines and
          dots). Cross-spectra obtained using a stimulus level of 70
          dB above threshold is also shown (dashed line) (from
          Møller, 1978b).

locked to the waveform of the sound stimulus (pseudorandom noise).
In a linear system that has a bandpass characteristic, such crosscorrelograms are valid estimates of the impulse response function
of the system under test. The Fourier transforms of these cross
correlograms consequently are an approximation of the frequency
transfer function of the system (Møller, 1974; O'Leary and Honrubia,
1975; Marmarelis and Marmarelis, 1978). In a system with a bandpass
characteristic, the derived frequency selectivity of the system is
a close approximation to what can be obtained using conventional
methods for determining the frequency transfer function, such as
single sinusoidal test signals.

The cross-correlation functions obtained from the responses of
single auditory nerve fibers using this method show damped oscillations, which is in agreement with the assumption that the auditory

periphery is a spectrum analyzer with a relatively narrow bandwidth
(Møller, 1977). Examination of the cross-correlation functions
obtained at different sound intensities reveals that the duration
of the damped oscillations decreases with increasing intensity
(Fig. 1). The decrease in duration corresponds to a broadening of
the frequency selectivity with increasing stimulus intensity. This
is further illustrated in Fig. 2, which shows a comparison between
frequency threshold curves and the Fourier transforms of cross-
correlation functions (cross-spectra) similar to those seen in
Fig. 1. Not only does the width of the tuning curves increase as a
function of the stimulus intensity, but also the center frequency
shifts downwards. The change in bandwidth (frequency difference
between points that are 10 dB below the maximum value) and the shift
in center frequency are illustrated in graphical form in Fig. 3.
These results are consistent for nerve fibers in the frequency range
with the center frequency from 1 kHz to 4 kHz. They show that the
auditory periphery cannot be regarded as a linear frequency analyzer, but that the frequency selectivity is a function of the
intensity of the noise used in these experiments.

The results using pseudorandom noise thus show that there is
a systematic shift in the peak frequency in the computed transfer
functions. This implies that the frequency of the largest response
shifts towards lower frequencies as the intensity is increased.
These results are in agreement with intracellular recordings from
single inner hair cells where pure tones at intensities above
threshold were used (Russel and Sellick, 1978, 1980) and from
certain experiments on cochlear mechanics (Rhode, 1971; Rhode and
Robles, 1974). Also experiments where recordings were made from
a large number of nerve fibers in the same animal when the stimulus
was the same pure tone at different intensities, show a similar
nonliniarity (Pfeiffer and Kim, 1974; Kim, et al., 1979; Kim, 1980).

The cross-spectra shown in Fig. 2 were normalized to the same
peak value. It may be assumed that the peak value of these cross-
spectra is a measure of the average degree of phase locking of the
discharges of the auditory nerve fibers. The effective stimulus is
the sound (pseudorandom noise) filtered by the spectral filter of
the auditory periphery. If the periphery of the auditory system
function in a manner similar to a linear spectrum analyzer, the
peak value of transfer functions such as those shown in Fig. 2
would increase proportionally with an increase in sound intensity.
The experiments described above show that this is not the case. On
the contrary, recordings from single nerve fibers show that the peak
value of the transfer function increases only up to a sound level
about 20 dB above threshold (Møller, 1978a). Above that level the
peak value reaches a plateau, as illustrated in Fig. 4, and maintains a nearly constant value throughout the intensity range that
was studied and comprises the larger part of the physiological
intensity range. The relationship between the stimulus sound inten-

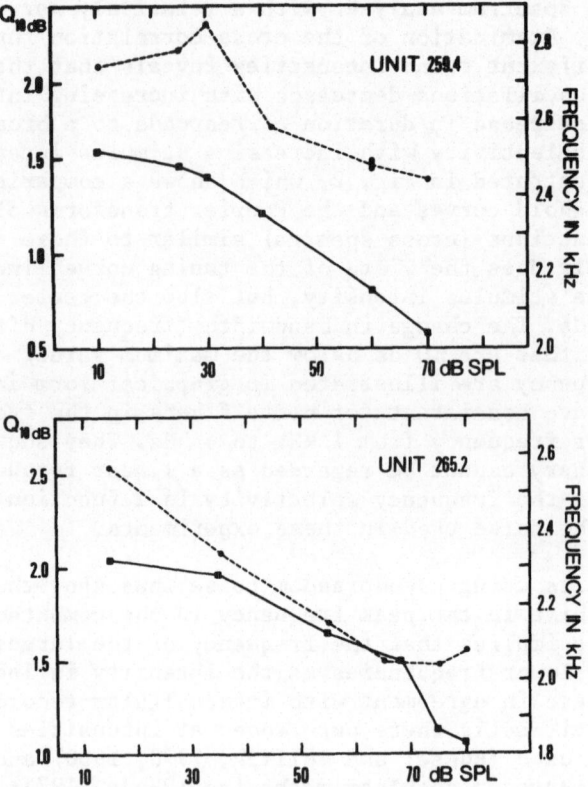

Fig. 3.   Relative width (dashes and circles) and the center frequency (solid lines and squares) of the transfer functions
of a single auditory nerve fiber determined using pseudorandom noise as stimuli. The relative width was measured
10 dB below the peak value ($Q_{10\ dB}$). These two values are
plotted against the stimulus intensity, given in dB SPL
measured in 1/3 octave frequency band (from Møller, 1977).

sity and the peak value of the cross-spectra has a more regular
course than the mean discharge rate (cf. Fig. 4).

The results shown above thus indicate that the frequency
selectivity of the peripheral auditory analyzer changes as a function of the stimulus sound intensity and that the (synchronized)
nerve impulse rate has a nearly constant value throughout most of
the physiological range. That shows that these nerve fibers can

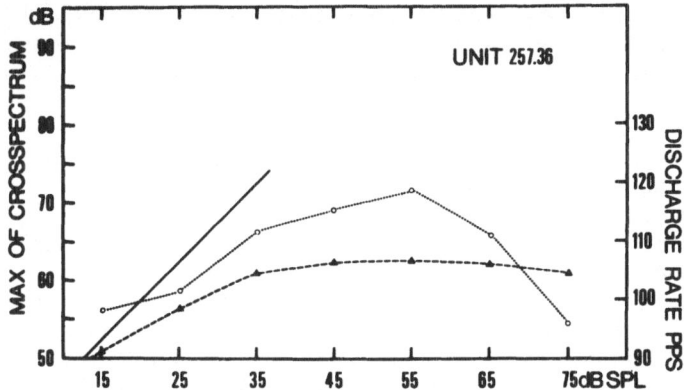

Fig. 4.   Typical average discharge rate (open circles and dashed
          lines) and relative peak value of the cross-spectra
          (triangles and dashed lines) as a function of the stimulus
          sound level (in dB SPL measured in a 1/3 octave band) for
          a nerve fiber in the auditory nerve of a rat (from Møller,
          1978a).

transmit temporal information about a sound even though their
average discharge rate has reached a saturation level. That indicates that the temporal pattern of the discharges may be the carrier
of information about frequency (spectrum) in the physiological
ranges of sound intensities and that frequency discrimination
according to the Place theory may occur only at low sound intensities and possibly in the high frequency range where phase locking
does not occur.

     When the damped oscillation in the cross-correlation functions
is examined in more detail, the fact emerges that the frequency of
this damped oscillation increases along the time scale (Møller and
Nilsson, 1978). The change in frequency of the damped oscillations
obtained from recordings from single nerve fibers of the auditory
nerve at three different intensities is presented in Fig. 5. This
change in frequency along the time scale of the damped oscillations
indicates that the auditory periphery behaves in a manner different
from that of ordinary bandpass filters composed of lumped components. In a system with lumped components, the frequency of the
damped oscillation is constant throughout its duration. The frequency modulation of the damped oscillatory response can be explained on the basis that the selectivity of the auditory frequency
analyzer has its origin in a system that consists of distributed

components. These results indicate that the frequency selectivity
of the peripheral auditory analyzer results from the transmission
line properties of the basilar membrane and not from a simple
resonator.

The results obtained using broadband noise at sound levels well
above threshold serve to question the importance of the spectral
selectivity in the auditory periphery as a basis for frequency
discrimination for complex sounds of physiological sound levels. If
the frequency discrimination in the auditory system relies on the
spectral selectivity of the auditory peripheral analyzer in the way
shown in the experiments described above, perception of pitch should
change with sound intensity. In addition the acuity of the spectrum
analyzer to separate spectral components would likely deteriorate
with increasing sound intensity. However, the results have not been
seen in psychoacoustic experiments.

Using the technique of recording from a large number of auditory nerve fibers in a single animal, a technique originally
developed by Pfeiffer and Kim (1975), Sachs and Young (1980) studied
the response pattern of several hundred nerve fibers to synthetic
vowels. They found that the discharge rate of nerve fibers with
different characteristic frequency reflected the spectrum envelope
of the sound for low stimulus intensities. When the stimulus intensity was increased to levels at or just below physiological
levels the discharge rate in the different nerve fibers no longer
reflected the spectrum of the sound. In further experiments, the
phaselocking to the time pattern of the stimulus sounds was studied
(Young and Sachs, 1980). It was found that the phaselocking was
maintained throughout the range of physiological sound levels,
despite the fact that average discharge rate had reached a saturation level.

These results, together with the results described above using
pseudorandom noise, indicate that the spectrum of a sound is not
conveyed to higher nervous centers of the auditory system by way of
the (average) discharge rate in different nerve fibers. It is more
likely that such information is carried in the time pattern of the
discharges of single auditory nerve fibers. Therefore, the conclusion may be that the spectral selectivity of the auditory peripheral
analyzer, when tested with broadband sounds, is different in many
ways from what it is in response to pure tones at threshold. The
above results indicate that spectral analysis is at least not the
only mechanism that enables the auditory system to discriminate
frequency. It is more likely that the spectral selectivity that
undoubtly exists in the auditory periphery serves to separate sound
with a wide spectrum into a (large) number of channels in which a
temporal analysis is performed. Such a separation in channels each
carrying information in a relatively narrow frequency band, would
facilitate temporal analysis as a basis for frequency discrimination

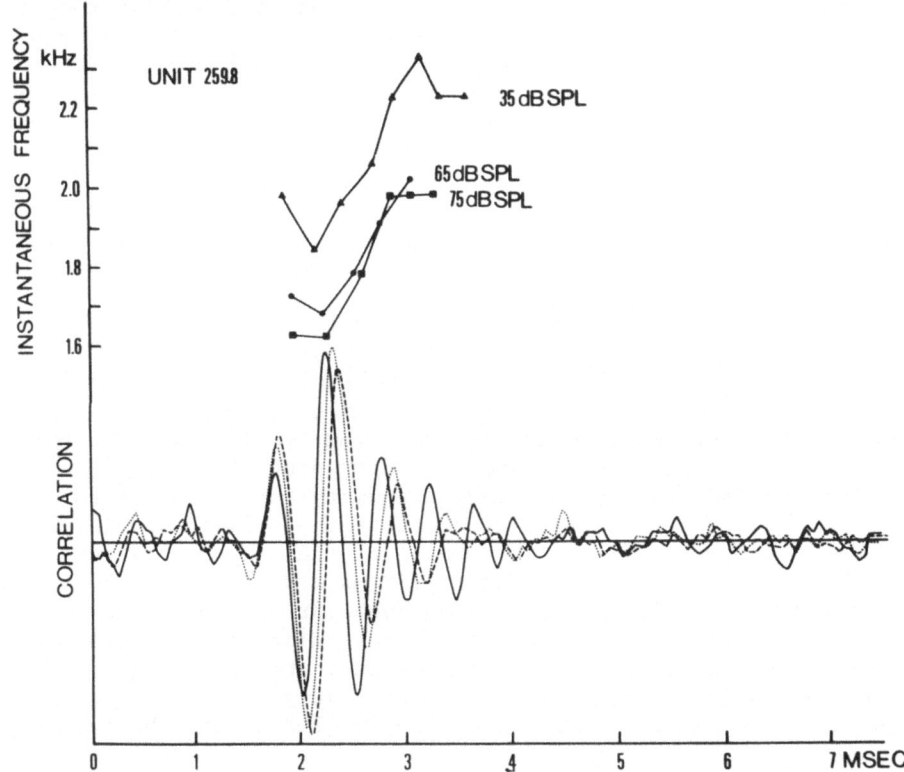

Fig. 5.   Lower graph: three cross correlation functions obtained
          from the response of an auditory nerve fiber in response
          to pseudorandom noise, solid line: 35 dB SPL, dotted line:
          65 dB SPL and dashed line: 75 dB SPL. The upper curves
          show the instantaneous frequency of the damped oscillation
          as a function of time. The instantaneous frequency is the
          inverse time between two zero crossings of a full wave of
          the oscillation, (from Møller and Nilsson, 1970).

of complex sounds.

     In applying the experimental results described above to
physiological sounds, it must be emphasized that both the synthetic
vowels and the pseudorandom noise serving as stimuli are steady
state sounds, while practically all natural sounds vary more or
less rapidly in both spectral composition and intensity. It is
easily demonstrated that even very simple time-varying sounds elicit
responses from neurons in the cochlear nucleus that are fundamen-
tally different from what would have been expected on the basis of
knowledge of the responses of the same nerve cells to sounds with
constant amplitude and spectrum. Relatively few experiments have

been reported where such sounds have been used, and most of the
published results originate from studies of the discharge pattern
of single nerve cells in the cochlear nucleus. These nerve cells
exhibit frequency threshold curves in response to tone bursts with
constant frequency that are not very different from those of single
auditory nerve fibers. When the stimulus is a tone whose frequency
is varied rapidly, the response patterns of the nerve cells in the
cochlear nucleus are markedly different from what would be expected
on the basis of their responses to tones with constant or slowly
varying frequency. Furthermore, the response patterns to tones with
varying frequency change in a systematic way as a function of the
rate of frequency change (Møller, 1969, 1971). A tone with rapidly
varying frequency gives rise to a response patterns that are more
localized in frequency than those of a tone with a slowly varying
frequency. The response pattern to a tone with rapidly varying
frequency is consequently different from what would have been expected on the basis of the frequency tuning properties of the cell
as it appears from the conventional frequency threshold curves
(see Fig. 6). Recent studies indicate that the primary auditory
fibers do not show a similar sharpening of the response pattern to
tones with rapidly varying frequency (Sinex and Geisler, 1980). Thus
the sharpening that is present in the cochlear nucleus is most
likely the result of the transformation of information that occurs
in the cochlear nucleus. Since the frequency threshold curves of
single neurons in the cochlear nucleus are very similar to those of
single auditory nerve fibers, it may be concluded that the transformation that takes place in the cochlear nucleus drastically modifies
the response to tones with rapidly varying frequency without markedly affecting the frequency selectivity to pure tones as it appears
in frequency threshold curves.

When bandpass filtered noise is used as a stimulus, the difference between the responses to steady state sounds and to sounds
with varying frequency become even larger than it was when tones
were used as the stimulus. Bandpass filtered noise with constant
center frequency is a less efficient stimulus than pure tones. The
firing rate of most cochlear nucleus units decreases when the bandwidth of the noise is increased (Greenwood and Goldberg, 1970).
Many nerve cells in the cochlear nucleus do not respond at all to
steady noise if the bandwidth of the noise is larger than a certain
value. However, when the center frequency of such noise is varied
at a certain rate, as illustrated in Fig. 7, units that did not
respond to bandpass filtered noise with slowly varying center frequency are seen to respond vigorously and with great frequency
selectivity to the same noise when its center frequency is varied
rapidly. Fig. 8 shows the relative height of the peaks of the
histograms of the responses to tones and noise bands as a function
of the rate of change in frequency of the tones and the center
frequency of the noise. These results show that the maximum relative increase in peak height has a smaller value for narrow band

noise than it has for a pure tone. It is clearly seen that these
units respond much better to sounds with changing frequency than
those with constant frequency.

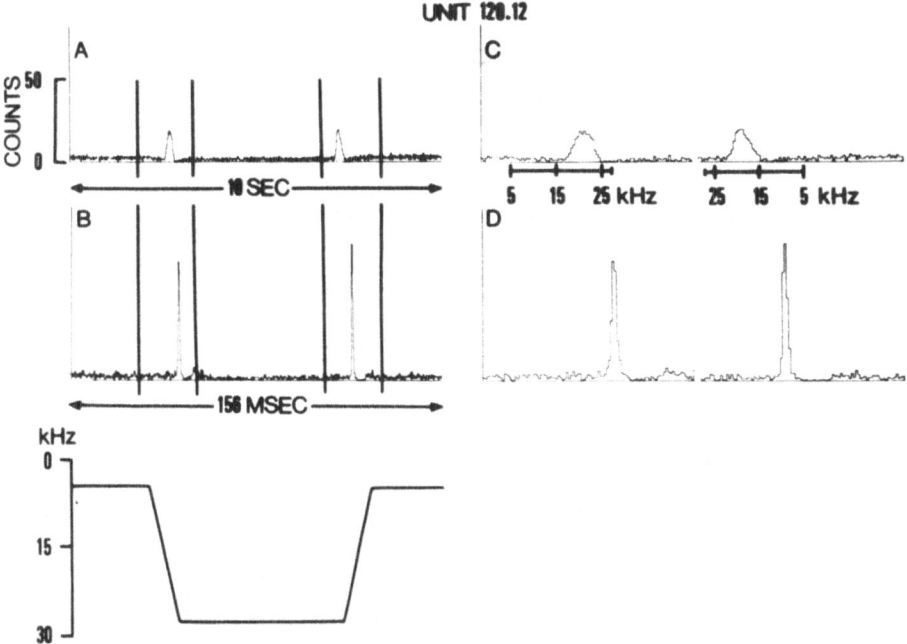

Fig. 6. Period histograms showing the distribution of discharges
         in response to trapezoidal frequency modulated tones (sweep
         tones) of different rates. The histograms show the distri-
         bution of nerve impulses over one cycle of the modulation.
         Upper row of histograms: sweep rate 0.1 Hz (period length
         10 second), corresponding to a rate of change in frequency
         of tone of 35 kHz/s. Left column of histograms: full sweep
         modulation; scheme of frequency change shown below. Right
         column of histograms: time scale expanded to show the part
         around the peaks in detail (as indicated by vertical lines
         in the histograms in the left column). Lower row of histo-
         grams: sweep rate 6.4 Hz (period length of 156 ms) cor-
         responding to a rate of change in frequency of 2.25 MHz/s.
         Each half of the histograms shows 100 bins corresponding
         to 16% of one full sweep cycle (from Møller, 1974b).

     Most natural sounds vary in amplitude more or less rapidly.
The discharge pattern of neurons in the cochlear nucleus typically
reproduces small changes in the intensity of a pure tone very well
(Møller, 1972). This is illustrated in Fig. 9 where a period histo-

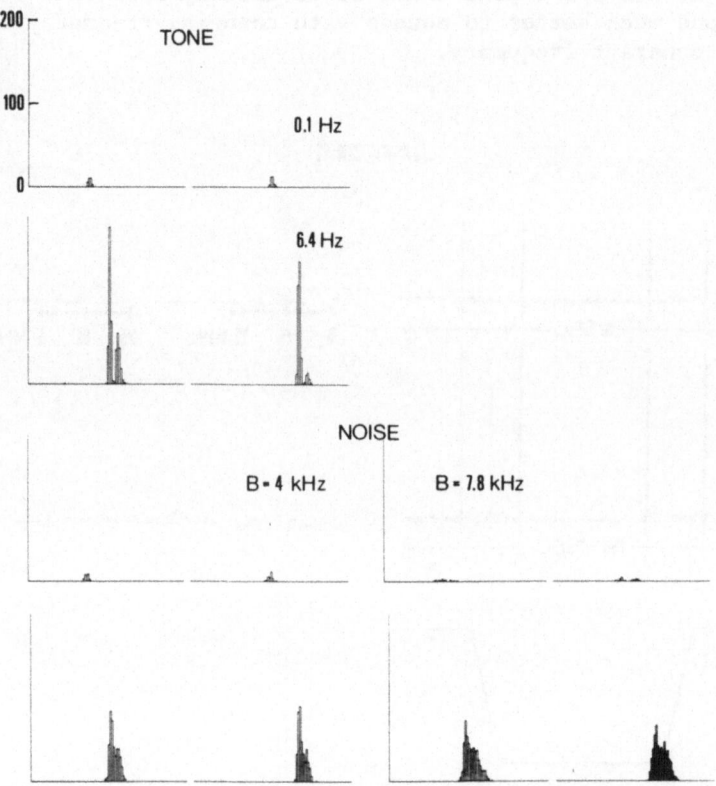

Fig. 7. Period histograms, of the responses to tones with varying
         frequency and to noise bands with varying center frequency.
         The histograms are similar to those of Fig. 6, right
         column and two different sweep rates, 0.1 and 6.4 Hz are
         shown. Results obtained using noise with two different
         bandwidths (4 and 7.8 kHz) are shown (from Møller, 1978c).

gram of the response to a sinusoidally amplitude modulated tone of
a unit in the cochlear nucleus is shown for two different modulation frequencies, 25 and 200 Hz. At the low modulation frequency
there is a relatively small modulation of the histogram, whereas
a modulation frequency of 200 Hz results in an almost 50% modulation of the histogram. The modulation of the intensity of the tone
was only $\pm$ 1 dB. The modulation of the period histograms presented
in Fig. 9 shows that a small change in the stimulus intensity gives
rise to a large modulation of the discharge rate of these neurons.
In many units, 2-4 dB modulation of the stimulus sound results in
a nearly 100% modulation of the discharge rate over sound intensities covering the physiological range. Fig. 9 also shows that the
degree of modulation of the discharge rate is different for different modulation frequencies. The modulation frequency at which

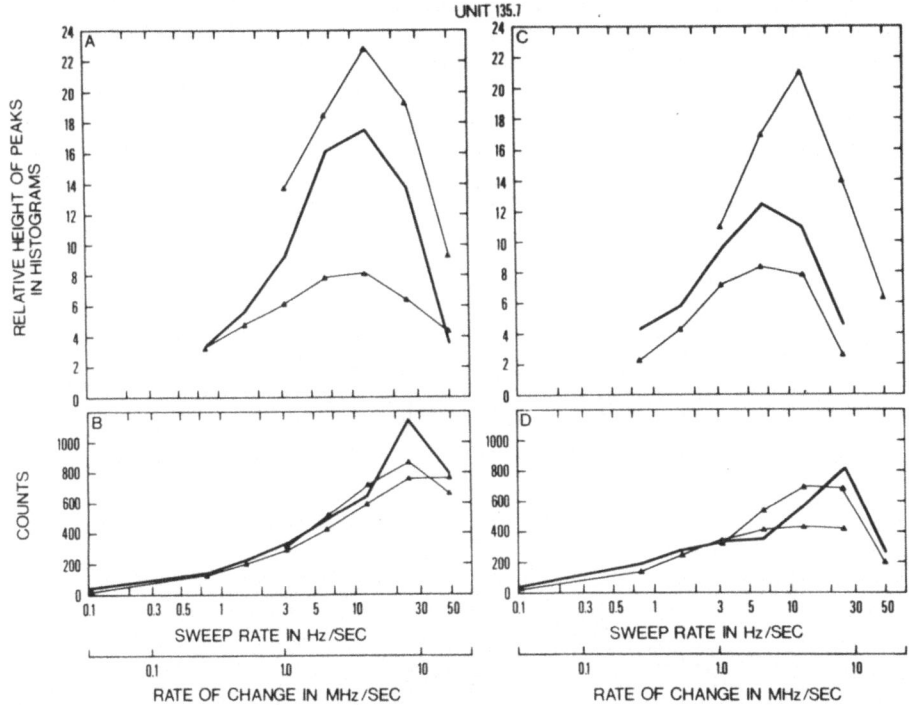

Fig. 8. Relative height of the histograms of the responses to tones
with changing frequency (heavy lines) and noise bands with
changing center frequency, (filled circles: bandwidth =
4 kHz, open circles: bandwidth = 7.8 kHz), shown as a function of rate of change (sweep rate, Hz/s, and rate of frequency change, MHz/s). The left hand graphs illustrate
increasing frequency (upgoing sweep) and the right hand
graph shows results obtained with increasing frequency
(from Møller, 1974a).

the cochlear nucleus neurons are most sensitive to amplitude modulation varies from unit to unit, but for most units it is in the range
of 50 to 300 Hz (Møller, 1972; Hirsch and Gibson, 1976). Thus, both
below and above that frequency, amplitude modulation gives rise to
less modulation of the discharge rate. It is also important to note
that the reproduction of small amplitude changes modulation gives
rise to a modulation of the discharge rate over a large range of
sound intensities (Møller, 1974c). This is illustrated in Fig. 10
which shows typical period histograms of the response from a nerve
cell in the cochlear nucleus to stimulation by a pure tone that was
sinusoidally amplitude modulated at two different modulation frequencies and for three different stimulus intensities. The mean
discharge rate of this unit reached a saturation level only about

10 dB above threshold (28 dB). However, it is obvious that despite
the fact that the mean discharge rate is nearly constant and
independent of the sound intensity, small, rapid changes in the
sound intensity amplitude modulation give rise to a large change
in the discharge rate provided that the intensity change occurs at
a certain rate. Thus, the slow (10 Hz) modulation results in a very
small modulation of the histograms at sound intensities more than
20 dB above threshold, whereas the fast (100 Hz) modulation of the
stimulus tone gives rise to a large modulation of the discharge
rate at all three intensities shown.

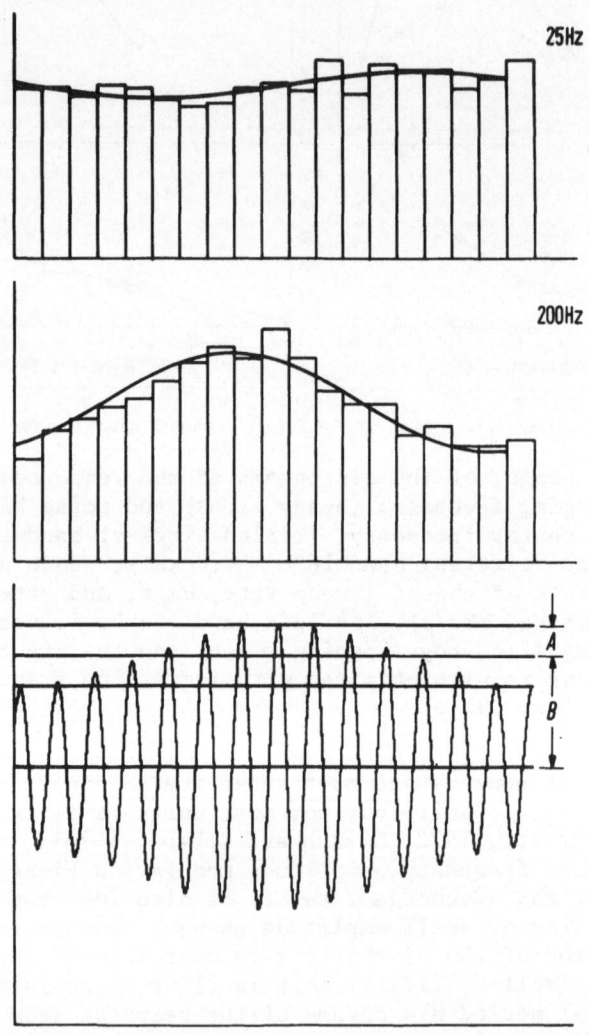

Fig. 9. Period histograms of the responses of a cell in the
         cochlear nucleus of a rat to sinusoidally (cont´d)

(Fig. 9. cont´d) amplitude modulated tones. The upper histograms were obtained at a modulation frequency of 25 Hz and
the lower at 200 Hz. The modulation depth was 20% in both
cases. The stimulus tone is seen below (from Møller, 1974b)

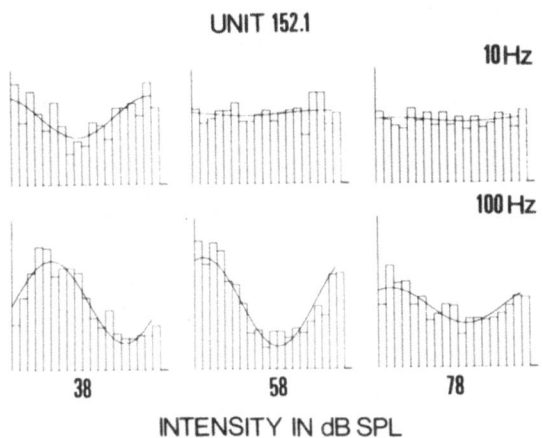

Fig. 10.   Period histograms of the responses of a cell in the
           cochlear nucleus of a rat to amplitude modulated tones
           at three different sound intensities. The modulation
           frequency was 10 Hz (upper row) and 100 Hz (lower row)
           and the modulation depth was 30%. The frequency of the
           tone was equal to the cells characteristic frequency
           (from Møller, 1974c).

      It is clear from the above experimental results that the
peripheral auditory system processes complex sounds in a way that
cannot be predicted on the basis of results obtained using simple
sounds such as pure tones and clicks. It is also evident that the
transformation in the cochlear nucleus of the information carried
by the fibers of the auditory nerve is substantial and the response
to time varying, complex sounds cannot be estimated either on the
basis of results obtained in studies where pure tones are used or

where steady state complex sounds are used. Results reported above
indicate that changes in both frequency and amplitude of sounds are
enhanced in the responses of single nerve cells in the cochlear
nucleus. Knowledge about the transformations that take place in the
more central auditory relay nuclei is sparse, but there is no reason
to believe that it is any less extensive than that demonstrated in
the cochlear nucleus. A few studies on coding of sounds, the amplitude or frequency of which change in single nerve cells in the
inferior colliculus, support that assumption. Thus, although frequency selectivity is a pronounced property of unit responses when
pure tones are used as stimuli, it may play a much less significant
role in processing of complex sounds than is indicated by the results
obtained using pure tone stimuli. The discharge pattern in response
to complex sounds in the physiological range of intensities seems
instead to be very sensitive to changes in the sound's frequency
and intensity. If the signal processing that takes place in the
auditory periphery is to be summarized briefly, the most adequate
description may therefore be that the system enhances changes in a
complex sound, rather than conveying spectrally analyzed information about the absolute distribution of energy within the audible
frequency range.

REFERENCES

Greenwood, D. D. and Goldberg, J. M., 1970, Response of neurons in
    the cochlear nuclei to variations in noise bandwidth and to
    tone-noise combinations, J. Acoust. Soc. Am., 47: 1022-1040.
Hirsch, H. R. and Gibson, M. M., 1976, Responses of single units
    in the cat cochlear nucleus to sinusoidal amplitude modulation
    of tones and noise: linearity and relation to speech percep-
    tion, J. Neurosci. Res., 2: 337-356.
Kiang, N. Y-S., Watanabe, T., Thomas, E. C. and Clark, L. F., 1965,
    "Discharge Patterns of Single Fibers in the Cat's Auditory
    Nerve", MIT Press, Cambridge, Mass.
Kim, D. O., 1980, Cochlear mechanics: implications of electro-
    physiological and acoustical observations, Hearing Res.,
    (in press).
Kim, D. O., Siegel, J. H. and Molnar, C. E., 1979, Cochlear non-
    linear phenomenon in two-tone responses, Scand. Audiol. Suppl.,
    9: 63-81.
Marmarelis, P. Z. and Marmarelis, V. Z., 1978, "Analysis of physio-
    logical systems. The white noise approach", Plenum Press, New
    York and London.
Møller, A. R., 1969, Unit responses in the cochlear nucleus of the
    rat to sweep tones, Acta Phys. Scand., 76: 503-512.
Møller, A. R., 1971, Unit responses in the rat cochlear nucleus to
    tones of rapidly varying frequency and amplitude, Acta Phys.
    Scand., 81: 540-556.
Møller, A. R., 1972, Coding of amplitude and frequency modulated

sounds in the cochlear nucleus of the rat, Acta. Phys. Scand.,
    86: 223-238.
Møller, A. R., 1974a, Coding of sounds with rapidly varying spectrum
    in the cochlear nucleus, J. Acoust. Soc. Am., 55: 631-640.
Møller, A. R., 1974b, Coding of amplitude and frequency modulated
    sounds in the cochlear nucleus, Acustica, 31: 202-299.
Møller, A. R., 1974c, Responses of units in the cochlear nucleus to
    sinusoidally amplitude modulated tones, Exp. Neurol., 45: 104-
    117.
Møller, A. R., 1977, Frequency selectivity of single auditory nerve
    fibers in response to broadband noise stimuli, J. Acoust. Soc.
    Am., 62: 135-142.
Møller, A. R., 1978a, Responses of auditory nerve fibers to noise
    stimuli show cochlear nonlinearities, Acta Oto-Laryngol., 86:
    1-8.
Møller, A. R., 1978b, Frequency selectivity of the peripheral audi-
    tory analyzer studied using broadband noise, Acta Phys. Scand.,
    104: 24-32.
Møller, A. R., 1978c, Frequency analysis in the periphery of the
    auditory system, Proc. of the Kybernetik Kongress, Munich,
    264-287.
Møller, A. R. and Nilsson, H. G., 1970, Inner ear impulse response
    and basilar membrane modelling, Acustica, 41: 258-262.
Pfeiffer, R. R. and Kim, D. O., 1974, Cochlear nerve fiber re-
    sponses: distribution along the cochlear partition, J. Acoust.
    Soc. Am., 58: 867-869.
Rhode, N. S., 1971, Observations of the vibration of the basilar
    membrane in squirrel monkeys using the Mossbauer technique,
    J. Acoust. Soc. Am., 49: 1218-1231.
O'Leary, D. P. and Honrubia, V., 1975, On-line identification of
    sensory systems using pseudorandom noise pertubations, Bio-
    phys. J., 15: 505-532.
Russel, I. J. and Sellick, P. M., 1978, Intracellular studies of
    hair cells in mammalian cochlea, J. Physiol. (London), 283:
    261-290.
Russel, I. J. and Sellick, P. M., 1980, Hearing Res., (in press).
Sachs, M. B. and Young, E. D., 1980, Effects of nonlinearities on
    speech encoding in the auditory nervy, J. Acoust. Soc. Am.,
    (in press).
Sinex, D. G. and Geisler, C. D., 1980, Auditory nerve-fiber re-
    sponses to frequency modulated tones. Third Midwinter Meeting
    of the Association for Research in Otolaryngology.
Young, E. D. and Sachs, M. B., 1979, Representation of steady-state
    vowels in the temporal aspects of the discharge pattern of
    populations of auditory nerve fibers, J. Acoust. Soc. Am.,
    66: 1381-1403.

ON PREDICTING THE RESPONSE OF AUDITORY NERVE FIBERS TO COMPLEX

TONES

Eric Javel

The Boys Town Institute for Communication
Disorders in Children
Omaha, Nebraska 68131, U.S.A.

INTRODUCTION

The response to sound of the mechanical and neural elements of
the peripheral auditory system is decidedly nonlinear. In recent
years quantitative studies of auditory responses to complex or
multicomponent tones have been performed with the result being that
the effects of the various nonlinearities, if not their causes, are
known in reasonable detail. However, despite the sizeable amount of
work, the way tones combine to produce responses in auditory nerve
fibers has remained unclear.

Presented here is a computational scheme which employs linear
combinations of basic response properties of auditory nerve fibers
to predict responses to arbitrary complex tones. The model predicts
both discharge rate and synchronization behavior. While it has been
developed to account for responses to low-frequency stimuli, it can
be expanded to accomodate responses to high-frequency tones as well.

RESULTS

Inputs to the Model

For a given signal consisting of n frequency components, the
model requires three inputs for each component. These are (1) the
behavior of discharge rate as a function of intensity, (2) the
behavior of discharge synchronization as a function of intensity
and (3) a description of the ability of each component to suppress
responses to other components.

Discharge rate and synchronization are described effectively by the logistic function y = 1 / (1 + exp(-ks)). For discharge rate and synchronization, respectively, the forms of the functions are

$$Rd = \frac{Rsat - Rspont}{1 + 10^{Sr(Tr-I)/20}} \quad \text{and} \quad V = \frac{Vmax}{1 + 10^{Sv(Tv-I)/20}}$$

For the equation describing discharge rate behavior, Rd is the expected "driven" discharge rate, Rsat is the saturation of maximum rate, Rspont is the spontaneous discharge rate, Sr is the slope of the function at its midpoint, Tr is the discharge rate treshold in decibels (dB), and I is the intensity of the tone in dB. The overall discharge rate is the "driven" rate plus the spontaneous rate, or R = Rd + Rspont. For the equation describing synchronization behavior, V is the expected synchronization coefficient, Vmax is the maximum synchronization, Sv is the slope of the function at its midpoint, Tv is the synchronization treshold in dB, and I is the intensity of the tone in dB. Minimum synchronization is zero.

These equations indicate monotonically-increasing, saturating responses. They are similar to those derived for discharge rate by Sachs and Abbas (1974), modified to accept intensity in dB rather than pressure units. Although the forms of the discharge rate and synchronization functions are similar, it is necessary to describe them separately because (a) synchronization typically possesses a lower threshold and slope than discharge rate and (b) synchronization behavior cannot be predicted from discharge rate behavior (McGee et al., 1979). The synchronization index we use is the vector strength measure of Goldberg and Brown (1969) which is equivalent to the magnitude of the coefficient at the stimulus frequency of Fourier transforms of period histograms.

The suppressive effect of a signal component on any other component is described by

$$S = 10 \, N \, \log \left(1 + 10^{(I-Q)/10} \right)$$

where S is the expected suppression in dB, N is the asymptotic slope of suppression growth (in dB suppression per dB increment in suppressor intensity), I is the intensity of the suppressor tone in dB, and Q is a factor related to suppression threshold. Q is given by

$$Q = Ts - (1/N) - 10 \, \log \left(1 + 10^{-1/10 \, N} \right)$$

where Ts is the suppression threshold in dB and N is as defined above. The suppression equations are modifications of those given

in Sachs and Abbas (1976), and they effectively describe our findings (Javel et al., 1978; Javel, 1979) that suppression behavior
depends only on the parameters of the suppressor tone, suppression
does not saturate, and suppression possesses a growth characteristic
which is ultimately linear in dB but which does not necessarily
have unity slope.

Outputs of the Model

The expected synchronization to a component in a complex input
signal is the synchronization generated at an "effective" intensity
which is lower than that actually used. The amount of the reduction
is equal to the sum, in dB, of the suppressive effects of the other
components, i.e.,

$$V\check{\ } (Fn, In) = V (Fn, In-S)$$

where $V\check{\ }(Fn, In)$ is the degree of synchronization at frequency Fn,
presented at intensity In along with other tones, and S is the sum
of the suppressive effects (in dB) on Fn of the other signal components. When suppression exists, then, synchronization is attenuated.
Since the response to each tone is suppressed by the other tones
present, suppression is mutual. That is, in a two-tone case $F2$
suppresses $F1$ at the same time that $F1$ suppresses $F2$.

The discharge rate contributed by a component to the overall
response is equal to its "unsuppressed" rate (the rate which would
have resulted had the tone been presented alone), multiplied by a
fraction which is the ratio of the "suppressed" synchronization
coefficient to the "unsuppressed" coefficient, i. e.,

$$Rd\check{\ } = (V\check{\ }/V)\ Rd$$

Alternatively this may be stated that the ratio of "suppressed" to
"unsuppressed" discharge rate is equal to the ratio of "suppressed"
to "unsuppressed" synchronization. Thus, suppression operates on
synchronization, which in turn operates on discharge rate.

Overall discharge rate in response to the complex tone is
simply the sum of the several driven rates, plus the spontaneous
rate.

Comparison of Predicted and Obtained Data

In Fig. 1 are shown the predictions of the model and data
obtained for overall discharge rate (part A), and synchronization
to F1 (part B) and F2 (part C), for a cat auditory nerve fiber.
The signal consisted of two tones at F1 = 5ØØ Hz and F2 = 4ØØ Hz.
Within a given sequence, F2 was fixed in intensity and F1 was varied
in intensity. The predictions for synchronization and for overall

discharge rate match those obtained reasonably well. Note that the model predicts that overall discharge rate when F2 is present at 9∅ dB will be greater than the maximum discharge rate for F1 presented alone, and that this is exactly what occured. Note also that the model predicts no suppression of overall discharge rate for this F2, and indeed no rate suppression resulted.

Utilizing input parameter values which we have found to be typical for two-tone situations in other cat auditory nerve fibers, we have been able to generate model predictions which range from considerable discharge rate suppression (usually in the case of F2 > F1) to no suppression of discharge rate (usually in the case of F2 < F1). Discharge synchronization, however, always exhibits suppression. The one parameter which affects the model output most drastically is N, the slope of the suppression growth function. Even though the way the model operates is the same for all stimulus configurations, the interplay among the parameters may cause very different modes of response to occur.

DISCUSSION

The model described here fits two-tone data well, but we have yet to test it for signals which consists of more than two tones or which consist of tones plus noise. For signals which possess a noise component, the suppression and discharge rate requirements of the model can easily be met, but establishing synchronization parameters would require techniques such as cross-correlation which are not with the present scope of the model. The same problem exists for signals which are too high in frequency to elicit phaselocked activity.

We have indicated that suppression operates on discharge synchronization and is not necessarily reflected in overall discharge rate. This implies that estimates of suppression magnitude obtained using discharge rate as an indicator are in error, and that considerable suppression may exist without it showing up in overall discharge rate. This can explain why some investigators have been unable to demonstrate suppression in the case of F2 < F1. As noted earlier, however, while suppression of discharge rate may not occur, suppression of synchronization always occurs.

In producing an output, our model distinguishes between the potential for excitation (Rd), and the excitation contributed by a suppressed component to the overall response (Rd⁻).This implies that the physical mechanism that generates suppression follows the mechanism that generates discharge rate, and that the two mechanisms are independent of one another. One interpretation is that the suppression mechanism resides in the hair cell and results from "competition" among signal components for representation at the hair

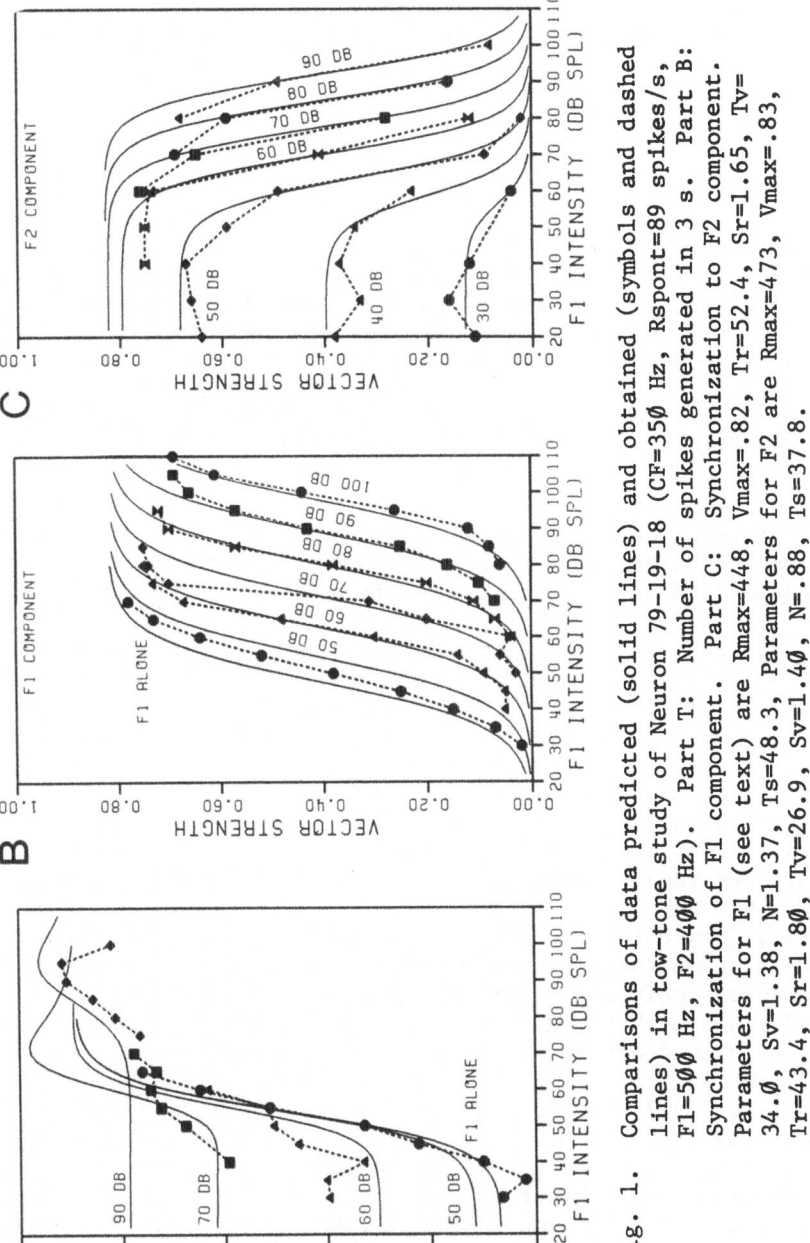

Fig. 1. Comparisons of data predicted (solid lines) and obtained (symbols and dashed lines) in tow-tone study of Neuron 79-19-18 (CF=350 Hz, Rspont=89 spikes/s, F1=500 Hz, F2=400 Hz). Part T: Number of spikes generated in 3 s. Part B: Synchronization of F1 component. Part C: Synchronization to F2 component. Parameters for F1 (see text) are Rmax=448, Vmax=.82, Tr=52.4, Sr=1.65, Tv= 34.0, Sv=1.38, N=1.37, Ts=48.3. Parameters for F2 are Rmax=473, Vmax=.83, Tr=43.4, Sr=1.80, Tv=26.9, Sv=1.40, N=.88, Ts=37.8.

cell output. This approach can explain why Rhode (1977), who measured suppression of basilar membrane motion, obtained suppression magnitudes considerably lower than those observed in either hair cells (Sellick and Russell, 1979) or auditory nerve fibers. While suppression of basilar membrane response exists, it is probably too small to be of consequence in affecting auditory nerve fibers' responses.

The operation of the model presented here can be compared to that derived by Rose et al. (1974). For several reasons, we have found Rose's assumptions and predictions to be not applicable to auditory nerve fibers. Among other arguments, his method of estimating "damping power" or suppression from pure-tone responses is invalid and his "cancellation" method for determining response dominance produces improper interactions among signal components.

ACKNOWLEDGMENTS

This work was supported by NINCDS grant NS-14880 and by a Biomedical Research Development Grant, both from NIH. The author acknowledges the help of JoAnn D. McGee and Edward J. Walsh, who assisted in data acquisition and analyses.

REFERENCES

Goldberg, J. M. and Brown, P. B., 1969, Response of binaural neurons of dog superior olivary complex to dichotic tonal stimuli: Some physiological mechanisms of sound localization, J. Neurophysiol., 32: 613-636.

Javel, E., 1979, Two-tone suppression in the auditory nerve of the cat: Suppression threshold and rate of growth, J. Acoust. Soc. Am., 66: S48.

Javel, E., Geisler, C. D. and Ravindran, A., 1978, Two-tone suppression in the auditory nerve of the cat: Rate-intensity and temporal analyses, J. Acoust. Soc. Am., 63: 1093-1104.

McGee, J., Walsh, E. J. and Javel, E., 1979, Discharge synchronization in auditory nerve and AVCN neurons, J. Acoust. Soc. Am., 65: S83.

Rhode, W. S., 1977, Some observations on two-tone interaction measured with the Mossbauer effect, in: "Psychophysics and Physiology of Hearing", E. F. Evans and J. P. Wilson, eds., Academic Press , New York.

Rose, J. E., Kitzes, L. M., Gibson, M. M. and Hind, J. E., 1974, Observations on phase-sensitive neurons of anteroventral cochlear nucleus of the cat: Nonlinearity of cochlear output, J. Neurophysiol., 37: 218-253.

Sachs, M. B. and Abbas, P. J., 1974, Rate versus level functions for auditory-nerve fibers in cats: Tone-burst stimuli, J. Acoust. Soc. Am., 56: 1835-1847.

Sachs, M. B. and Abbas, P. J., 1976, Phenomenological model for two-tone suppression, J. Acoust. Soc. Am., 60: 1157-1163.

Sellick,P. M. and Russell, I. J., 1979, Two-tone suppression in cochlear hair cells, Hearing Res., 1: 227-236.

EFFECTS OF MASKING NOISE ON THE REPRESENTATION

OF VOWEL SPECTRA IN THE AUDITORY NERVE

Herbert F. Voigt, Murray B. Sachs and Eric D. Young

Dept. of Biomedical Engineering
The Johns Hopkins School of Medicine
Baltimore, Maryland, U.S.A.

INTRODUCTION

Steady-state vowels can be discriminated in background noise
at signal to noise ratios of less than −15 dB (Dewson, 1968). Thus,
the information relating to the spectrum of a vowel stimulus which
is contained in the firing patterns of the ensemble of auditory-
nerve fibers must remain when the vowels are presented in background
noise. Spectral features may be encoded in auditory-nerve discharge
patterns in at least two ways: One in terms of the distribution of
average discharge rate as a function of fibers´ characteristic
frequencies (CF) (rate-place representation, Sachs and Young, 1979),
and the other in terms of temporal or phase-locked responses
(temporal-place representation, Young and Sachs, 1979). We have
shown previously (Young and Sachs, 1979) that the temporal-place
code provides a more robust representation of vowel formant frequencies than does the rate-place code. In this paper we shall compare
the effects of broad-band noise on these two representations of the
spectrum of the vowel /ɛ/.

METHODS

Large numbers of auditory-nerve fibers were studied in cats
anesthetized with pentobarbital. The CF of each fiber was determined and responses to the synthesized, steady-state vowel /ɛ/,
presented alone and in the presence of noise, were recorded. The
spectrum of /ɛ/ is shown in Fig. 1; the fundamental frequency of
the vowel is 112 Hz. The first formant frequency is 0.512 kHz and
is located between the fourth and fifth harmonics of the fundamental; the second formant is at 1.792 kHz and the third formant

is at 2.432 kHz. Vowel duration was 400 ms; stimuli were presented once per second until 1000 spikes were recorded during the stimulus interval. A rate-place profile was obtained by plotting each fiber's average discharge rate (measured over the last 380 ms of the vowel stimulus) as a function of the CF of the fiber.

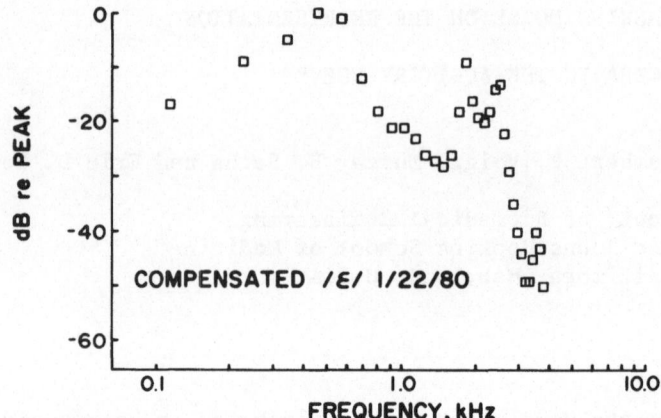

Fig. 1.   Amplitude spectrum of the vowel /ɛ/; measured near the cat's eardrum during the experiment. The spectrum of /ɛ/ has been compensated for the human external ear transfer function (Wiener and Ross, 1946).

Period histograms were computed as estimates of a fiber's instantaneous discharge rate as a function of time through one cycle of the vowel. The discrete Fourier transform of the period histogram was computed. The amplitude (spikes/s) of each harmonic component in the transform was used as a measure of the fiber's response to the stimulus component at the corresponding frequency. We have previously shown that the following temporal-place profile provides a useful representation of the response of the population of auditory-nerve fibers to a vowel: For each harmonic of the vowel the magnitude of the Fourier transform component at that harmonic is averaged over all the fibers with CFs within half an octave of the harmonic frequency. This measure, called the "average localized synchronized rate," (ALSR) is plotted as a function of harmonic frequency. Note that it is a combination of rate, place and temporal information about the stimulus spectrum.

In order to study the effects of broad-band masking noise on these two representations of vowel spectra, responses to the vowel

were recorded in a silent background and in the presence of continuous broad-band masking noise. The level of the vowel was constant
at 67 dB SPL and the level of the masking noise was varied, producing signal-to-noise ratios of 26, 16, 6 and -4 dB. Signal-to-noise
ratio was measured in the band below 3.0 kHz.

RESULTS

    Fig. 2 illustrates the data obtained from one unit. The CF of
the fiber was 740 Hz. The left column shows post-stimulus time (PST)
histograms and the right column shows period histograms derived
from the same data. The top row shows the responses to the vowel
alone. The PST histogram shows an increase in average rate elicited
by the vowel. The period histogram shows a phase-locked response
dominated by the fifth harmonic of the vowel. At a signal-to-noise
ratio of 16 dB, the change in average rate produced by the vowel
is reduced. The rate response disappears completely at -4 dB. However, there is a clear phase-locked response in the period histogram even at the lowest signal-to-noise ratio. Similar phase-locked
responses to vowel fundamental frequencies have been shown by Kiang
and Moxon (1974).

    In Fig. 3 we compare the rate-place profile with the temporal-
place representation of the ALSR. In the left column the average
discharge rates of individual fibers are plotted as a function of
their CFs. Units with spontaneous rates greater than 1 spike/s are
plotted as X´s and low spontaneous rate units (less than 1 spike/s)
are plotted as squares. The solid line is a moving window weighted
average of the rates of the high spontaneous units. In the rate
profile for the vowel alone (upper left), there is at most a weak
indication of peaks at the formant frequencies (arrows). There
are no signs of formant-related peaks at either of the signal-to-
noise ratios shown (6 dB and -4 dB), because most of the units´
discharge rates are driven to their maximum values by the noise.

    The temporal-place representations for these same three condi-
tions are shown on the right. In each plot there are three prominent
peaks which correspond to the first three formants of the vowel
(arrows). The peaks stand out at both signal-to-noise ratios. There
is another peak in the ALSR for the vowel alone at frequencies
corresponding to the 9th and 10th harmonics of the fundamental
frequency. This peak does not correspond to a peak in the vowel
spectrum (Fig. 1), and is most likely a distortion product related
to other vowel components, similar to the component at the second
harmonic of the first formant discussed in our earlier paper (Young
and Sachs, 1979). Notice that adding a broad-band noise background
to the vowel causes a reduction in the size of this peak.

CONCLUSION

The temporal-place representation of vowel spectra is superior
to the rate-place representation at moderate to high vowel levels
(Young and Sachs, 1979). In addition it retains information about
vowel spectra in the presence of background noise. The rate-place
representation does not reflect the formant structure of the vowel
even at moderate signal-to-noise ratios.

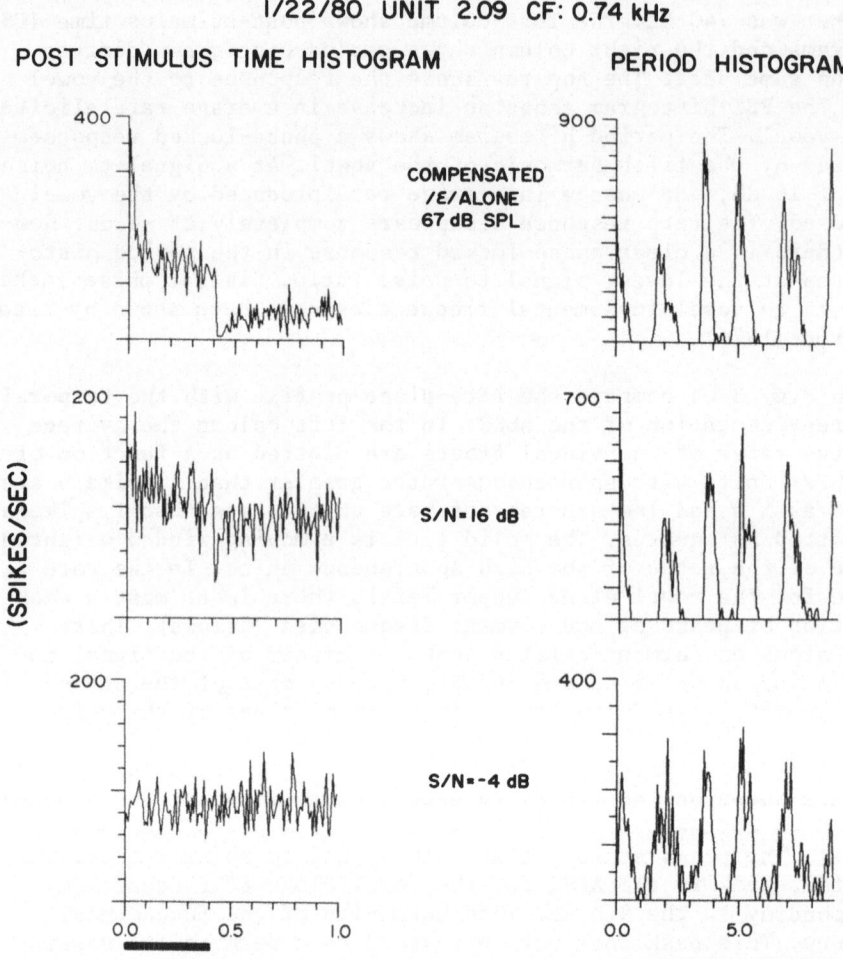

Fig. 2.   Post stimulus time (PST) histograms (left column) and period
          histograms (right column) from one of 190 units recorded
          on 1/22/80. The period histogram abscissa is one cycle of
          the vowel.

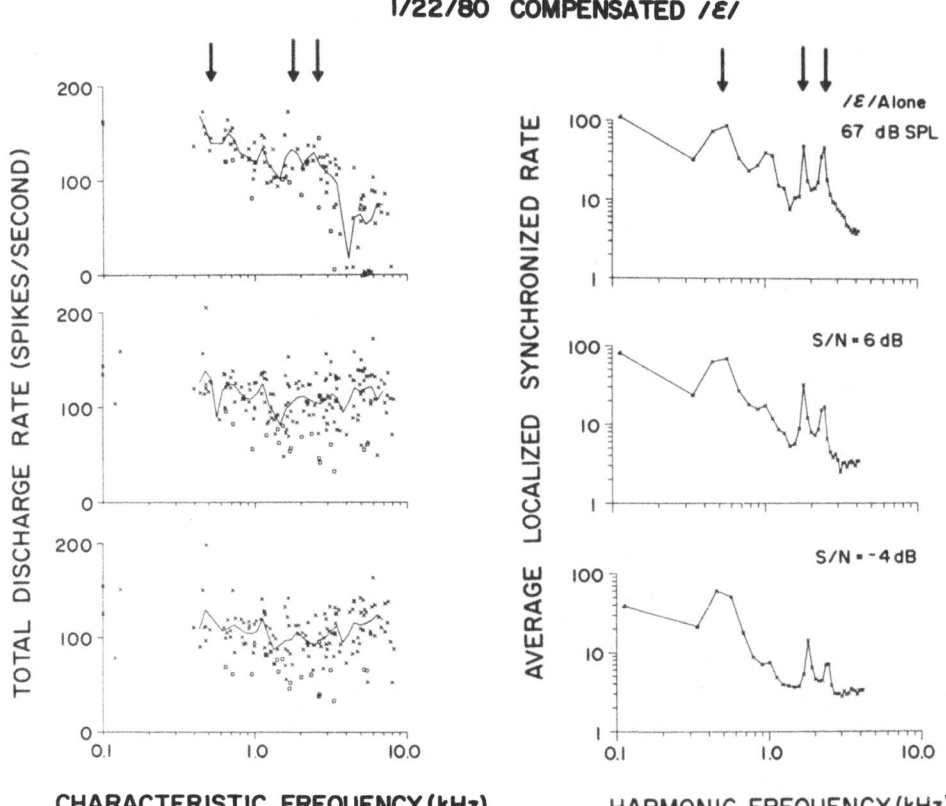

Fig. 3. Rate-place representation (rate profiles, left column) and
         temporal-place representation (ALSR, right column) derived
         from vowel alont (top row) and vowel-in-noise (middle and
         bottom rows) conditions. Arrows point to formant frequencies
         of /ɛ/. In the rate profiles, units with spontaneous rates
         greater thal 1 spike/s are plotted as X´s; low spontaneous
         units (less than 1 spike/s) are plotted as squares.

REFERENCES

Dewson, J. H., 1968, Efferent olivocochlear bundle: Some relation-
    ships to stimulus discrimination in noise, J. Neurophysiol.,
    31: 122-130.
Kiang, N. Y. S. and Moxon, E. C., 1974, Tails of tuning curves of
    auditory-nerve fibers, J. Acoust. Soc. Am., 55: 620-630.
Sachs, M. B. and Young, E. D., 1979, Encoding of steady-state vowels
    in the auditory nerve: Representation in terms of discharge
    rate, J. Acoust. Soc. Am., 66: 470-479.

Wiener, F. M. and Ross, D. A., 1946, The pressure distribution in
    the auditory canal in a progressive sound field, J. Acoust.
    Soc. Am., 18: 401-408.
Young, E. D. and Sachs, M. B., 1979, Representation of steady-state
    vowels in the temporal aspects of the discharge patterns of
    populations of auditory-nervy fibers, J. Acoust. Soc. Am., 66:
    1381-1403.

NEURONAL CIRCUITS IN THE DORSAL COCHLEAR NUCLEUS

Kirsten K. Osen and Enrico Mugnaini*

Anatomical Institute, University of Oslo
Oslo 1, Norway
*Department of Biobehavioral Sciences, University
of Connecticut
Storrs, 06268, Connecticut, U.S.A.

The dorsal cochlear nucleus (DCN) is of interest from several points of view, not least from its large variation among species. In rodents and carnivores the nucleus is multilayered with many structural similarities with the cerebellar cortex (Lorente de Nó, 1933, 1979; Mugnaini et al., 1980a,b,c). In primates and particularly man, it is less clearly stratified (Moore and Osen, 1979; Moore, 1980). In whales it is either rudimentary (Osen and Jansen, 1965) or absent (Ogawa and Arifuko, 1948). The cytoarchitecture of the DCN has been described by several authors (Lorente de Nó, 1933, 1979; Osen, 1969; Brawer et al., 1974; Kane, 1974a,b; Moore and Osen, 1979; Moore, 1980; and others), but up to now the intrinsic neuronal circuitry of the nucleus has been poorly understood and nothing is known about its functional role in audition.

The present account is a survey of our published as well as ongoing observations on the DCN in the mouse, rat, guinea pig, and cat (Mugnaini et al., 1980a,b,c). These studies include sections from normal and lesioned animals stained with procedures such as Golgi, Nissl, Woelke, Nauta, Fink and Heimer, Koelle, Protargol, HRP, immunohistochemical methods for GAD, semithin sections, and electron microscopy. This description will be limited to the DCN of the cat which has been studied most extensively, but it applies largely to other subprimate mammals as well.

In the cat the long axis of the DCN is tilted with its dorsal end 45° medially and 15° caudally from the vertical direction. The axis is slightly curved, expecially at its dorsal and ventral ends (Fig. 1A). The descending branches of the cochlear nerve fibers,

after having penetrated the posteroventral cochlear nucleus, bend
laterally and rostrally to enter the DCN at its caudal margin
(Fig. 1B,E). Here they are ordered cochleotopically across the long
axis of the nucleus with the basal (high frequency) fibers situated
dorsally and the apical (low frequency) fibers ventrally (Fig. 1B).

The superficial molecular layer of the DCN is encircled by a
granule cell layer (Fig. 1C) that spreads rostrally over the surface
of the ventral cochlear nucleus (VCN)(Mugnaini et al., 1980c). Granule cells are also found in the various layers of the DCN. The thin

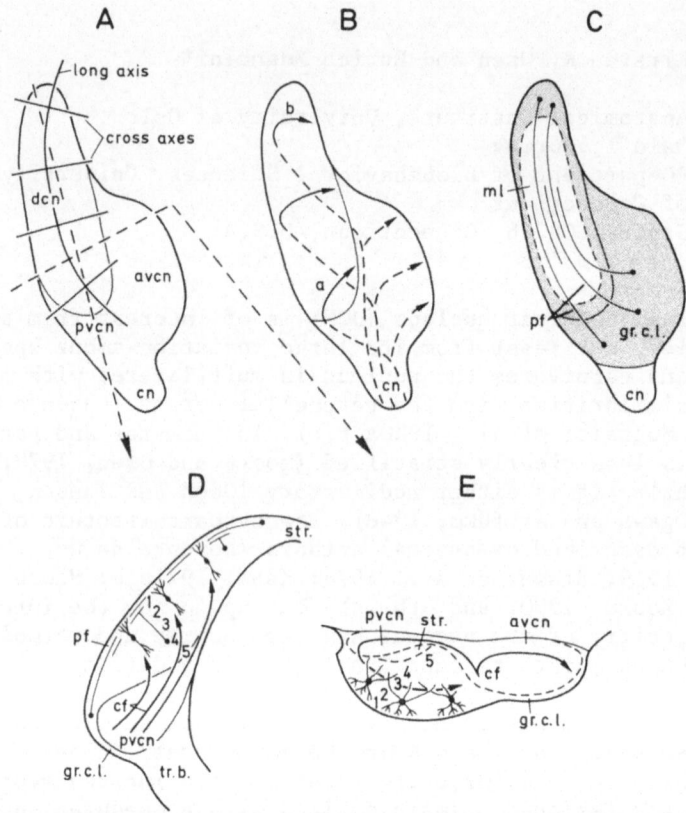

Fig. 1. The cat cochlear nuclei in surface view (A-C) and in sec-
         tions cut parallel to (D) and across (E) the long axis of
         the DCN, as indicated in A. B shows the orientation of
         primary cochlear fibers with apical (a) fibers ventrally
         and basal (b) fibers dorsally. C shows the extent of the
         superficial granule cell layer (gr.c.l., rastered) with
         granule cells giving origin to parallel fibers (pf) of the
         molecular layer (ml). D and E show the relation (cont´d)

(Fig. 1 cont´d) of parallel fibers and cochlear fibers (cf)
to the apical and basal pyramidal cell dendrites, respectively. Abbr.: avcn, anteroventral cochlear nucleus; cn,
cochlear nerve; pvcn, posteroventral cochlear nucleus; str.,
acoustic striae; tr.b., trapezoid body; 1-5, layers of the
DCN.

and mostly unmyelinated granule cell axons converge toward the molecular layer where they run in parallel to each other and to the long
axis of the nucleus (Fig. 1 C,D) constituting a large proportion of
the molecular layer neuropil.

The main target of both the parallel fibers and the primary
afferents in the DCN are the bipolar pyramidal cells (Fig. 2). The
perikarya of these cells are situated in a staggered row deep to the
molecular layer, in layer 2 of the DCN. They are the principal neu-

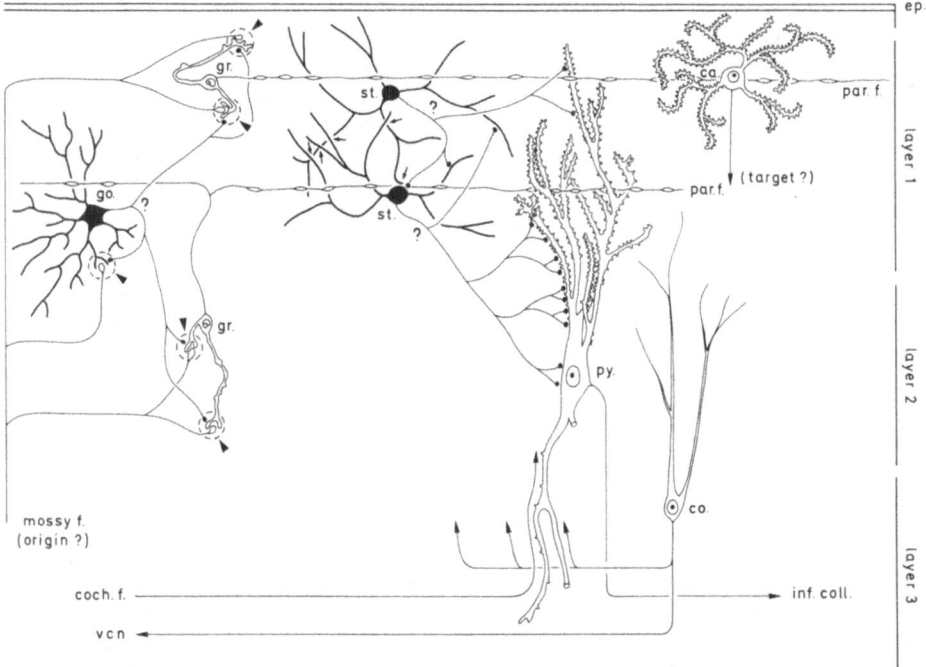

Fig. 2.   Diagram of layers 1, 2 and 3 of DCN in a plane parallel to
          its long axis. Abbr.: ca., cartwheel cell; co., corn cell;
          coch.f., cochlear nerve fiber;go., Golgi cell; gr., granu-
          lar cell; inf. coll., inferior colliculus; par.f., parallel
          fiber; py., pyramidal cell; st., stellate cell. Glomeruli
          are indicated by arrowheads, gap junctions by arrows.

rons of the nucleus and project to the contralateral inferior
colliculus (Osen, 1972; Adams, 1979). Their relatively smooth and
sparsely branched basal dendrites occupy layer 3 of the DCN, and
synapse with the primary afferents. Their spiny and heavily branched
apical dendrites, on the other hand, enter the molecular layer and
synapse with the parallel fibers. The distinct orientation of the
two types of afferents are reflected upon the apical and basal
dendritic fields. Both of these are clearly anisotropic with the
larger diameter oriented across the long axis of the DCN. The basal
field is, however, considerably wider and more flattened in the
cross sectional plane than is the apical field (Fig. 1 D,E). Whereas
individual apical dendritic arbors of neighbouring cells interdigitate with each other amply, the basal arbors interdigitate almost
exclusively with arbors of cells located in the same cross sectional
plane of the DCN, i.e., in the same isofrequency plane. The basal
dendrites, thus, seem to maintain the cochleotopy of the primary
fiber system, while the apical dendrites with the parallel fiber
input apparently serve a different function.

We have evidence that fibers from several sources reach the
granule cell system. As first described by Kane (1974a), the granule
cell dendrites participate in glomerular synaptic arrays (Fig. 2,
arrowheads) that are centered on a large mossy fiber terminal with
the characteristics of excitatory boutons (Mugnaini et al., 1980b).
Some of these fibers are probably collaterals of cholinergic olivocochlear fibers (McDonald and Rasmussen, 1971). Most of them, however, are of unknown origin (Kane, 1977). Mossy fibers also establish
contact with dendrites and bodies of slightly larger neurons, the
so-called Golgi cells (Mugnaini et al., 1980b) (Fig. 2). These may
give rise to smaller boutons of the inhibitory type which are found
peripherally in both the granule and Golgi cell glomeruli. As in
the cerebellum, thus, the Golgi cells are probably inhibitory interneurons controlling the mossy fiber input to granule cells.

Similar to the parallel fiber – Purkinje cell contacts in the
cerebellum, the cochlear granule cell axons form excitatory type of
asymmetric synapses with rounded vesicles on the spines of the
apical pyramidal cell dendrites. The cochlear parallel fibers also
supply the molecular stellate cells which are local interneurons
like their counterparts in the cerebellum (Lorente de Nó, 1979).
These cells may give rise to at least the majority of the inhibitory
type of small boutons with pleiomorphic vesicles which synapse on
perikarya and dendritic shafts of the pyramidal cells (Mugnaini et
al., 1980a). Our immunohistochemical studies, now in progress, show
that these boutons contain GAD. The stellate cells, therefore, may
be GABA-ergic neurons which, by providing a feed-forward inhibition
to the pyramidal cells, modulate the parallel fiber input to the
latter. The stellate cells are interconnected by gap junctions and
may, therefore, tend to fire synchronously (Mugnaini et al., 1980a).

The transmitter of the cochlear parallel fibers is still unknown. It may be glutamate as in the cerebellum (Bird et al., 1978).

A third target of parallel fibers in the molecular layer of the DCN are the cartwheel neurons of Brawer et al. (1974) (Fig. 2). Their dendrites are spiny like the apical pyramidal cell dendrites, but in contrast to the latter cells, the cartwheel neurons do not have basal dendrites penetrating deeper layers of the nucleus. These neurons, therefore, may not receive primary afferents. Judged by location and input the cartwheel neurons could represent the "build-up units" and the pyramidal cells the "pauser units" of Godfrey et al. (1975), assuming that the primary afferents are responsible for the first response and the parallel fiber input for the build up pattern. Other candidates for "build-up units" may be pyramidal cells that are dislocated in the molecular layer and have a predominant apical dendritic arbor. The projection of the cartwheel neurons is not defined.

Besides the excitatory (presumably glutamatergic (Bird et al., 1978)) primary afferents, the basal dendrites of the pyramidal cells also receive input from other sources. Some of these afferents may originate from trapezoidal or strial fibers descending to layer 3 of the DCN from the superior olivary complex (Borg, 1973; Elverland, 1977) and the inferior colliculus (Noort, 1969; Kane, 1977; Kane and Finn, 1977). The effect of these tracts is not known. The basal dendrites may also receive input from other neurons within the DCN, particularly the corn cells (Lorente de Nó, 1933, 1979) whose perikarya are situated in layer 3 (Fig. 2). Like other neuronal elements of this layer, the corn cell dendrites spread out in the cross sectional plane of the DCN while they have a restricted distribution in the opposite direction. According to Lorente de Nó (1979) the corn cell axon projects to the VCN after having given off collaterals to neighbouring isofrequency planes within the DCN, thus establishing either lateral inhibition or excitation. The transmitter substance is not known. Interestingly, the corn cells seem to be restricted to the same region of the DCN as the "on-type-S units" of Godfrey et al. (1975). Layers 2-5 of the DCN also contain other types of neurons which are not included in this survey. The DCN also receives afferent fibers from the VCN (Lorente de Nó, 1933, 1976, 1979). Thus, there are short loops between the two parts of the cochlear nuclear complex, but the details of these connections are unknown.

The species variations of the DCN mentioned in the beginning, are to a large extent bound to a different development of the granule cell system. This is well developed in rodents and carnivores and disappears gradually or is transformed in primates (Moore, 1980) and is nearly absent in man (Moore and Osen, 1979). So far we have no comprehensive answer to the question of the functional role of this cerebellar-like neuronal circuitry in audition, and why it changes so conspicuously during phylogeny.

ACKNOWLEDGMENTS

This work was supported in part by NIH grant No. 09904 (To E. Mugnaini). We wish to thank Anne-Lise Dahl, Kari Øztürk, Margrethe Lynnebakken, Einar Risnes and Tove Eliassen for careful assistance.

REFERENCES

Adams, J. C., 1979, Ascending projections to the inferior colliculus, J. Comp. Neurol., 183: 519-538.

Bird, S. J., Gulley, R. L., Wenthold, R. J. and Fex, J., 1978, Kainic acid injections result in degeneration of cochlear nucleus cells innervated by the auditory nerve, Science, 202: 1087-1089.

Borg, E., 1973, A neuroanatomical study of the brainstem auditory system of the rabbit. II. Descending connections, Acta Morphol. Neerl.-Scand., 11: 49-62.

Brawer, J. R., Morest, D. K. and Kane, E. C., 1974, The neuronal architecture of the cochlear nucleus of the cat, J. Comp. Neurol., 155: 251-300.

Elverland, H. H., 1977, Descending connections between the superior olivary and cochlear nuclear complexes in the cat studied by autoradiographic and horseradish peroxidase methods, Exp. Brain Res., 27: 397-412.

Godfrey, D. A., Kiang, N. Y. S. and Norris, B. E., 1975, Single unit activity in the dorsal cochlear nucleus of the cat, J. Comp. Neurol., 162: 269-284.

Kane, E. C., 1974a, Synaptic organization in the dorsal cochlear nucleus of the cat: A light and electron microscopic study, J. Comp. Neurol., 155: 301-330.

Kane, E. C., 1974b, Patterns of degeneration in the caudal cochlear nucleus of the cat after cochlear ablation, Anat. Rec., 179: 67-92.

Kane, E. S. 1977, Descending inputs to the cat dorsal cochlear nucleus: an electron microscopic study, J. Neurocyt., 6: 583-605.

Kane, E. S. and Finn R. C., 1977, Descending and intrinsic inputs to dorsal cochlear nucleus of cats: a horseradish peroxidase study, Neurosci., 2: 897-912.

Lorente de Nó, R., 1933, Anatomy of the eighth nerve. III. General plan of structure of the primary cochlear nuclei, Laryngoscope, 43: 327-350.

Lorente de Nó, R., 1976, Some unresolved problems concerning the cochlear nerve, Ann. Otol. Rhinol. Laryngol. 85: 2-28, Suppl. 34.

Lorente de Nó, R., 1979, Central representation of the eighth nerve, in: "Ear diseases, deafness and dizziness", V. Goodhill, ed.,

Harper Row, Hagerstown.

McDonald, M. D. and Rasmussen, G. L., 1971, Ultrastructural characteristics of synaptic endings in the cochlear nucleus having
acetylcholinesterase activity, Brain Res., 28: 1-18.

Moore, J. K., 1980, Loss of lamination as a phylogenetic process
in the primate dorsal cochlear nucleus, J. Comp. Neurol.,
(in press).

Moore, J. K. and Osen, K. K., 1979, The cochlear nuclei in man,
Am. J. Anat., 154: 393-418.

Mugnaini, E., Dahl, A-L., Osen, K. K. and Fiori, M. G., 1980a,
Stellate cells of the molecular layer in the dorsal cochlear
nucleus of rat, guinea pig and cat: Synaptic connections and
mutual coupling by gap junctions, Am. J. Anat., (in press).

Mugnaini, E., Osen, K. K., Dahl, A.-L., Friedrich, Jr., V. L. and
Korte, G., 1980b, Fine structure of granule cells and related
interneurones (termed Golgi cells) in the cochlear nuclear
complex of cat, rat and mouse, J. Neurocytol., (in press).

Mugnaini, E., Warr, W. B., Osen, K. K., 1980c, Distribution and
light microscopic features of granule cells in the cochlear
nuclei of cat, rat and mouse, J. Comp. Neurol., (in press).

Noort, J. van, 1969, The structure and connections of the inferior
colliculus. An investigation of the lower auditory system,
Thesis, van Gorcum and Co. N. V., Leiden.

Ogawa, T. and Arifuku, S., 1948, On the acoustic system in the
cetacean brains. Sci. Rep. Whales Res., Inst. Tokyo, 2:1-20.

Osen, K. K., 1969, Cytoarchitecture of the cochlear nuclei in the
cat, J. Comp. Neurol., 136: 453-484.

Osen, K. K., 1972, Projection of the cochlear nuclei on the inferior
colliculus in the cat, J. Comp. Neurol., 144: 355-372.

Osen, K. K. and Jansen, J., 1965, The cochlear nuclei in the common
porpoise, Phocaena phocaena., J. Comp. Neurol., 125: 223-258.

THE INTERNAL ORGANIZATION OF THE DORSAL COCHLEAR NUCLEUS

Eric D. Young and Herbert F. Voigt

Dept. of Biomedical Engineering
The Johns Hopkins University School of Medicine
Baltimore, Maryland 21205 U.S.A.

The dorsal cochlear nucleus (DCN) is distinguished from other parts of the cochlear nucleus by the intricacy of its neural circuitry (Lorente de No, 1933; Kane, 1974). The response properties of neurons in the DCN reflect this morphological complexity by differing markedly from those of auditory nerve fibers and other cochlear nucleus neurons. One striking feature of DCN responses is the extent to which they are dominated by inhibition which is weakened or abolished by anesthesia (Evans and Nelson, 1973). In this paper, we will present evidence that some of the inhibition observed in DCN responses is generated by inhibitory interneurons within the DCN.

Response Types in the DCN

The response properties of single cells in the DCN of unanes-thetized, decerebrate cats can be divided into two classes, Type II/III and Type IV (Evans and Nelson, 1973; Young and Brownell, 1976). The principal properties of these two types are illustrated in Fig. 1. Type II/III cells are characterized by excitatory respon-ses to best frequency (BF) tone bursts at all levels above tresh-old. The data in Figs. 1A and 1C are typical of about 2/3 of Type II/III units. These units (called Type II/III-LS below) have no spontaneous activity; they typically respond to BF tones with high discharge rates, but give relatively little if any response to broad band noise (70% do not respond to noise at all). The remaining Type II/III units, those with significant spontaneous activity (Type II/III-HS), will not be discussed in this paper.

127

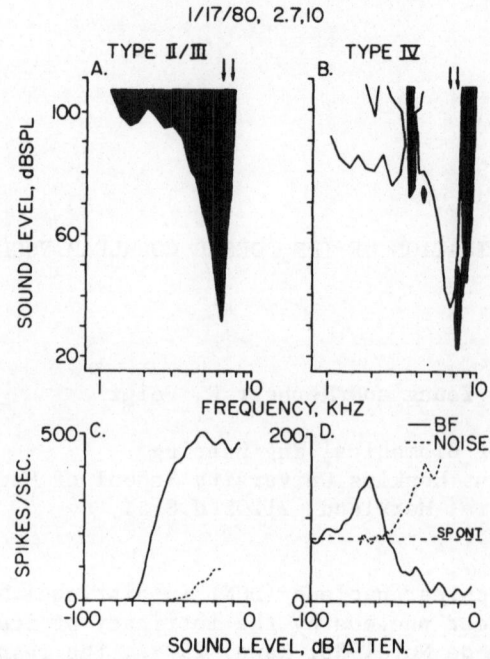

Fig. 1.   A, B. Response map of a Type II/III-LS (A) and a Type IV (B)
          unit, recorded simultaneously with two electrodes. Black
          areas are excitatory (>20% increase in rate and >2.5
          spikes/s); unshaded areas are inhibitory (>20% decrease
          in rate). Inhibitory regions could not be mapped for the
          Type II/III-LS unit since it had no spontaneous activity.
          Arrows point to BF of the Type II/III-LS unit (left arrow,
          6.51 kHz.) and Type IV unit (right arrow, 7.41 kHz.).
          C, D. Discharge rate during 200 ms stimulus versus sound
          level for the same Type II/III-LS (C) and Type IV (D) units.
          Solid lines show BF tone responses; dotted lines show noise
          responses. Spontaneous rate of Type IV unit shown by dashed
          line in D.

     Type IV units usually give excitatory responses to BF tones
within 20 to 30 dB of threshold, but give strongly inhibitory responses to tones at or near BF at higher sound levels. Scattered
excitatory areas are sometimes observed at higher levels (e.g. at
3.5 and 8.5 kHz. in Fig. 1B). The transition from excitation to
inhibition is sharp in Type IV units; this is illustrated by the
strongly nonmonotonic rate versus level function in Fig. 1D (solid
line). Unlike Type II/III-LS units, Type IV units almost always have

spontaneous activity and give excitatory responses to broad band
noise that are about as strong as their responses to BF tones
(dotted curve in Fig. 1D).

Relationship between Response Types and Morphological Cell Types

Fig. 2 is a schematic drawing showing a simplified summary of
our current understanding of the internal wiring diagram of the DCN
(Lorente de No, 1933; Brawer et al., 1974; Osen, 1969; Kane, 1974).
Most axons that leave the DCN through the dorsal acoustic stria
(DAS) originate in fusiform (F) or giant (G) cells (Adams and Warr,
1976). The table at top right in Fig. 2 shows that most (34/43)units
giving Type IV responses could be antidromically stimulated from
the DAS whereas few Type II/III units, especially those without
spontaneous activity, could be stimulated. Thus, Type IV units
account for most of the output activity originating in the DCN that
is carried through the DAS to more central parts of the auditory
system. Since they rarely can be antidromically stimulated from the
DAS, Type II/III-LS units can be associated with DCN interneurons.

The interneuronal constituents of the DCN are small (S) and
granule (g) cells. Their axons terminate mostly within the DCN;
one type of small cell in the deep DCN has been described as
sending a collateral of its axon to the anteroventral cochlear
nucleus (AVCN; Lorente de No, 1933). The table at top left in Fig.
2 shows that a few units giving Type II/III responses could be
antidromically stimulated from the region of the posterior AVCN
through which the DCN/AVCN intranuclear association pathway runs.
Thus, at least some Type II/III responses are recorded from small
cell interneurons in the DCN.

Interactions between Type II/III and Type IV Cells

If Type II/III responses are recorded from interneurons whose
axons terminate on principal cells giving Type IV responses, the
question of the nature of this interaction arises. The response
properties of the two types of units suggest that Type II/III-LS
responses are recorded from inhibitory interneurons. This statement
is based on the fact that Type II/III-LS units tend to discharge
strongly in response to those stimuli which are inhibitory to Type
IV units and vice versa (Young and Brownell, 1976). Type II/III-LS
units are not spontaneously active and do not respond to broadband
noise, whereas Type IV units are spontaneously active and respond
strongly to noise. Type IV units tent to respond to BF tones only
at low sound levels; Type II/III-LS units respond strongly to tones
at moderate and high levels and have BF tone tresholds that are
higher than those of Type IV units (median difference 9 dB). The
treshold difference is illustrated by the response maps in Figs.
1A and 1B which were simultaneously obtained from a pair of units.
The Type II/III-LS unit (Fig. 1A) has a BF threshold about 10 dB

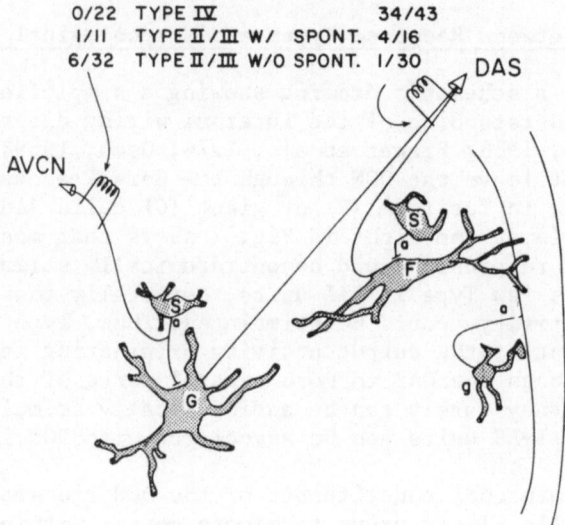

NUMBER OF UNITS WHICH COULD
BE ANTIDROMICALLY STIMULATED

0/22	TYPE IV	34/43
1/11	TYPE II/III W/ SPONT.	4/16
6/32	TYPE II/III W/O SPONT.	1/30

Fig. 2.   Simplified wiring diagram of the DCN, showing internal
         connections and outputs. F - fusiform cell; G - giant cell;
         S - small cell; g - granule cell. The DAS contains axons
         from fusiform and giant cells plus some from superficially
         located small cells (Adams and Warr, 1976) and from outside
         the DCN (not shown). Axons of certain small cells in deep
         DCN bifurcate, sending one branch to AVCN and one to DCN
         neuropil (shown terminating on a giant cell, although
         detailed evidence is lacking; Lorente de No, 1933). Table
         above shows the number of units antidromically stimulated
         from the AVCN (left) or DAS (right)/ the number of units
         tested

higher than that of the Type IV unit (Fig. 1B). As a consequence
of their higher thresholds, the excitatory areas of Type II/III-LS
units response maps frequently overlap almost exclusively inhibitory
areas of the response maps of simultaneously isolated Type IV cells
with similar BFs (Voigt and Young, 1980; e. g. the maps in Fig. 1).
Thus, the response properties of Type II/III-LS units are exactly
the properties needed to provide the inhibitory input to Type IV
units.

     More direct information about Type II/III-Type IV interactions
can be obtained by examining crosscorrelograms of simultaneously
recorded spike trains from pairs of cells (Moore et al., 1970).
Three examples of crosscorrelograms are shown in the right column
of Fig. 3. These were obtained from pairs consisting of a Type II/
III-LS and a Type IV unit which were simultaneously recorded with

two electrodes. The correlograms show the average discharge rate in
the Type IV unit given that a spike occurred in the Type II/III-LS

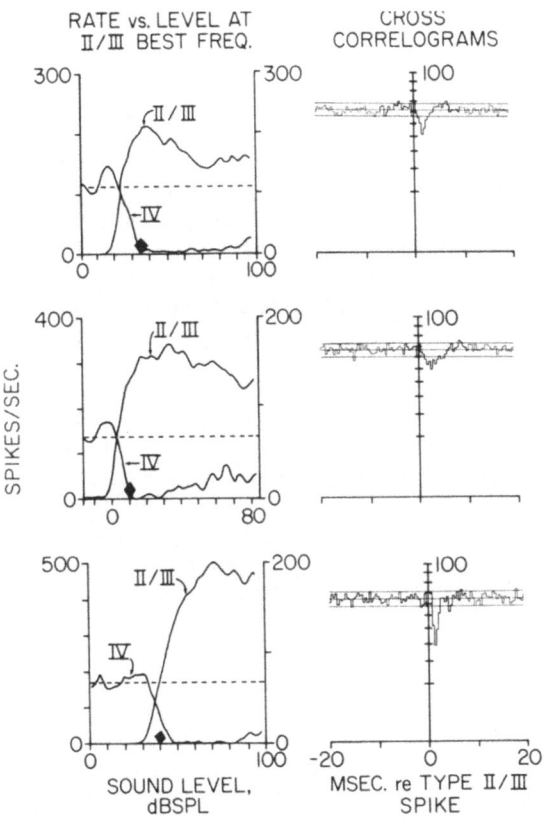

Fig. 3. Left column: Discharge rate versus sound level for three
         different Type II/III-LS, Type IV pairs. Stimuli were
         200 ms tones at Type II/III-LS BF (2.57, 5.42, and 6.51
         kHz., top to bottom; Type IV BFs were 3.06, 4.05, and
         7.41 kHz.). Left ordinate is for Type II/III-LS data and
         right ordinate is for Type IV data. Type IV spontaneous
         rate shown by dashed lines.
         Right column: Crosscorrelograms of spike trains from the
         same three pairs, computed from activity evoked by 50 s
         tones at Type II/III-LS BF and level indicated by diamond
         on abscissa of corresponding rate plot. Abscissa is time
         of occurrence of spike in Type IV unit relative to spike
         in Type II/III-LS unit. Each bin is 0.4 ms. Ordinate scale
         is Type IV discharge rate, plotted on square root coordi-
         nates. Horizontal lines show average long-delay correlation
         (average Type IV rate) and 95% confidence limits (cont'd)

(Fig. 3 cont'd) on null hypothesis of independent spike
trains. Bottom two plots show data from same pair of units
as Fig. 1.

unit at time zero (see Voigt and Young, 1980 for details of computation). In each case, there is a significant decrease in the response probability of the Type IV unit for a few milliseconds after
spikes in the Type II/III-LS unit; this is indicated by the through
in the crosscorrelogram just to the right of the origin. A trough
like this is expected in the crosscorrelogram of two units connected
by an inhibitory synapse (Moore et al., 1970). Similar inhibitory
troughs were observed in the crosscorrelograms of most (12/17)Type
II/III-Type IV pairs studied with a single electrode (Voigt and
Young, 1980); in most of those pairs, the Type II/III element had
no spontaneous activity (11/12). A similar result holds for pairs
studied with two electrodes (and therefore physically further apart),
although results are not yet complete.

The left column of Fig. 3 shows plots of discharge rate versus
sound level for the same three pairs of units. The stimuli were
200 ms tone bursts at the BF of the Type II/III-LS unit. In each
case, there is a close correspondence between the treshold of the
Type II/III cell and the beginning of the sharp decline in discharge
rate in the Type IV cell. This sort of reciprocal relationship
between Type II/III-LS and Type IV response is always observed in
pairs with inhibitory troughs in their crosscorrelograms, but is
frequently not observed when the crosscorrelograms are flat.

One final piece of evidence for an inhibitory connection
between Type II/III-LS and Type IV units is provided by an observation reported previously (Voigt and Young, 1980). In one case, the
responses to stimuli in a portion of the inhibitory area of a Type
IV unit became excitatory when a simultaneously isolated Type II/
III-LS unit (which had shown an inhibitory trough in its cross-
correlogram with the Type IV cell) exhibited a sustained injury
discharge and disappeared. The frequency region in which the transformation occurred was the area occupied by the response map of the
Type II/III-LS unit.

REFERENCES

Adams, J. C. and Warr, W. B., 1976, Origins of axons in the cat's
        acoustic striae determined by injection of horseradish per-
        oxidase into severed tracts, J. Comp. Neurol.,170: 107-121.
Brawer, J. R., Morest, D. K. and Kane, E. C., 1974, The neuronal
        architecture of the cochlear nucleus of the cat, J. Comp.
        Neurol., 155: 251-300.
Evans, E. F. and Nelson, P. G., 1973, The responses of single
        neurones in the cochlear nucleus of the cat as a function of

their location and the anaesthetic state, Exptl. Brain. Res.
17: 402-427.

Kane, E. C., 1974, Synaptic organization in the dorsal cochlear
    nucleus of the cat: A light and electron microscopic study,
    J. Comp. Neurol., 155: 301-329.

Lorente de No, R., 1933, Anatomy of the eighth nerve. III General
    plan of the structure of the primary cochlear nuclei, Laryn-
    goscope, 43: 327-350.

Moore, G. P., Segundo, J. P., Perkel, D. H. and Levitan, H., 1970,
    Statistical signs of synaptic interaction in neurons, Biophys.
    J. 10: 876-900.

Osen, K. K., 1969, Cytoarchitecture of the cochlear nuclei in the
    cat, J. Comp. Neurol.,136: 453-482.

Voigt, H. F. and Young, E. D., 1980, Evidence of inhibitory inter-
    actions between neurons in the dorsal cochlear nucleus, J.
    Neurophysiol. (in press).

Young, E. D., 1980, Identification of response properties of ascend-
    ing axons from dorsal cochlear nucleus. Brain Res. (in press).

Young, E. D. and Brownell, W. E., 1976, Responses to tones and
    noise of single cells in dorsal cochlear nucleus of unanesthe-
    tized cats, J. Neurophysiol., 39: 282-300.

SESSION IV
CENTRAL AUDITORY MECHANISMS A
Chairmen: J. E. Hind and M. Molnár

FUNCTIONAL ORGANIZATION OF THE INFERIOR COLLICULUS

J. Syka, R. Druga* , J. Popelář and B. Kalinová

Institute of Experimental Medicine
Czechoslovak Academy of Sciences
128 08 Prague 2, U nemocnice 2
*Dept. of Anatomy, Medical School, Charles University
Prague, Czechoslovakia

INTRODUCTION

The basic subdivision of the inferior colliculus (IC) in
mammals into three main parts – the central nucleus, the pericentral
nucleus and the external nucleus (Fig. 1) is known since the pio-
neering work of Ramón y Cajal (1911). Recent anatomical and electrophysiological studies, beginning with the work of Rose et al. (1963)
up to the more recent results by Aitkin et al. (1975) have demonstrated that these three nuclei may have a different function. For
example Aitkin et al. (1978) demonstrated that the external nucleus
of the IC has an extensive somatosensory input and that neurones in
this part of the IC may belong to a system which integrates the
information about the space parameters impinging upon different
sensory organs. Major interest has been concentrated, however, upon
the morphology and function of the central nucleus of the IC. This
essentially oval nucleus, which forms the central part of the IC,
consists of two subdivisions which have been found to be similar in
many species of mammals – the smaller dorsomedial division, consisting mainly of large cells, receiving fibres from the auditory
cortex and the larger ventrolateral division, which has a pronounced
laminar arrangement of cell dendrites and axons (Morest 1964, Rockel
and Jones 1973, Fitzpatrick 1975). Laminae, which provide a basis
for the tonotopic organization of neurones in the nucleus have been
demonstrated in the cat by Merzenich and Reid (1974). The present
report summarizes our data about the organization of the central
nucleus of the IC in the rat, with respect to the ascending and
descending inputs to this structure, its tonotopic arrangement and
the influence of auditory deprivation on the development of inputs

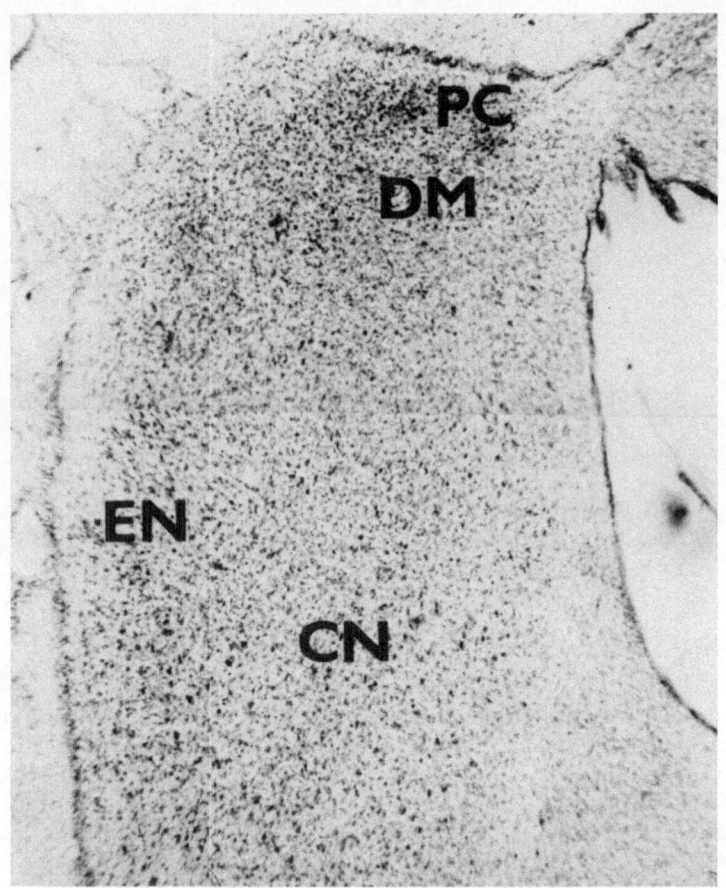

Fig. 1.   A coronal section through the posterior third of the inferior colliculus in the rat showing central (CN), pericentral (PC) and external (EN) nuclei. DM, dorsomedial
part of the central nucleus of IC. x 45.

to the IC. In addition, the properties of responses of IC neurones
to binaural acoustic stimuli will be compared in some laboratory
species.

METHODS

     Experiments were performed on hooded rats, pigmented guinea-
pigs and rabbits. Rats were anaesthetized with sodium pentobarbital
(30 mg/kg), guinea-pigs and rabbits with urethane (20 % solution
0.8 ml/100g), both injected intraperitoneally. The inferior colliculus was penetrated under different angles either through the
overlying cortex or through the cerebellum. Extracellular unit
activity was recorded by means of stainless steel microelectrodes

or glass micropipettes (filled with 3 M KCl), both having in impedance of 1-5 MΩ. Sound stimuli at frequencies ranging from 0.1 kHz
to 50 kHz were generated with special sealed piezoelectric loudspeakers (Salava et al., 1979). The sound pressure acting upon the
ear drum was controlled. Binaural stimuli were delivered with differences in time and intensity at 5 μs or 1 dB steps.

In one series of experiments, the external auditory meatus in
the ratlings was ligated 10 days after birth. After several months,
0.3 μl of 30 % solution of horseradish peroxidase Sigma VI were
injected into the IC ipsilateral or contralateral to the previously
ligated external auditory meatus. In the second group of monaurally
deprived rats, the ligature was removed, the external meatus cleaned
and the unit activity was recorded in the IC.

The horseradish peroxidase was injected in 15 non-deprived rats
into different sites of the IC. Small amounts (0.1 - 0.2 μl) were
slowly ejected under pressure from a 1 μl Hamilton microsyringe.
Rats were sacrificed after 24 - 48 hours, the brain removed and
processed according to the original method of Graham and Karnovsky
(1966). Section were studied under both bright and dark field
illumination. In 5 rats, the auditory cortex was destroyed by elec-
trolytical lesions. After 5 days, the rats were sacrificed and after
fixation blocks of the brain stem were impregnated according to the
method of Nauta and Gygax (1954).

RESULTS

Tonotopic Organization of the Central IC Nucleus in the Rat

The characteristic frequencies in the central nucleus of the
IC for penetrations oriented caudally and ventrally succesively
increased from 0.5 kHz to 50 kHz. In Fig. 2 are shown typical
changes in the characteristic frequency (CF) in two parallel penetrations oriented ventrally. The finding is essentialy similar to
the results obtained by Merzenich and Reid (1974) in the cat. Reconstruction of the tonotopic arrangement of the IC, based on data
pooled from several penetrations revealed isofrequency layers of
neurones, which formed sheets inclined along the sagittal section
from the dorsocaudal to the ventrorostral side, with the lowest
characteristic frequencies found dorsally and the highest ventrally.
Fig. 3 demonstrates the reconstruction in the sagittal section
from 6 tracks which penetrated the IC under different angles. The
isofrequency layers were also tilted down at the most medial and
lateral sides, when viewed on the frontal section of the IC. An
approximately 2 kHz change in the characteristic frequency was
obtained during 100 μm movement of the electrode tip in a caudal
and ventral direction (i.e. across the isofrequency layers); the
largest extent of layers was found for frequencies from 10 to 15kHz.

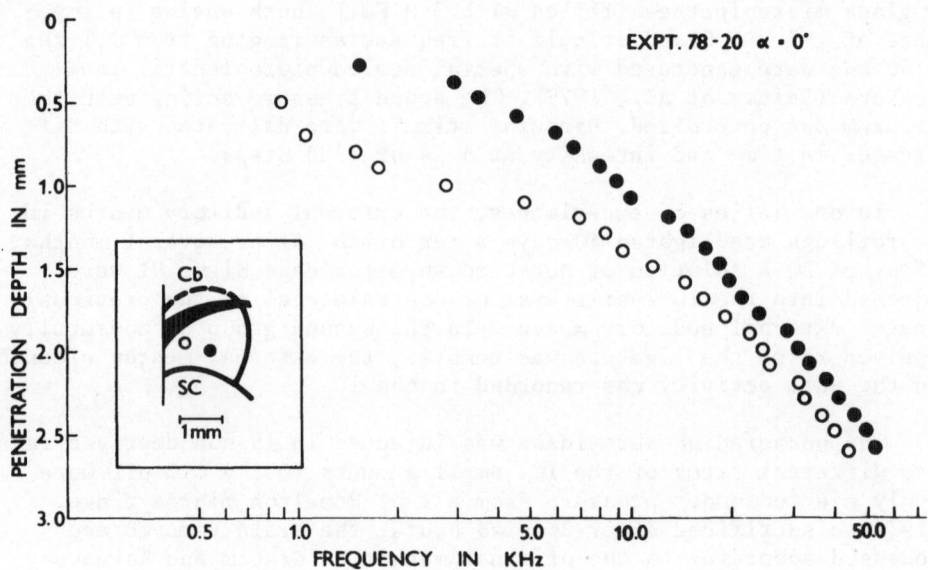

Fig. 2. Distribution of CF in two penetrations through the IC in
the rat. Penetrations were oriented perpendicularly to the
dorsal surface of the IC in the dorso-ventral direction.
The inset shows the localization of penetrations on the
dorsal surface of the IC. Ordinate - depth from the dorsal
surface in mm. Abscissa-frequency in kHz. SC = superior
colliculus; Cb = cerebellum.

    The thresholds for individual neurones at their characteristic
frequencies were measured for reconstructing the threshold of
audibility in the hooded rat. Fig. 4 summarizes the data from 67
neurones recorded in normal non-deprived rats. Lowest thresholds
were observed in neurones with the CF between 5 - 10 kHz. In view
of these findings we estimated the approximate extent of individual
isofrequency layers in the central nucleus of the IC, which may
roughly indicate the number of neurones with similar frequencies.
The comparison of the numbers indicating the extent of individual
layers and the threshold of audibility in the hooded rat has shown
good correspondence; i.e. the lower the threshold, the larger is
the extent of the layer for the individual frequency and, consequently, the larger is the number of neurones with this CF in the
central nucleus of the IC.

Ascending and Descending Projections to the IC in the Rat

    The organization of ascending and descending inputs to the
central nucleus of the IC in the rat was studied with the aid of

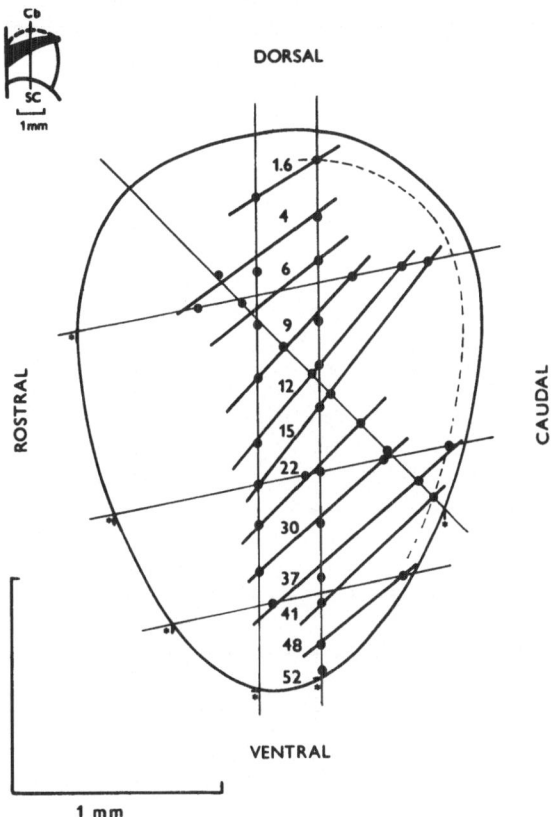

Fig. 3. Sagittal section through the IC in the rat with indicated
         orientations of isofrequency layers. Reconstruction from
         6 tracks, which penetrated the IC under different angles.
         Lesions at the end of tracks are indicated with asterisks.
         Numbers in the middle – frequency in kHz. The inset pictu-
         re shows the orientation of the sagittal section, SC = su-
         perior colliculus, Cb = cerebellum.

horseradish peroxidase (HRP) injections. Labelled cells were found
in all subcollicular auditory nuclei (with exception of the medial
nucleus of the trapezoid body), in the contralateral inferior colliculus and in the ipsilateral auditory cortex. Table 1 presents
the summary of results of HRP injections in 15 animals. In the table, the number of crosses indicates the relative extent of labelling
in the nuclei ipsilateral and contralateral to the side of injection. The more powerful inputs to the IC came from all cochlear
nuclei on the contralateral side, from the ipsilateral lateral
superior olivary nucleus, dorsal nuclei of the lateral lemniscus of

Table 1. Distribution of HRP Labelled Neurones in the
Auditory System of the Rat after Injection of
the HRP into the Central Nucleus of the IC.

	Ipsilateral	Contralateral
DCN	+	+ + +
AVCN	+	+ + +
PVCN	+	+ + +
LSO	+ +	+
MSO	+	0
LTB	+	+
VTB	+	+
MTB	0	0
SPN	+ +	+
DNLL	+ + +	+ + +
VNLL	+ + +	0
IC		+ + +
Cortex	+ + +	0

DCN, dorsal cochlear nucleus; AVCN, anteroventral cochlear
nucleus; PVCN, posteroventral cochlear nucleus; LSO,
lateral superior olivary nucleus; MSO, medial superior
olivary nucleus; LTB, lateral nucleus of the trapezoid
body; VTB, ventral nucleus of the trapezoid body; MTB,
medial nucleus of the trapezoid body; SPN, superior
paraolivary nucleus; DNLL, dorsal nucleus of the lateral
lemniscus; VNLL, ventral nucleus of the lateral lemniscus;
IC, inferior colliculus. Number of crosses indicates
the density of HRP labelled neurones.

both sides, ipsilateral ventral nucleus of the lateral lemniscus,
contralateral inferior colliculus and the ipsilateral auditory
cortex.

Almost all of the major cell types within the subcollicular
nuclei were labelled with HRP. An exception were octopus cells in
the posteroventral cochlear nucleus and the large spherical cells
in the anteroventral cochlear nucleus, which were not labelled.
Typical labelled cells in the dorsal cochlear nucleus, auditory
cortex and the supraolivary nucleus are shown in Fig. 5. Labelling

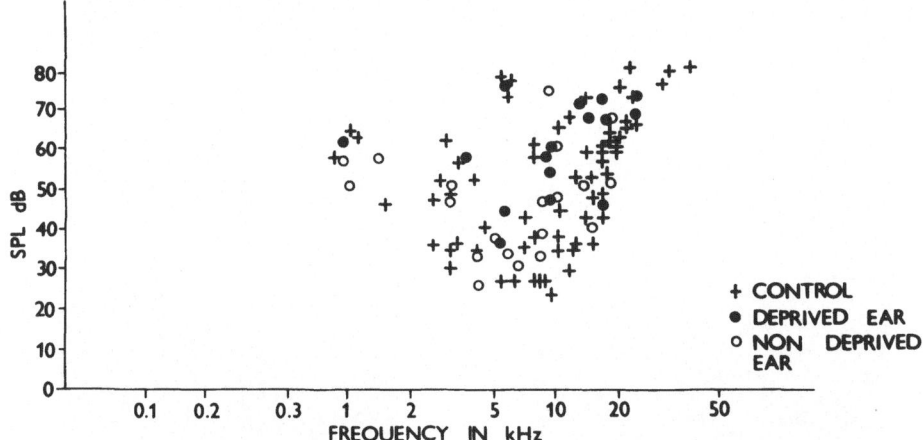

Fig. 4.   Plot of characteristic frequencies of neurones in the in-
          ferior colliculus of the rat. Crosses, control non-deprived
          animals; open circles, CF of neurones in monaurally depri-
          ved animals, driven from the non-deprived ear; full circ-
          les, CF of neurones, stimulated from the deprived ear.

in the auditory cortex was confined to the pyramidal cells in
layer V.

     Injections of small amounts of HRP into different parts of
the IC resulted in tonotopic labelling of cells within restricted
regions of the cochlear nuclei, lateral superior olivary nucleus
and dorsal nucleus of the lateral lemniscus. The injections into
the dorsal, low frequency part of the IC produced labelling of
cells within the ventral regions of cochlear nuclei, whereas the
injections into the ventral high frequency part resulted in the
labelling of cells in the dorsal regions of cochlear nuclei.

     The tonotopic arrangement of inputs to the central nucleus
of the IC in the rat was less evident, when the labelling after
injections into restricted parts of the IC was analyzed in the
auditory cortex. These results led us to investigate the tonotopic organization of cortical neurones in the rat, which has not
been studied before. The microelectrodes were inserted perpendicularly into the temporal cortex, in the region indicated by the
HRP labelling. In the upper part of Fig. 6 the extent of the area
labelled with the HRP injection into the IC is depicted. The results of mapping of the area within the square are shown in the
lower left side of the figure. Similar data from another animal
(right side of the figure) demonstrate the course of isofrequency
contours, with the representation of high frequencies in the ro-

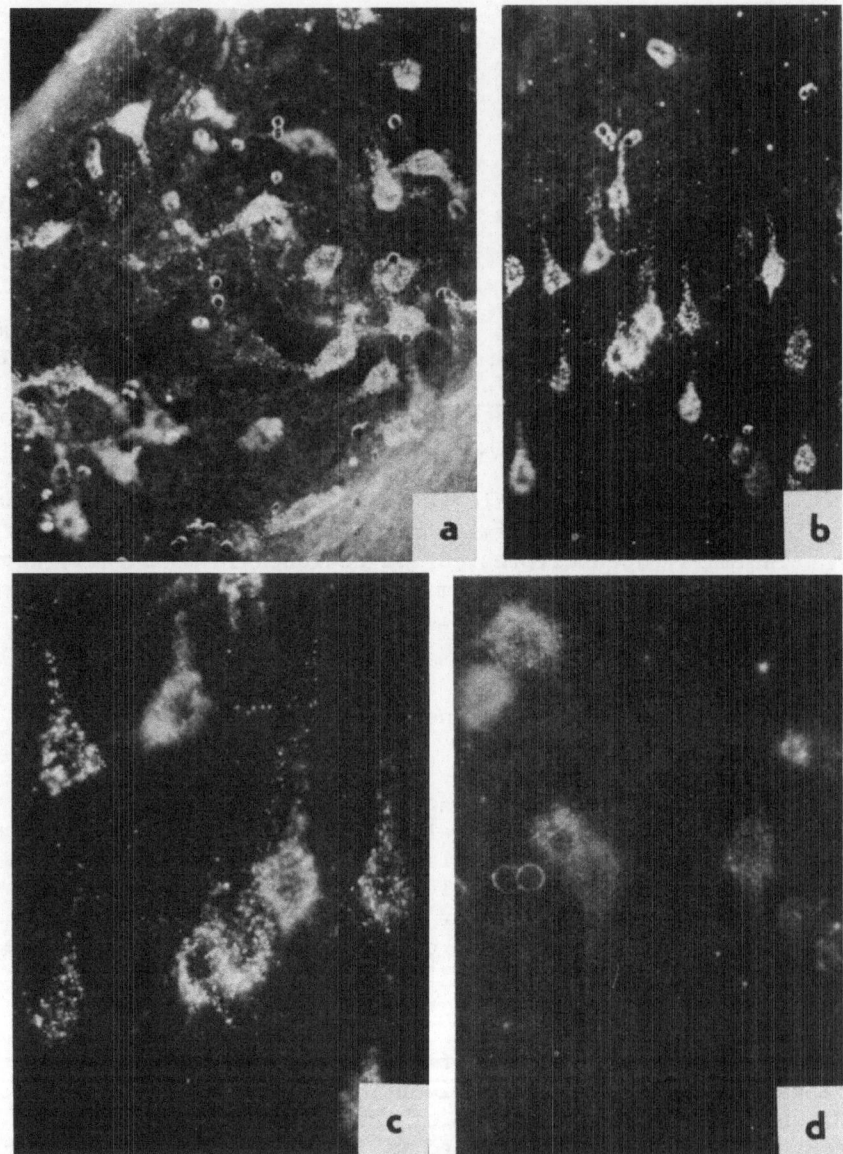

Fig. 5. a) Labelled neurones in the dorsal cochlear nucleus after
injection of HRP into contralateral central nucleus of IC.
Magnif. x 265. b) Labelled neurones in layer V of ipsilateral auditory cortex. x 265. c) Heavily labelled perikarya and apical dendrites of pyramidal neurones in layer
V of ipsilateral auditory cortex. x 640. d) HRP positive
cells in ipsilateral superior paraolivary nucleus (SPN).
x 640.

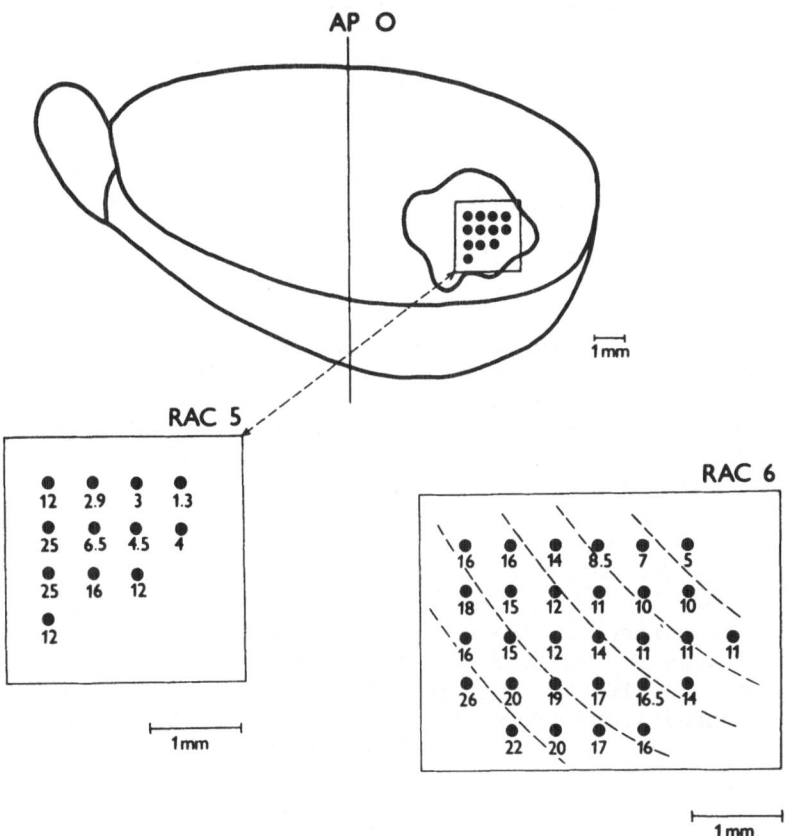

Fig. 6. Localization of labelled neurones in the cortex of the rat
after injection of the HRP into the IC. The extent of the
HRP labelled area in the rat cortex is shown in the lateral
view. Numbers in squares represent CF in kHz of neurones
found during perpendicular penetrations of the microelectrode through the cortex at the place marked by a dot.
Two typical results from rats RAC 5 and RAC 6 are shown.

stral and ventral, and low frequencies in the dorsal and caudal
region of the auditory area. We have not investigate the total HRP-
labelled area up to now; probably, the auditory cortex in the rat
may be subdivided into primary and secondary areas as in other
mammals.

The diffuse labelling of neurones in the auditory cortex in
the rat might be, at least partially, a consequence of localized
termination of the descending fibres in the IC, which may not be
tonotopically organized. The auditory cortex was, therefore, destroyed in 5 rats with electrolytic lesions and the localization

of fibre and terminal degeneration in the inferior colliculus was
studied in sections stained by the method of Nauta and Gygax. The
extent of lesions in all animals roughly corresponded with the extent of the HRP labelling in the auditory cortex after the injections into the IC. Fig. 7 summarizes the findings of terminal, preterminal and fibre degenerations in the IC. Degenerative changes in
the ipsilateral IC we found in the external nucleus, in the dorsomedial division of the central nucleus and in the lateral parts of
the laminated ventrolateral division of the central nucleus. Degenerations on the contralateral side were sparser and essentially
restricted to the dorsomedial division of the central nucleus. In
the lower part of Fig. 7 are shown micrographs of degenerated fibres
and the preterminal and terminal fragments in the IC. The results
of degeneration studies are thus in good agreement with the data
obtained in the cat by Diamond et al. (1969). In contrast to the
results of HRP injections, the degeneration studies demonstrated the
descending input to the IC coming from the auditory cortex of both
sides.

Effects of Auditory Deprivation in the Rat

Rats with unilateral ligature of the external auditory meatus
were investigated approximately 6 months after the operation. In
one group of rats electrical responses of IC units were recorded,
while in a second group HRP was injected into the IC. Preliminary
data were obtained in 4 rats in each of these groups.

Thresholds at characteristic frequencies for 20 neurones a-
coustically stimulated from the non-deprived contralateral ear did
not essentially differ from the thresholds found in control non-
deprived animals (Fig. 4, open dots). In 15 neurones stimulated
from the deprived ear (after the ligature had been removed and the
auditory meatus cleaned), the thresholds were higher than in the
control animals (Fig. 4, full dots). The increased thresholds were
not observed, when the neurones were stimulated from the non-deprived
ear (in the case of neurones, which are excited from both ears). We
may thus consider the increase in the thresholds to be caused probably by the damage of peripheral conduction mechanisms.

In two rats, 0.3 µl of HRP was injected into the central part
of the IC ipsilateral to the deprived ear. The distribution of labelled cells and their quantity in the nuclei which project to the
IC were essentially the same as in control non-deprived animals.
In two rats which had received the HRP injections into the IC contralateral to the deprived ear, we observed a proportional decrease of
labelled cells in all nuclei projection to the IC, including the
auditory cortex. These preliminary data, which need further verification, suggest that the metabolic processes subserving the retrograde

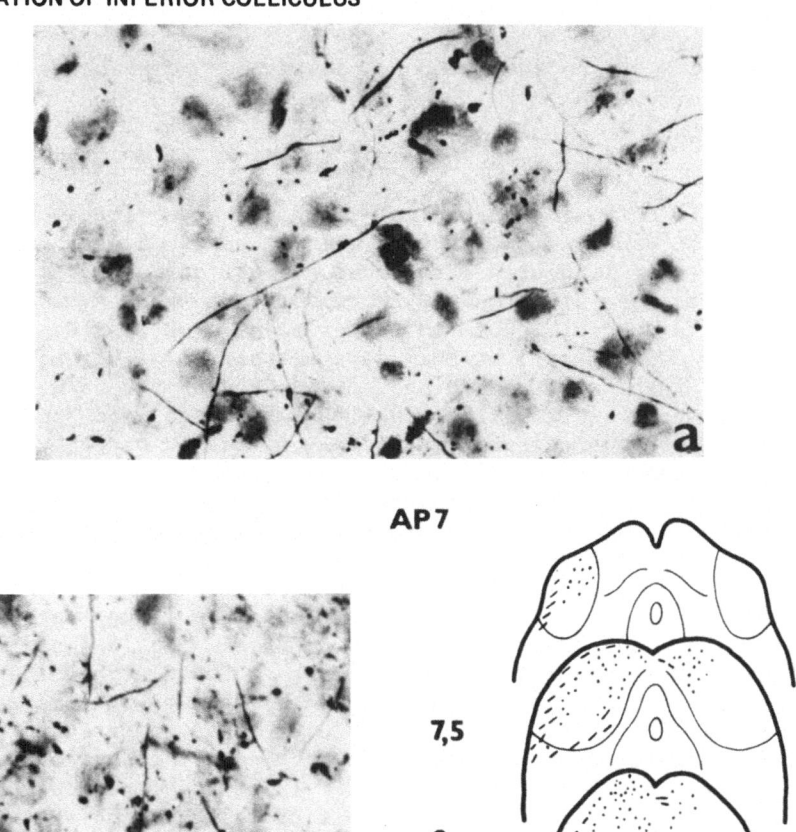

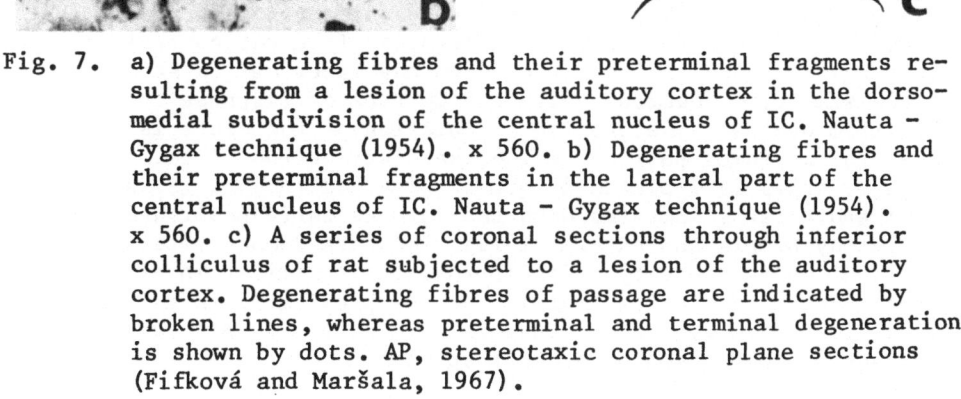

Fig. 7. a) Degenerating fibres and their preterminal fragments re-
         sulting from a lesion of the auditory cortex in the dorso-
         medial subdivision of the central nucleus of IC. Nauta –
         Gygax technique (1954). x 560. b) Degenerating fibres and
         their preterminal fragments in the lateral part of the
         central nucleus of IC. Nauta – Gygax technique (1954).
         x 560. c) A series of coronal sections through inferior
         colliculus of rat subjected to a lesion of the auditory
         cortex. Degenerating fibres of passage are indicated by
         broken lines, whereas preterminal and terminal degeneration
         is shown by dots. AP, stereotaxic coronal plane sections
         (Fifková and Maršala, 1967).

transport of HRP into the auditory pathway may be changed after
auditory deprivation.

Responses of IC Neurones to Monaural and Binaural Tonal Stimuli

In addition to the anatomical connections and functional properties of IC neurones in the rat, we have obtained data concerning
the responses of IC neurones to tonal stimuli on other species.
These studies were performed in two laboratory species, i.e. in
rabbits (Syka et al., 1980) and in guinea-pigs. Responses of IC
neurones to 200 ms tonal stimuli were classified either as "sustained" or "on" responses, according to the shape of the PSTH. The
types of responses classified usually as "chopped" and "paused"
(Pfeiffer, 1966) were found very rarely in the IC of rabbits and
guinea-pigs and were classified as "sustained" responses. In Fig. 8,
in the upper part, the distribution of the two types of neuronal
responses is shown in four laboratory species (the data for the
cat were taken from Rose et al., 1963 and those for the rat from
Marusyeva, 1971). In rabbits (Syka et al., 1980) and in guinea-pigs,
we found an approximately similar distribution of the response types, i.e. 85 % and 83 % "sustained" responses, and 15 % and 17 %
"on" responses respectively. In contrast to the distribution of PSTH
types in rabbits and in guinea-pigs, both types of responses were
equally distributed among investigated neurones in cats and rats
(Fig. 8). However, the difference in the occurrence of the two types of responses between the groups of laboratory species may be
caused by different types of anaesthesia: whereas rabbits and guinea-
pigs were anaesthetized with urethane, cats (Rose et al., 1963) and
rats (Marusyeva, 1971) were under barbiturate anaesthesia.

The responses of IC neurones to binaural tonal stimuli were
divided into three main types, based on the classification proposed
by Goldberg and Brown (1969). The EE type is characterized by excitatory responses to monaural stimulations of both ears, the binaural response may be either higher than any of the monaural responses (EE - facilitation) or lower, in the case of binaural occlusion (EE - occlusion). The EI type of binaural response is characterized by monaural excitation (usually from the contralateral
ear); the response is inhibited, when the other ear is stimulated
at the same time. Neurones with the EO type of response are driven
only from one ear, whereas the stimulation of the other is ineffec-
tive. In the lower part of Fig. 8 the distribution of the binaural
response types is depicted in guinea-pigs and rabbits (Syka et al.,
1980) and this is compared with the data recently published for
the cat by Semple and Aitkin (1979). The data demonstrate that the
distribution response types is approximately similar in all these
three laboratory species (rabbit, guinea-pig, cat). In guinea-pigs,
the EI response types were found in a relatively small number of
neurones in comparison with other species.

INFERIOR COLLICULUS

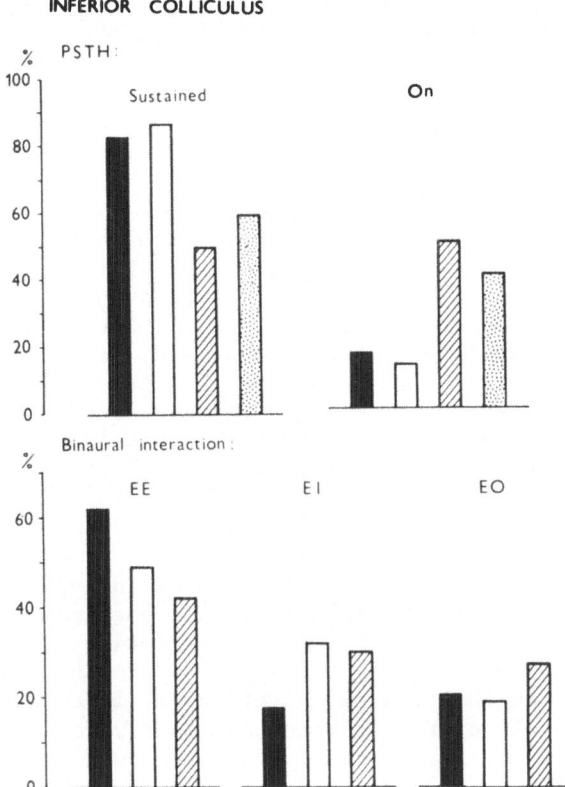

Fig. 8. Distributions of types of responses to binaural acoustical
stimuli in IC neurones of guinea-pigs (n=68); rabbits
(Syka et al., 1980; n=97) and cats (Semple and Aitkin,
1979; n=553). Bottom: distribution of two basic types of
PSTH - "sustained" and "on" responses in guinea-pigs
(n=277); rabbits (Syka et al. 1980; n=151), cats (Rose
et al., 1963) and rats (Marusyeva, 1971).

DISCUSSION

In summary, the data characterize some basic parameters of
responses of the inferior colliculus neurones, the tonotopic arrangements of the central nucleus of the IC and the organization of
ascending and descending inputs to the IC. With respect to many
structural and functional parameters, the IC in the rat is similar
to the IC in the cat. Isofrequency layers described in the rat correspond to the similar organization in the cat (Merzenich and Reid,
1974; Semple and Aitkin, 1979). The difference mainly concerns the

span of characteristic frequencies: The lowest CF in the rat was
around 1 kHz, whereas in the cat CF lower than 100 Hz may be found.
Similarly, higher values of high frequency CF were in the rat in
comparison with the cat. The maximum sensitivity, i.e. the lowest
point in the audibility curve is also shifted towards the lower
frequencies in the cat; in the rat, the maximum sensitivity was
found to be from 5 to 10 kHz.

The tonotopical arrangement of the central nucleus of IC in
the rat was briefly described by Clopton and Winfield (1973).
Similarly, the afferents to the IC were traced in the rat by Beyerl
(1978). Our data are essentially in agreement with the result of
Beyerl (1978); in addition to his data we also described the labelling of neurones in the nuclei of the lateral lemniscus. Similar
distribution of HRP labelled cells in the nuclei of the lateral
lemniscus was observed in the cat by Roth et al. (1978) and Adams
(1979). With respect to the density of labelled cells in the ipsilateral dorsal nucleus of the lateral lemniscus our data agree with
those by Roth et al. (1978), demonstrating the equal heavy labelling
of cells in dorsal nuclei of the lateral lemniscus bilaterally. In
the olivary complex of rat, the small size of individual nuclei
limits the possibility to differentiate the labelling of cells according to the tonotopic principle, with exception of the large
lateral superior olivary nucleus. In contrast to the data published
by Beyerl (1978), we observed relatively heavy labelling of cells
in the contralateral inferior colliculus, mostly confined to the
central nucleus of the IC. In the cortex, we found HRP labelled
cells ipsilaterally in a large area, which is quite similar to the
area given by Beyerl (1978, Fig. 7). Only pyramidal cells in layer
V were labelled. To our surprise cells in the contralateral side of
the cortex were not labelled, although the degeneration studies by
Diamond et al. (1969) in the cat indicated pathways descending to
the IC of both sides. In order to exclude the possible difference
between species, the lesions were performed in the cortical area,
corresponding to HRP labelling. The results were essentialy the
same as those found in the cat by Diamond et al. (1969), i.e. terminal and fibre degenerations were found in the dorsomedial division of the IC on both sides. It is possible that our injections
of HRP and probably also those by Beyerl (1978), did not ensure
sufficient spreading of HRP into the dorsomedial division so that
the labelling resulted only in the uptake of HRP in the larger
ventrolateral division of the central nucleus of the IC. It remains
to be explained, however, why the HRP labelled cells in the cortex
are not distributed tonotopically, in the case of small injections
into different parts of the IC. Further experiments will be necessary
to explain this problem.

The types of binaural interactions in neurones of the IC were
found to be similarly distributed in three different laboratory
species (guinea-pig, rabbit, cat) with the exception of a relatively

small number of neurones with the EI type of interaction in guinea-pig. The importance of the classification of IC neurones according to the binaural response properties was recently stressed by the finding that binaural response classes in the cat are distributed differentially throughout the central nucleus of the IC (Roth et al., 1978; Semple and Aitkin, 1979). For example, EO units were concentrated caudally, ventrally and laterally, while the EI units were mostly found rostrally. It is probable that this type of space arrangement will also be present in the IC of other species. In connection with the clustering of neurones with the same type of binaural interaction (which probably receive their input from different subcollicular nuclei) it is of interest that the effects of electrical stimulation of the IC in an awake cat are dependent on the spatial localization of the tip of the stimulating electrode in a similar way (Syka and Radil-Weiss, 1971). Thus electrical stimulation of the ventral and caudal point of the IC evoked predominantly isolated movements of the contralateral pinna, whereas bilateral pinna movements were elicited from other parts of the IC. The similarity between the distribution of monaurally sensitive units and the point which evokes isolated contralateral pinna movements speaks in favour of identical functional organization of the central nucleus of the IC with respect to space perception in a similar way as the organization of the superior colliculus.

REFERENCES

Adams, J. C., 1979, Ascending projections to the inferior colliculus, J. Comp. Neur., 183: 519–538.

Aitkin, L. M., Webster, W. R., Veale, J. L. and Crosby, D. C., 1975, Inferior colliculus. I. Comparison of response properties of neurons in central, pericentral and external nuclei of adult cat, J. Neurophysiol., 38: 1196–1207.

Aitkin, L. M., Dickhaus, H., Schult, W. and Zimmermann, M., 1978, External nucleus of inferior colliculus: auditory and spinal somatosensory afferents and their interactions, J. Neurophysiol., 41: 837–847.

Beyerl, B. B., 1978, Afferent projections to the central nucleus of the inferior colliculus in the rat, Brain Research, 145: 209–223.

Clopton, B. M. and Winfield, J. A., 1973, Tonotopic organization in the inferior colliculus of the rat, Brain Research, 56: 355–358.

Diamond, I. T., Jones, E. G. and Powell, T. P. S., 1969, The projection of the auditory cortex upon the diencephalon and brain stem in the cat, Brain Research, 15: 305–340.

Fifková, E. and Maršala, J., 1967, Stereotaxic atlases for the cat, rabbit and rat, in: Bureš, J., Petráň, M., Zachar, J., "Electrophysiological methods in biological research", Academia, Prague.

FitzPatrick, K. A., 1975, Cellular architecture and tonotopic orga-
    nization of the inferior colliculus of the squirrel monkey,
    J. Comp. Neur., 164: 185-208.
Goldberg, J. and Brown, P. B., 1969, Response of binaural neurons
    of dog superior olivary complex to dichotic tonal stimuli: Some
    physiological mechanisms of sound localization, J. Neurophy-
    siol., 32: 613-636.
Graham, R. C. and Karnovsky, M. J., 1966, The early stages of absor-
    ption of injected horseradish peroxidase in the proximal tubules
    of mouse kidney: ultrastructural cytochemistry by a new tech-
    nique, J. Histochem. Cytochem., 14: 291-302.
Marusyeva, A. M., 1971, Temporal characteristics of the auditory
    neurons in the inferior colliculus, in: "Sensory processes at
    the neuronal and behavioral level", G. V. Gersuni, ed., Academ-
    ic Press, New York.
Merzenich, M. M. and Reid, M., 1974, Representation of the cochlea
    within the inferior colliculus of the cat, Brain Research, 77:
    397-415.
Morest, D. K., 1964, The laminar structure of the inferior colli-
    culus of the cat, Anat. Rec., 148: 314.
Nauta, W. J. H. and Gygax, P., 1954, Silver impregnation of degenera-
    ting axons in the central nervous system. A modified techni-
    que, Stain Technol., 29: 91-93.
Pfeiffer, R. R., 1966, Classification of response patterns of spike
    discharges for units in the cochlear nucleus: Tone burst sti-
    mulation, Exp. Brain Res., 1: 220-235.
Ramón y Cajal, S., 1911, Histologie du système nerveux de Í homme
    et des vertébrés, Vol. II. Consejo Superior de Investigaciones
    Cientificas Instituto Ramón y Cajal, Madrid, 1955.
Rockel, A. J. and Jones, E. G., 1973, The neuronal organization
    of the inferior colliculus of the adult cat, I. The central
    nucleus, J. Comp. Neur., 147: 11-60.
Rose, J. E., Greenwood, D. D., Goldberg, J. M. and Hind, J. E.,
    1963, Some discharge characteristics of single neurons in the
    inferior colliculus of the cat. I. Tonotopical organization,
    relation of spike counts to tone intensity, and firing
    patterns of single elements, J. Neurophysiol., 26: 294-320.
Roth, G. L., Aitkin, L. M., Andersen, R. A. and Merzenich, M. M.,
    1978, Some features of the spatial organization of the central
    nucleus of the inferior colliculus of the cat, J. Comp. Neur.,
    182: 661-680.
Salava, T., Syka, J. and Popelář, J., 1979, Sealed sound system
    for small animal in auditory research, Physiol. Bohemosl.,
    28: 271.
Semple, M. N. and Aitkin, L. M., 1979, Representation of sound
    frequency and laterality by units in central nucleus of cat
    inferior colliculus, J. Neurophysiol., 42: 1626-1638.

Syka, J. and Radil-Weiss, T., 1971, Electrical Stimulation of the
    tectum in freely moving cats, Brain Research, 28: 567-572.
Syka, J., Radionova, E. A. and Popelář, J., 1980, Discharge characteristics of neuronal pairs in the rabbit inferior colliculus, in preparation.

INTEGRATION AND SEGREGATION OF INPUT TO THE CAT INFERIOR COLLICULUS

M. N. Semple and L. M. Aitkin

Departments of Psychology and Physiology
Monash University
Clayton, Australia 3168

The axons of cells from several nuclei on both sides of the brainstem converge at the midbrain to provide input to the central nucleus of the inferior colliculus (ICC) (Woolard and Harpman, 1940; Roth, Aitkin, Andersen and Merzenich, 1978; Adams, 1979). These converging afferents must terminate in an orderly fashion since, in common with other primary auditory nuclei, ICC displays a precise spatial representation of tonal frequency (Rose, Greenwood, Goldberg and Hind, 1963; Merzenich and Reid, 1974; Semple and Aitkin, 1979). However, recent studies of the organization of this nucleus have revealed, in addition, an apparent relationship between unit location and binaural response properties, which may reflect a partial segregation within ICC of the ascending brainstem projections (Roth et al., 1978; Semple and Aitkin, 1979). The present study was designed to investigate further the possibility that ascending afferents might be segregated within ICC of the cat.

One of the ascending inputs to ICC is provided by the dorsal acoustic stria (DAS), a tract originating in the contralateral dorsal cochlear nucleus (DCN). In 19 adult cats initially anaesthetised with sodium pentobarbital (40 mg/kg) and maintained with ketamine hydrochloride (to effect, usually about 10 mg), we examined the effects in ICC of applying electrical stimulation to the medial, caudal surface of the contralateral DCN, where DAS arises. Electrical stimuli consisted of single 0.1 ms square pulses applied via a concentric bipolar stimulating electrode (DC impedance $\doteq$ 150 Ω). In addition, auditory stimulation was provided in the form of tone pips (duration 300 ms, rise and fall 5 ms) presented at 1/s via sealed systems incorporating probe microphone assemblies. Bilateral cochlear microphonic recordings were examined periodically during each experiment to provide information about the state of the periph-

eral auditory system. Responses in ICC to either electrical or auditory stimulation were recorded with tungsten microelectrodes.

From a sample of 285 single units recorded in ICC, 153 responded
to electrical stimulation of the contralateral DCN. It could be argued that some of these units might have been activated indirectly
through pathways other than DAS; hence, we decided to consider in
detail the properties of those units responding at shortest latency
and with lowest threshold, believing that these units, at least,
must have received direct input from the contralateral DCN. The resultant sample consisted of 33 units responding to electrical stimulation at latencies <2.0 ms and with thresholds <2.0 V.

All 33 units were located laterally in ICC (in sagittal planes
between 4.1 and 5.1 mm from the midline) and all but two had best
frequencies to auditory stimulation of >6.4 kHz (indicative of ventral disposition). Nine units responded to contralateral monaural
stimulation only (EO); 18 were excited contralaterally and inhibited
ipsilaterally (EI); and the remaining 6 units exhibited more diverse
binaural interactions.

It is consistent with the known physiology of DCN units that
most ICC units apparently receiving direct input from the contralateral DCN were excited by contralateral tones, but it was not expected that such a large proportion of the sample would exhibit ipsilateral inhibition. Previous reports of weak crossed inhibitory
influence observed with some units recorded in DCN (Webster and
Riisik, 1975; Young and Brownell, 1976) would not appear to account
for the remarkably sensitive inhibitory effects observed in the majority of EI units of the present study. Figure one provides an example of such a unit. It can be seen (Fig. 1A) that response of unit
78-11-41 to contralateral monaural tones (20 kHz, 40 dB SPL) consists
of an initially vigorous discharge adapting to a lower level of firing, which is maintained for the duration of the 300 ms tone burst.
When the same stimulus is presented simultaneously on the ipsilateral
side (Fig. 1B) the later spikes are completely inhibited.

The response of unit 78-11-41 to electrical stimulation of DCN
is shown in Fig. 1C. The lower trace includes several deflections
from the baseline: the first biphasic transient component is the
artifact of the shock stimulus; the second major component (marked
by an arrow) is the response of unit 78-11-41 and the later spikes
are from a second unit which was apparently unresponsive to tonal
stimulation. It is evident that the discharge of unit 78-11-41 is
superimposed on additional evoked activity; the upper trace of Fig.
1C shows the features of this characteristic field response recorded
at the same site after unit 78-11-41 had disappeared. Three components
are evident: a "shock" artifact (S), "tract" (T) activity (presynaptic axons) and a "response" (R) component (post-synaptic neurones).

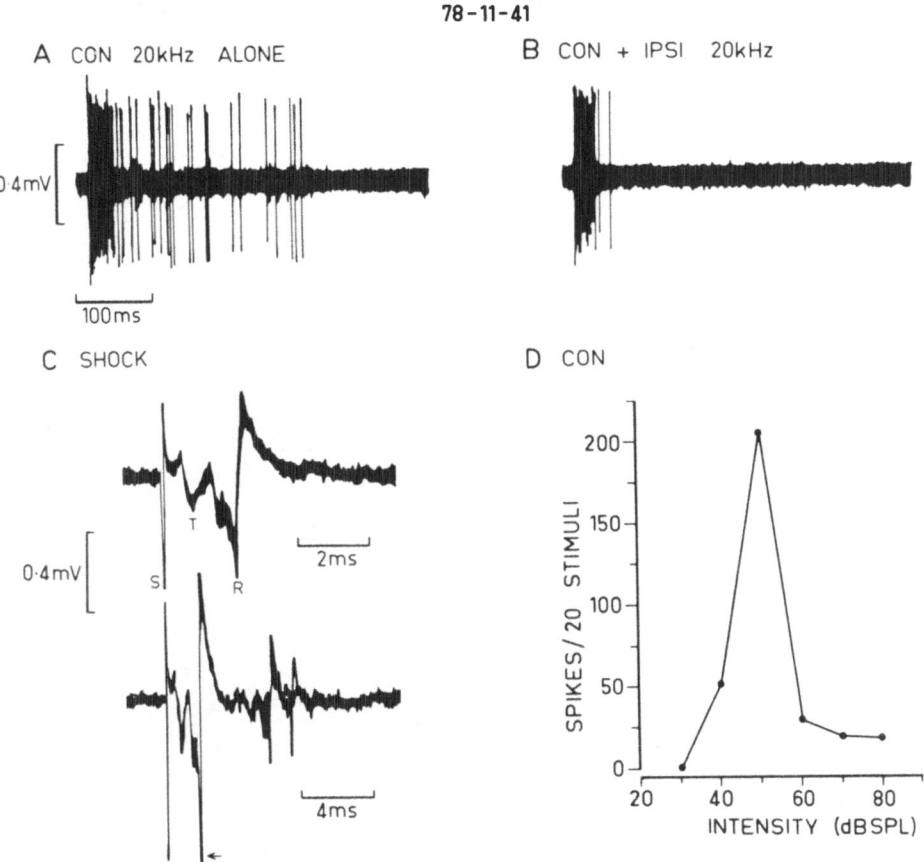

78-11-41

A CON 20kHz ALONE

B CON + IPSI 20kHz

0·4mV

100ms

C SHOCK

D CON

0·4mV

S   R

T

2ms

4ms

SPIKES / 20 STIMULI

INTENSITY (dB SPL)

Fig. 1.   Unit 78-11-41. Response to contralateral monaural tones (A)
          and to binaural tones (B) photographed from the oscilloscope.
          A single best frequency (20 kHz) tone burst at 40 dB SPL and
          300 ms duration was used in both cases. C: upper trace shows
          field response to 3V shock following the disappearance of
          Unit 41 (T: tract wave, R: post-synaptic response wave;
          S: stimulus artefact). Lower trace: Unit 41 (arrow) fires
          at peak of R wave. A second unit discharges approximately
          4 ms later. D: spike count/intensity function for Unit 41
          measured with 20 kHz contralateral tone pips.

     Another feature of unit 78-11-41 was that its discharge rate
was nonmonotonically related to the contralateral tonal stimulus.
As intensity was increased above threshold the spike output rose to
a maximum at 50 dB SPL; further increase in stimulus strength produced a reduction in firing (Fig. 1D). Sufficient information pertaining to the relationship between discharge and stimulus intensity

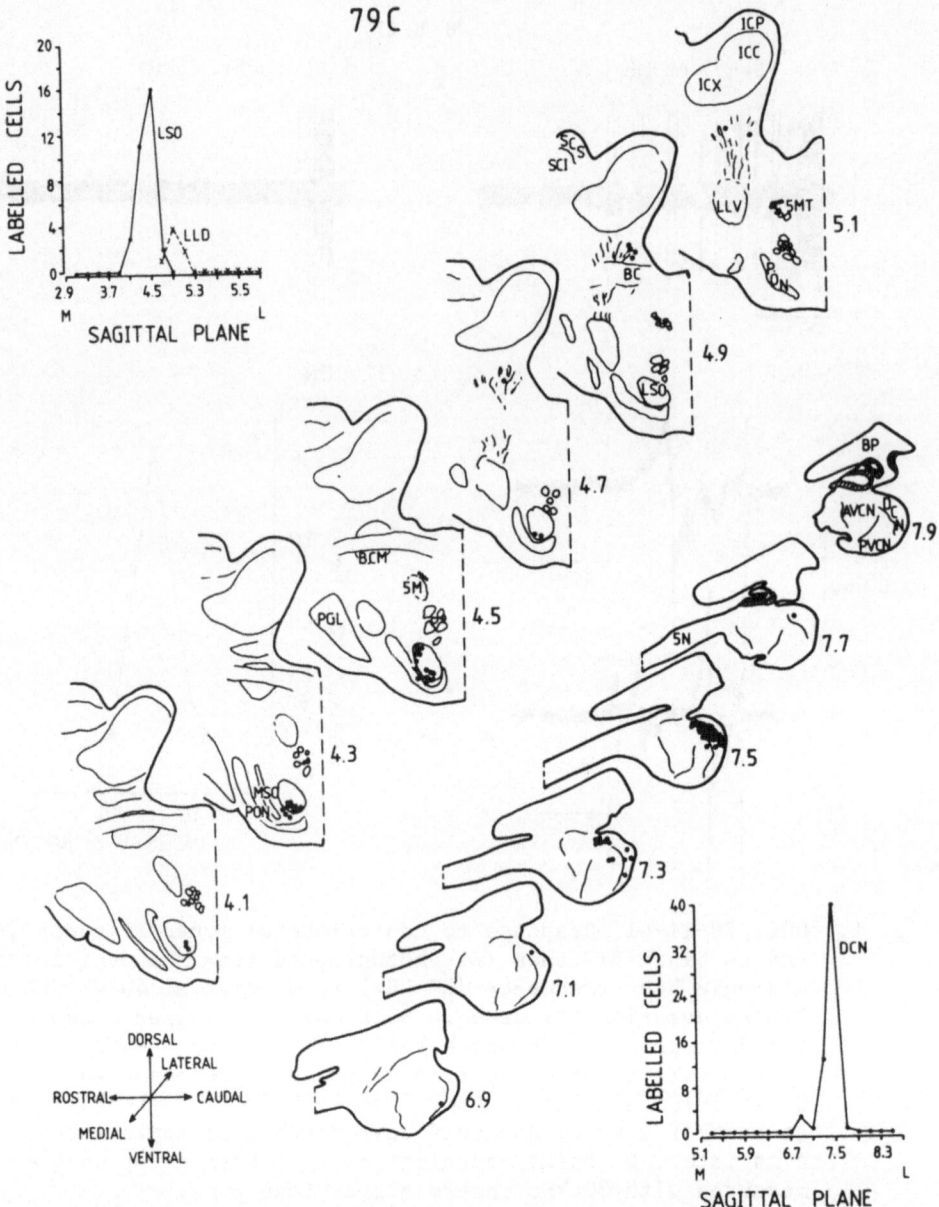

Fig. 2   Reconstructions of sagittal sections showing distribution
of HRP-labelled cells in the lateral superior olive (LSO),
dorsal cochlear nucleus (DCN) and dorsal nucleus of the lateral lemniscus (LLD) on the side contralateral to electro-
phoretic injection from a recording micropipette in ICC of
cat 79C. ICP, ICX: pericentral and external nuclei (cont´d)

(Fig. 2 cont`d) of inferior colliculus; SCS, SCI: superficial
and intermediate layers of superior colliculus; LLV; ventral
nucleus of lateral lemniscus; 5MT: motor nucleus of trigeminal
nerve; PON: peri-olivary nucleus: BC: brachium conjunctivum;
BCM: marginal nucleus of brachium conjunctivum; PGL: lateral
pontine grey; MSO: medial superior olive; BP: brachium pontis;
AVCN, PVCN: anteroventral and posteroventral cochlear nuclei;
5N: trigeminal nerve. Numbers appended to sections: sagittal
planes in mm. Insets: numbers of labelled cells in LSO, LLD
(upper left) and DCN lower right plotted as a function of
sagittal plane from medial (M) to lateral (L).

was obtained for 23 of the 33 units in our restricted sample. Of
these, 20 displayed nonmonotonic functions at best frequency with
maxima <60 dB SPL. Comparison of cochlear microphonic recordings
indicated a minimum separation of 45 dB, so it seems unlikely that
acoustic crossover could have produced these nonmonotonic functions.

The considerable nonmonotonicity of these functions is quite
consistent with responses recorded in DCN (Greenwood and Maruyama,
1965) but, as previously indicated, the sensitivity of ipsilateral
inhibitory effects is not predictable on this basis. It seems that,
in addition to a projection from DCN, some units in ventrolateral
ICC may receive input from a source which provides an inhibitory
influence.

A likely candidate for this source of ipsilateral inhibiton has
been revealed in a second series of experiments utilizing the retrograde tracer horseradish peroxidase (HRP). Anaesthetic conditions
and auditory stimulation were the same as in the electrical stimulation study, but single unit responses to tones were recorded in
ICC with micropipettes (5-10 μm tip) containing 10% HRP in normal
saline. In small regions where physiological characteristics appeared
fairly uniform, HRP was injected by low current iontophoresis to
label the sources of afferent input. Histochemical procedures used
to identify HRP label were identical to those described by Lane
(1978). Although these experiments are still in progress, useful data
have been obtained from 7 adult cats. Typically, labelling has been
observed in only two or three of the major projecting nuclei, with
different combinations of nuclei labelled by injections into different regions of ICC.

It was pertinent to the electrical stimulation study that when
HRP was injected into a discrete region of ventrolateral ICC characterized by EI and EO single unit properties (best frequency 7.5 kHz),
label was observed in the contralateral lateral superior olivary
complex (LSO) and in contralateral DCN. Fig. 2 demonstrates the pattern of labelling observed in this experiment. Reconstructions of
parasagittal sections at intervals of 0.2 mm are shown at the level
of the superior olivary complex (4.1 to 5.1) and of DCN (6.9 to 7.9).

Filled circles represent single labelled cells; labelling occurs in
LSO, DCN and sparsely in the dorsal nucleus of the lateral lemniscus
(LLD). These were the only regions on either side of the brainstem
or midbrain in which labelling was detected. The number of labelled
cells is plotted separately as a function of the sagittal plane over
which LSO and DCN extend. It is evident that only a very restricted
region of each nucleus contains labelled cells.

It would appear from the present results that ascending input
might remain partially segregated within ICC but that fine patterns
of integration also exist. In particular, the DAS projection is
restricted to ventrolateral parts of the contralateral ICC and appears to be integrated with input from LSO.

ACKNOWLEDGMENTS

The authors would like to thank Judy Sack and Jill Poynton for
illustrations and photography, and Pam Ward who typed the manuscript.
This study was supported by grants from the Australian Research
Grants Committee.

REFERENCES

Adams, J. C., 1979, Ascending projections to the inferior colliculus,
     J. Comp. Neur., 183: 519-538.
Greenwood, D. D. and Maruyama, N., 1965, Excitatory and inhibitory
     response areas of auditory neurons in the cochlear nucleus,
     J. Neurophysiol., 28: 863-892.
Lane, J. K., 1978, A protocol for horseradish peroxidase HRP histo-
     chemistry, in: "Neuroanatomical Techniques", Society for Neuro-
     science, Bethesda, Maryland, 69-61.
Merzenich, M. M. and Reid, M. D., 1974, Representation of the cochlea
     within the inferior colliculus of the cat, Brain Res., 77: 397-
     415.
Rose, J. E., Greenwood, D. D., Goldberg, J. M. and Hind, J. E., 1963,
     Some discharge characteristics of single neurons in the inferior
     colliculus of the cat. I. Tonotopic organization, relation of
     spike-counts to tone intensity, and firing patterns of single
     elements, J. Neurophysiol., 26: 294-320.
Roth, G. L., Aitkin, L. M., Andersen, R. A. and Merzenich, M. M.,
     1978, Some features of the spatial organization of the central
     nucleus of the inferior colliculus of the cat, J. Comp. Neur.,
     182: 661-680.
Semple, M. N. and Aitkin, L. M., 1979, Representation of sound fre-
     quency and laterality by units in the central nucleus of the
     cat inferior colliculus, J. Neurophysiol., 42: 1626-1639.
Webster, W. R. and Riisik, D. M., 1975, Binaural interaction in the
     cochlear nucleus of the cat, Proc. Aust. Physiol. Pharmacol.
     Soc., 6: 174-175.

Woollard, H. H. and Harpman, J. A., 1940, The connexions of the
    inferior colliculus and of the dorsal nucleus of the lateral
    lemniscus, J. Anat. (Lond.), 74: 441-458.
Young, E. D. and Brownell, W. E., 1976, Responses to tones and noise
    of single cells in dorsal cochlear nucleus of anaesthetised
    cats. J. Neurophysiol., 39: 282-300.

SOME FACETS OF THE ORGANIZATION OF THE PRINCIPAL DIVISION OF

THE CAT MEDIAL GENICULATE BODY

L. M. Aitkin, M. B. Calford, C. E. Kenyon and
W. R. Webster

Departments of Physiology and Psychology
Monash University
Clayton, Victoria, Australia

INTRODUCTION

It is our purpose in this paper to review knowledge accumulated about the medial geniculate body - the auditory nucleus of the thalamus - by presenting new findings pertinent to older questions. Two major divisions of the cat medial geniculate body (MGB) have long been recognized: a region of medium-sized, closely-packed cells forming the dorsal and lateral sector of this part of the thalamus, and a medial division populated by larger, more loosely-packed neurons. These divisions are usually referred to as the principal division (Mgp) and magnocellular (Mgm) divisions, respectively (Rose and Woolsey, 1949, 1958).

Studies of the MGB by Morest, using the Golgi technique (Morest 1964, 1965a) have revealed that Mgp has two subdivisions, the dorsal and ventral divisions, separable by virtue of their cellular packing and cellulo-dendritic arrangements. The ventral division contains neurons with tufted dendrites arranged in laminae in the lateral part (pars lateralis, LV) or in coils ventromedially (pars ovoidea, OV), while neurons of the dorsal division have radiate dendrites whose dendritic fields are larger in the dorsal nucleus (D) than in the deep dorsal nucleus (DP). A number of recent studies have shown that the various nuclei within Mgp derive their afferents from different midbrain sources and, in turn, project in contrasting ways upon the auditory cortex (Andersen et al., 1980; Colwell and Merzenich, 1975; Kudo and Niimi, 1978; Morest 1965b; Niimi and Naito, 1974; Oliver and Hall, 1978; Raczkowski, Diamond and Winer, 1976; Sousa-Pinto, 1973).

163

Physiological studies have, in some rexpect, lagged behind the increasing body of structural and connectional information about Mgp. Some information exists pertaining to the tontopic organization and tonal discharge characteristics of units in the ventral division (Aitkin and Webster, 1972). However, tone-evoked discharges have proven difficult to elicit in the dorsal nucleus of the barbiturate and nitrous oxide anesthetized animal (Aitkin, 1973; de Ribaupierre, Toros and de Ribaupierre, 1975) although it is known that unit responses in this region may be evoked by click stimuli in the chloralose - anesthetized cat (Altman et al., 1970).

In order to expand our knowledge of sensory processing in MGB, we have carried out a number of related studies of this nucleus. In this paper we will attempt to answer the following questions:

(1)   What are the differences in responses to tones of units in the dorsal and ventral divisions?

(2)   What can be deduced about the disposition of isofrequency contours in the ventral division by means of microelectrode penetrations of varying trajectory?

(3)   What are the details of the connections between discrete, physiologically - defined sectors of the ventral and dorsal divisions, and the inferior colliculus, as determined by microinjections of horseradish peroxidase.

Some aspects of the first part of this study have been previously published in abstract form (Calford and Webster, 1979) and will be dealt with relatively briefly in this paper.

METHODS

The basic operative, stimulation and recording procedures used in this study were very similar to those described in some detail in a previous publication from this laboratory (Semple and Aitkin, 1979) and only new details will be elaborated here. Anesthesia was induced in 34 cats by intraperitoneal injections of sodium pentobarbital (40 mg/kg) and supplemented by intramuscular doses of ketamine hydrochloride (10 mg). An occipital craniotomy was performed and a portion of the occipital cortex (not impinging on the auditory cortex) was removed to expose the midbrain and posterior thalamus. In most experiments electrodes were directed into MGB under visual control, usually in lateral-to-medial or dorsocaudal-to-ventrorostral directions.

Unit discharges were recorded with either tungsten-in-glass microelectrodes or glass micropipettes with tip diameters of 3-10 μm filled with 2M NaCl containing pontamine sky blue dye (impedances 1-2 MΩ). In later experiments micropipettes were filled with a 10 %

solution of horseradish peroxidase HRP; Sigma Type VI in 0.9% sterile saline (impedances 10-20 MΩ). Excellent spike isolation was
obtained with the three types of microelectrode, although better
stability appeared to be provided with the tungsten electrodes.

Horseradish Peroxidase Injections

Careful assessments were made of the threshold best frequencies
(BF) of single units in the 18 experiments in which micropipettes
filled with HRP were used. It was usually possible with vertical
penetrations to describe a region in ventral MGB in which BF varied
over only a small range. With 5 lateromedial penetrations a larger
portion of the cochlear representation was revealed electrophysiologically over a distance of 1-2 mm, and in these cats HRP was effluxed
along the entire cochlear map.

A cathodal current of 0.5 - 1μA, passed for 3-5 min, produced
an HRP efflux adequate to label neurons in the inferior colliculus.
However, the number of labeled cells did not appear to be related
in any simple way to the measured current passed. The enzyme was
usually deposited along a track length of 0.5-1mm with currents of
this size passed every 100μm. Retrograde transport was allowed to
continue for 16-24 hr., after which time a lethal dose of anesthetic
was administered and cats were perfused transcardially with 1-2 l of
warm (38°C) 0.9% saline, followed by a similar volume of cold (4°C)
2% glutaraldehyde in 0.1 M phosphate buffer (pH 7.2). The brain was
removed and relevant blocks were immersed overnight in 30% sucrose
dissolved in refrigerated, buffered glutaraldehyde, followed by 30%
sucrose in buffer alone (2-3 days). Blocks were then cut at 50μm on
a freezing microtome, and alternate sections were stained with thi-
onine or reacted with tetramethylbenzidine according to the method
of Lane (1978). The latter sections were counterstained with neutral
red.

The positions of HRP-labeled cells in the inferior colliculus
were determined with high-power light microscopy and located on low
power outline drawings of the midbrain.

RESULTS

Responses to Tones of Neurons in the Principal Division

Unit responses to tones in the ventral division of MGB have
been documented in a number of previous studies (Aitkin and Webster,
1972; Aitkin and Prain, 1974; Gross et al., 1974; Kallert, 1974;
Rouiller et al., 1979).

The results of the present study conform to the earlier observations. Most ventral division units were sharply-tuned with well-
defined BFs, exhibited binaural interaction and responded only at

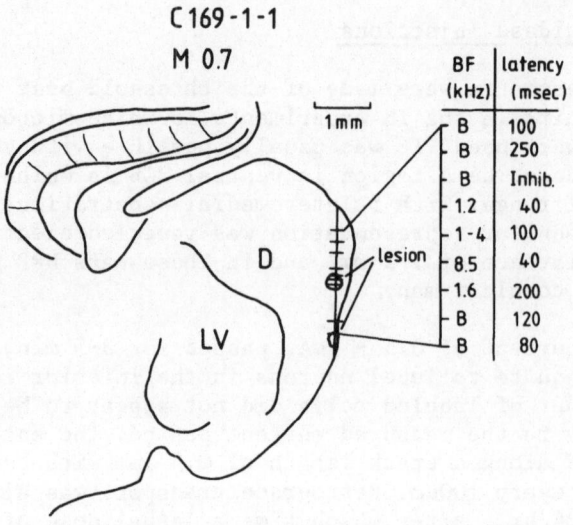

C 169-1-1
M 0.7

BF (kHz)	latency (msec)
B	100
B	250
B	Inhib.
1.2	40
1-4	100
8.5	40
1.6	200
B	120
B	80

Fig. 1. Micropipette track through caudal MGB(D) in experiment C169
shown in a schematic sagittal section located 0.7 mm medial
to the lateral edge of the MGB (M 0.7). Cross bars along
electrode track correspond to unit recording sites whose
best frequencies (BFs) and first spike latencies are shown
on the right. Two lesions are shown as irregular circles.
B: broad tuning; Inhib.: inhibitory response; D, LV: dorsal
division and pars lateralis of MGB.

the onset of the tone. The prevalence of onset responses in the
ventral division may be due to a powerful and widespread post-synaptic inhibition generated within MGB (Aitkin and Dunlop, 1969; Etholm,
1975); this inhibition presumably eliminates spikes occurring during
the latter part of the stimulus relayed from the inferior colliculus.

In contrast to the ventral division, little information exists
about discharge patterns in the dorsal division, but many units in
its caudal sector have been sampled in the present study. These are
characteristically labile and broadly-tuned with ill-defined BFs,
and tend to fire at much longer latencies than units in ventral MGB.

A representative track through the caudal dorsal division
(Fig. 1) reveals these traits, and the long latencies of 5 of these
units (at least 100 ms) are suggestive of a rebound from inhibition
(Aitkin and Dunlop,1969). Thus, the initial acoustic event for some
dorsal units may be inhibitory. The broad frequency response exem-
plified by most of these units precludes the investigation of tonotopic organization in the dorsal division. This property is in division units, with their associated orderly cochleotopic organization.

Electrodes passing through both Mgp (D) and Mgp (V) usually
first encountered units behaving in the way described above (Fig. 1).
In some penetrations labile, long latency units with broad frequency
responses were succeeded by more secure responses with latencies of
20-50 ms. The occurrence of the latter units usually signalled the
close proximity of sharply-tuned ventral division neurons with latencies of 8-20 ms. It was possible in some experiments to correlate
the physiological boundary between MGB (D) and MGB (V) responses
with a clear cytoarchitectonic border. Unfortunately, it was not
always possible to obtain clear tone-evoked responses in the dorsal
division - in these cases it is possible that unit responses were
depressed by the anaesthetic level.

Tonotopic Organization of MGB (LV)

We have examined further aspects of the distribution of frequency
-sensitive elements in MGB (LV) by comparing the BFs of units isolated in both lateral to medial penetrations and in obliquely-angled
tracks running from dorso-caudal to ventro-rostral. These have been
made with micropipettes, in contrast to the larger metal electrodes
used in previous studies, and probably provide unit recordings from
a more representative sample of neuronal soma diameters than have
been made previously in this laboratory.

Best frequencies assessed during 3 latero-medial penetrations
at the junction of the caudal and middle thirds of MGB, obtained
from pars lateralis (Fig. 2A) characteristically showed a regular
increase from low to high as the micropipette proceeded medially.
Low octaves (below 2 kHz) appeared compressed into the first 0.5 mm
of tissue relative to the somewhat more expanded representation of
the higher octaves medially. Such findings confirm previous studies
(Aitkin and Webster, 1972; de Ribaupierre and Toros, 1976) and suggest a disposition of iso-frequency contours of the sort shown in

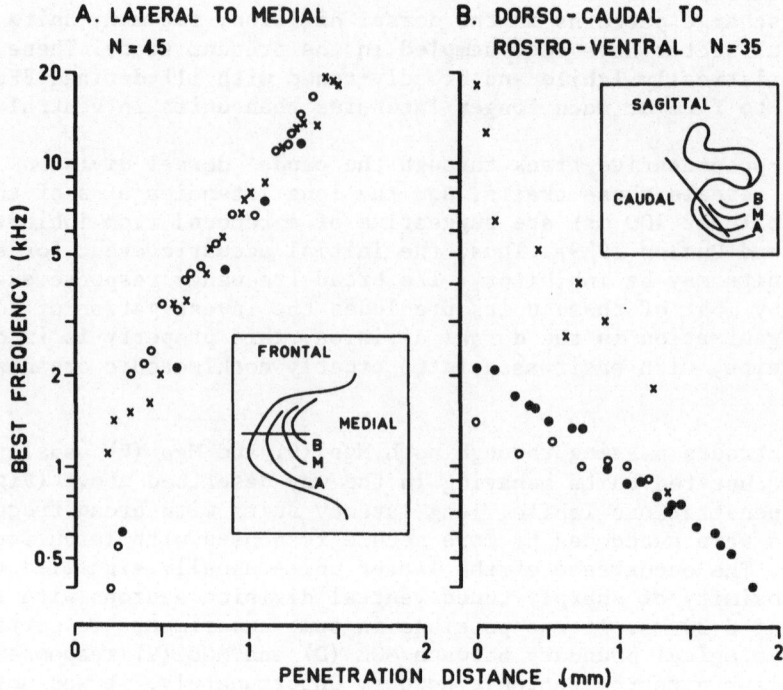

Fig. 2.   Best frequencies of 45 units recorded in 3 lateral-to-medial
          penetrations (A) and of 35 units isolated in 3 dorso-caudal
          to rostro-ventral penetrations (B), plotted against pene-
          tration distance in mm. Symbols are used only to distin-
          guish different experiments. Insets: schematic illustrations
          of suggested laminar dispositions, related to the apex,
          middle and base (A, M & B, respectively) of the cochlear
          partition.

Fig. 2A; inset. These have a similar orientation to the cellulo-
dendritic laminae observed in Golgi preparations of MGB (LV) by Morest (1965a).

     Penetrations made through MGB (LV) in the vertical plane pro-
vided less stereotyped sequences of BF which depended on the angle
to the vertical plane, and on the caudo-rostral and latero-medial
points of entry into the ventral division. An inclination of approx-

imately 30° caudal to the vertical plane was commonly used, and three
representative tracks are summarized in Fig. 2B, where descending
sequences of BF were observed. With caudal entry points lateral
tracks usually began with lower best frequencies than did more medial
penetrations (e. g. crosses versus filled circles in Fig. 2B); incli-
nations greater than 45° to the frontal plane often sampled relatively constant BFs. Results such as these suggest the isofrequency
contours schematically depicted in the inset of Fig. 2B.

Taken together, our data provide a picture of a cochlear representation in MGB (LV) where low frequencies are located laterally,
ventrally and caudally while hight frequencies occur medially, dorsally and rostrally.

Collicular Input to MGB (LV) Traced with HRP

The projection of the inferior colliculus upon the medial geniculate body has previously been studied in the cat using large
lesions of the inferior colliculus (Moore and Goldberg, 1963; Powell
and Hatton, 1969) or with relatively gross injections of anterograde
tracer into that structure (Andersen et al., 1980). Aside from the
brief report of Kudo and Niimi (1978), the fine details of the topographic connections between these nuclei remain to be described.

We have paid close attention to the labeling patterns in the
ipsilateral ICC following small extrusions of HRP into MGB (LV).
Although no data will be presented, it should be noted that most,
but not all, injections resulted in a very much smaller number of
labeled neurons in the contralateral ICC. When labeled neurons were
found contralaterally, their distribution mirrored that found in the
ipsilateral ICC.

It is appropriate to begin with an example of a lateral-to-medial penetration in which HRP was extruded along a discrete cylinder
containing a major portion of the cochlear representation. One such
experiment is shown in Fig. 3 (AWK-3) in which the micropipette
entered MGB (LV) at the junction of the caudal and middle thirds of
this region. Small clusters responded at 400-500 Hz and subsequently
8 units and many unit clusters were isolated, providing a regular
BF sequence from 1 to 15 kHz (Fig. 3A, B). A positive current of 500
μA (3 minutes on, 2 minutes off) was applied every 100 μm as the
pipette was withdrawn. Cells labeled following retrograde transport
of HRP to ICC were distributed across the latero-medial extent of
dorsal ICC (Fig. 3C) and were in greatest numbers medially and rostrally (Fig. 3C, 2.7 - 3.15; Fig. 3D). Thus, a latero-medial (apex
to base) injection in MGB LV provided a latero-medially orientated
labeled strip in ICC. Such a strip would be expected to encompass
a wide frequency range (Semple and Aitkin, 1979). The apparent high
BF focus in ICC is probably attributable to a large initial efflux

of HRP than that occurring at the end   lateral part   of the injection.

A contrasting experiment in which 0.5 µl of a 30 % solution of HRP was injected (with a 10 µl microsyringe) into the lateral sector of Mgp – containing units with low best frequencies – is shown in Fig. 4 (78 – G). The distribution of retrograde labeling within ICC was very consistent from section to section although the cell numbers varied as a function of rostro-caudal location in ICC. In all sections a swathe of labeled neurons formed a crescent along the dorsolateral boundary of ICC, particularly at rostral levels (Fig. 4,

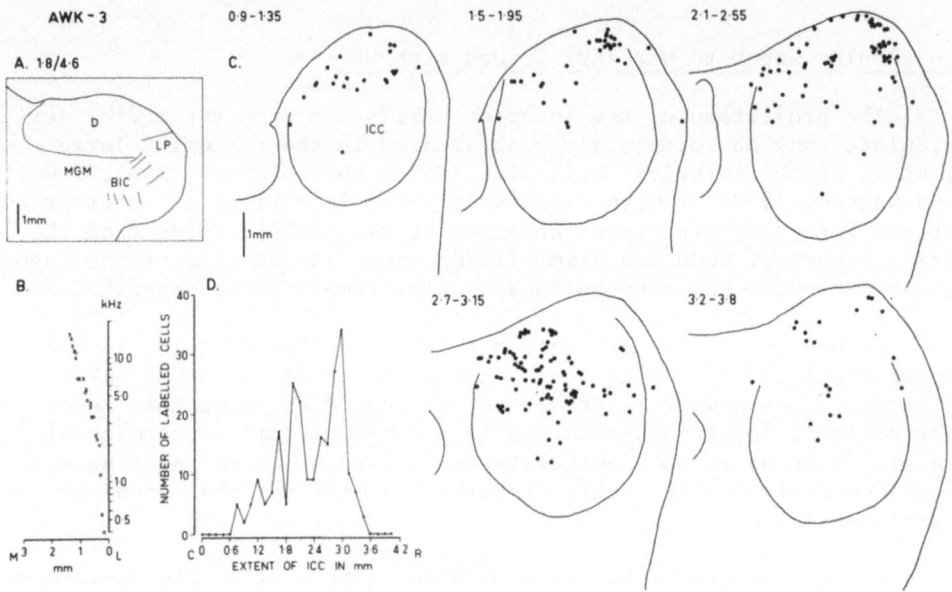

Fig. 3.   A: Micropipette track through pars lateralis (LP) in experiment AWK-3 shown in a schematic frontal section taken 1.8 mm rostral to the caudal tip of MGB. Total caudo-rostral extent of MGB = 4.6 mm. MGM: magnocellular division of MGB; BIC: brachium of inferior colliculus.
B: Best frequencies of units (open circles) and clusters (crosses) isolated along the above track, plotted against lateral (L) to medial (M) penetration distance in mm (B).
C: Schematic outline drawings summarizing the distribution of labeled cells (dots) in sections located at the stated distances from the caudal boundary of the inferior colliculus (0.9-1.35 etc.).
D: Plot of total number of labeled cells against (cont'd)

(Fig. 3 cont´d) caudal (C) to rostral (R) extent of ICC in mm.

1.7, 2.0), populating a region known electrophysiologically to contain low BFs (Semple and Aitkin, 1979). This swathe of cells is largely orthogonal to that in Fig. 3 shown to result from an injection over a wide cochlear segment, a fact which might be predicted from the nature of the frequency representations in the two structures.

Seven injections were made where the micropipette traversed MGB (LV) from dorsocaudal to ventrorostral. In these, HRP was extruded along a track segment containing a restricted cochlear representation. In cat AWK 10 (Fig. 5) a regular descending sequence of BFs was encountered in MGB (LV) beginning dorsally at 2.1 kHz and ending with 0.5 kHz BF (Fig. 5A). An efflux of HRP was made along the track between 0.5 and 0.75 kHz BF locations. Labeled cells in ICC were mostly located between 5 and 6 mm from the midline (Fig. 5B) and formed a three-dimensional slab extended in the dorso-ventral plane (Fig. 5E, F) and tilted in the horizontal plane to occupy a more dorsal position medially (Fig. 5, G, H).

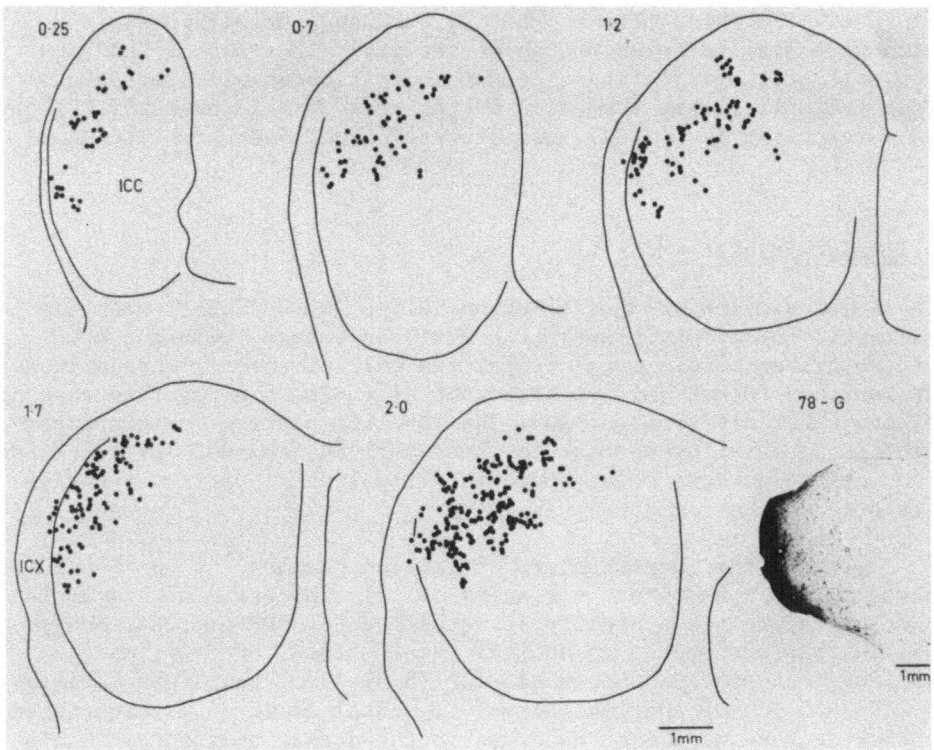

Fig. 4. Representative frontal sections from caudal (cont´d)

(Fig. 4 cont´d) (top left) to rostral (bottom, middle)
through ICC at stated distances (mm) from caudal boundary,
in experiment 78-G. Bottom right: Frontal section through
ipsilateral MGB reacted for HRP but not counterstained.

The higher BF injection of cat C219 (Fig. 6) was made between
2.7 and 7.0 kHz along a track which later exhibited a reversal of BF
(2.6 to 14.9 kHz) at the presumed border between MGB (LV) and MGB (OV)
(Fig. 6A). As in Fig. 5, labeled neurons were confined to sagittal
planes between 5 and 6 mm from the midline (Fig. 6B). However, relative to the low BF injection of Fig. 5, the slab was displaced
ventrally within ICC and tilted from dorso-caudal to ventrorostral
(Fig. 6D, E).

These experiments are preliminary, but certain features of the
pattern of connections between MGB (LV) and the ipsilateral ICC are
already apparent. First, labeled cells were found in the lateral part
of ICC and were always located in regions which could be predicted,
given the BF of the injection site, from the known tonotopic organization of ICC. Secondly, a small slab or patch in MGB (LV) received
input from a small patch in ICC. Thus the connections between ICC and
MGB (LV) are essentially point-to-point. One qualifying comment
should also be made, however: not all ICC neurons within a given
patch were labeled, even though it is likely that all of the axon
terminals at the injection site have been flooded with retrograde
tracer. Finally, very few cells in the pericentral nucleus (ICP) or
the external nucleus (ICX) were labeled when injections were made
in MGB (LV).

Collicular Input to MGB (D)

A large microsyringe injection (0.5 µl of 30 % HRP) was made in
the caudal dorsal division in cat MC-1. An intense reaction with
tetramethylbenzidine revealed that the bulk of the retrograde tracer
was confined to MGB (D) but that some diffusion had occurred rostrally into other divisions of MGB. The labeling pattern in this experiment is clearly in contrast to those derived from MGB (LV) injections,
with labeled neurons being found in virtually all parts of the ipsilateral auditory midbrain (Fig. 7).

Owing to the large numbers of labeled neurons, it was found
convenient to summarize the results of this injection with symbols
denoting the relative density of occurrence of HRP neurons. Heavy
labeling was restricted to nucleus sagulum (SAG, SP 7.0), medium
labeling found in ICX and caudal ICP (5.0 - 7.0) and light labeling
in ICC (4.0 - 6.0) and the lateral tegmental field (6.0). Scattered
labeled neurons were also observed in the dorsal and ventral nuclei
of the lateral lemniscus (LLD and LLV, 5.0 - 6.0).

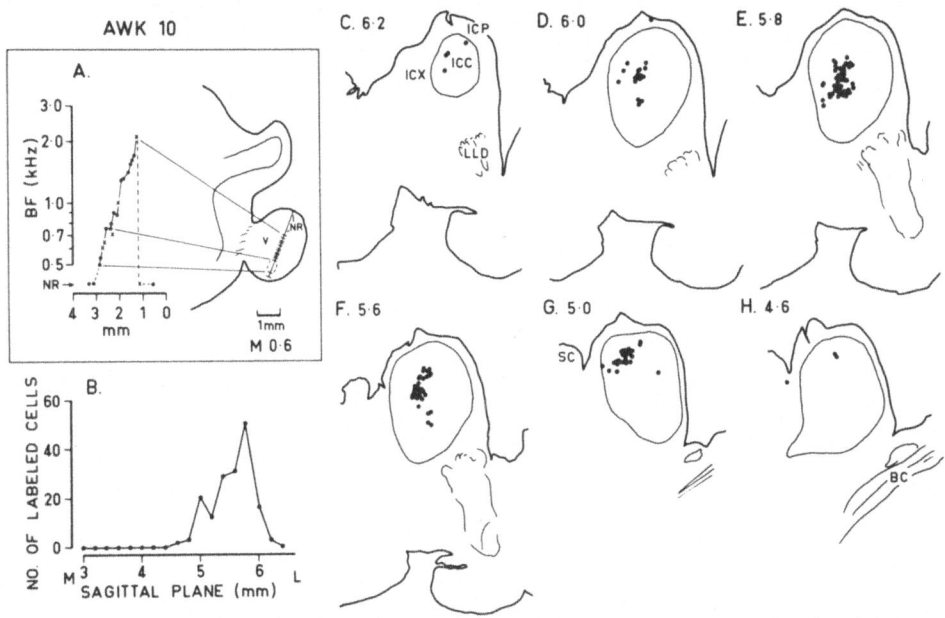

Fig. 5.   A: Micropipette track through MGB (V) of cat AWK-10, shown
in a schematic sagittal section 0.6 mm medial to the lateral
edge of MGB (M 0.6). Dotted line near end of electrode
track: location of HRP extrusion; NR: no response; other
nomenclature as for Fig. 3.
B: Number of labeled cells in ICC plotted against sagittal
plane, from medial (M) to lateral (L).
C-H: Outline drawings summarizing the distribution of labeled cells in representative sagittal sections through
ICC at the stated sagittal planes (6.2, etc.). SC: superior
colliculus; LLD: dorsal nucleus of lateral lemniscus. 1 mm
calibration in A applies to all sections.

This experiment thus identified the various midbrain candidates
which could supply important connections to MGB (D). We have continued this study using micropipettes filled with HRP which were
first used to record response patterns in caudal MGB (D) and, subsequently, to extrude HRP over discrete areas of interest. In one
such experiment (MC-2, Fig. 8) 7 units with complex properties were
recorded in caudal MGB (D) Fig. 8A. Three offset responses, one
inhibitory unit and three long-latency discharges to pure tones were
observed, some of which were very broadly tuned.

A small deposition of HRP along a restricted portion of this
track led to a highly discrete labeling pattern in the midbrain -
a narrow band of darkly-labeled cells placed very laterally in
nucleus sagulum (Fig. 8, B-D). This discrete injection result con-

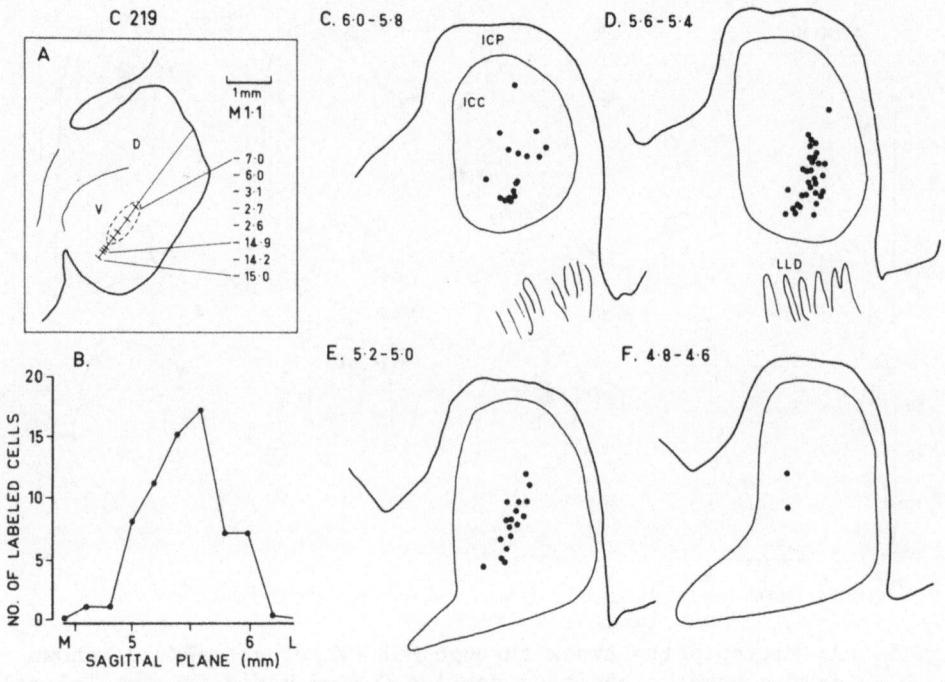

Fig. 6.   A: Micropipette track through MGB (D) and (V) in cat C219,
          shown in schematic sagittal section 1.1 mm medial to the
          lateral edge of MGB.
          B: Number of labeled cells in ICC plotted against sagittal
          plane.
          C-F: Outline drawings summarizing the distribution of la-
          beled cells in representative sagittal sections through
          ICC. Other nomenclature as for Fig. 5.

firms that the heavily-labeled cells of nucleus sagulum in Fig. 7
are indeed an important source of afferents to MGB (D).

    Subsequent experiments in progress suggest that the connections
between the auditory midbrain and MGB (D) may be quite complex but
that the major midbrain relay, ICC, is likely to play very little
part in this pathway.

DISCUSSION

    We have presented data pertaining to the organization of the
principal division of the medial geniculate body. Our results have
confirmed certain earlier observations and also fill some gaps in
our knowledge of the relationships between the auditory thalamus
and midbrain.

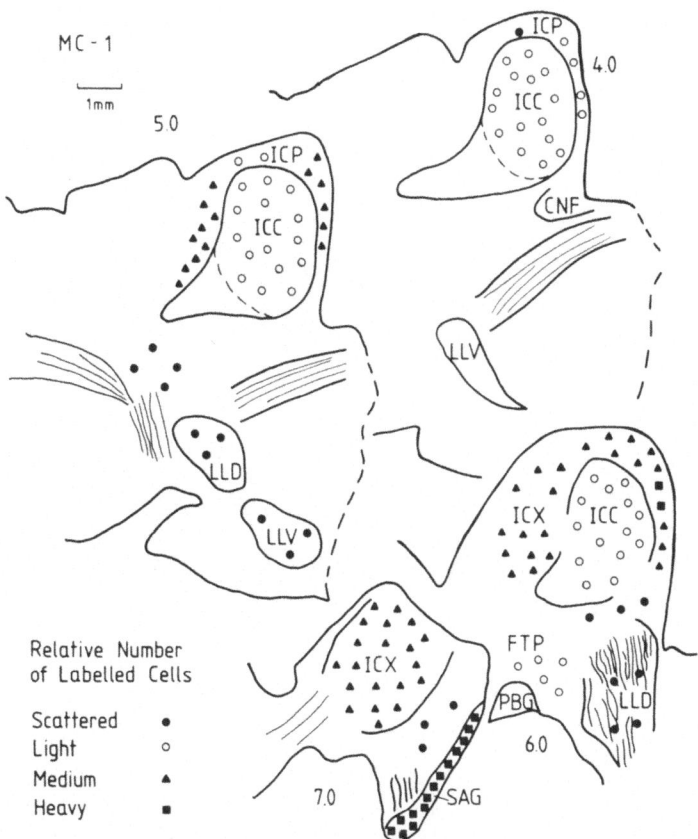

Fig. 7.   Relative distribution of labeled cells in the auditory midbrain following an 0.5 μ l injection of 30 % HRP into the
caudal tip of the MGB in cat MC-1. LLV: ventral nucleus of
lateral lemniscus; CNF: cuneiform area; FTP: paralemniscal
tegmental field; PBG: parabigeminal nucleus; SAG: nucleus
sagulum; ICP: pericentral nucleus of inferior colliculus;
ICX: external nucleus of inferior colliculus. Other nomenclature as in Fig. 5.

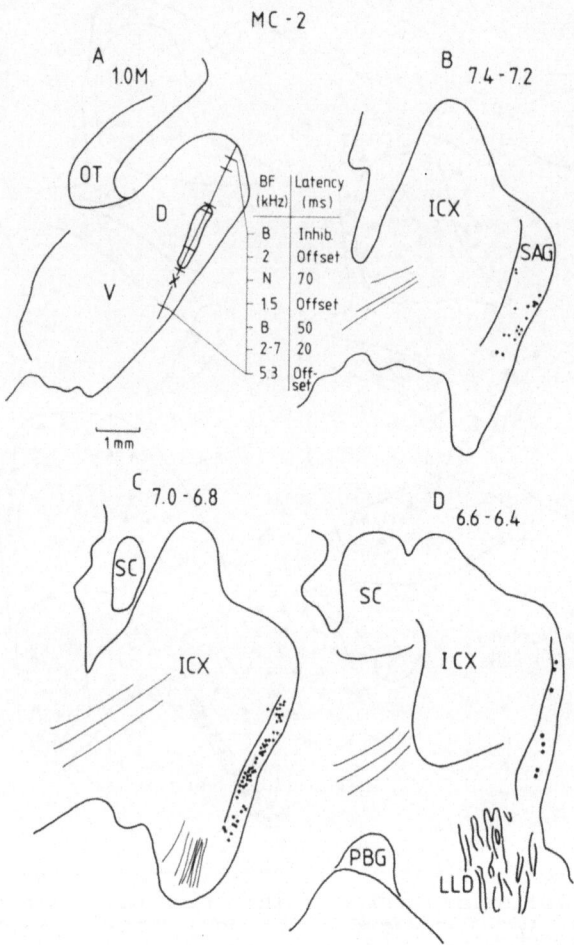

Fig. 8. A: Micropipette track and extrusion site in MGB (D) in cat
MC-2, illustrated by a sagittal outline drawing 1.0 mm
medial to the lateral edge of MGB. OT: optic tract; N:
response to white noise only; Offset: response (cont´d)

(Fig. 8 cont´d) to the offset of the tone pip.
B-D: Outline drawings of sagittal sections through the auditory midbrain at the stated sagittal planes. Nomenclature as for Figs. 5 and 7.

Dorsal Division of MGB

We have found that units in the dorsal division of MGB respond to tones in a labile fashion, often firing at long latencies to a broad range of tonal frequencies. In some experiments units were isolated in MGB (D) which could not be driven in any obvious way by tones, and other units only responded to complex sounds. It is possible that the level of barbiturate anesthesia may be an important determinant here; anesthesia has been shown to depress firing in other divisions of MGB (Aitkin and Prain, 1974). In support of this suggestion, most of the responsive units in the present study were obtained later in experiments, when ketamine, rather than pentobarbital, may have been the dominant anesthetic. A depressing action of barbiturate anesthesia is also suggested by the fact that MGB (D) units in chloralose-anesthetised animals often respond to acoustic stimuli (Altman et al., 1970).

The lability of MGB (D) unit responses may also be related to the nature of the connections this area receives from the auditory midbrain. In a detailed study of the tree shrew, Tupaia glis, Oliver and Hall (1978) have used both anterograde and retrograde tracing methods to define auditory colliculogeniculate connections. Their results suggest that MGB (D) receives most of its input from nucleus sagulum, with additional light projections from the dorsal nucleus of the lateral lemniscus and from the midbrain tegmentum. We have reached a similar conclusion with regard to nucleus sagulum, and Morest has shown that fibers reach MGB (D) as part of a lateral tegmental system medial to the brachium of the inferior colliculus (Morest, 1965b). In an anterograde study of the cat by Andersen and his colleagues (1980) evidence has been provided for a projection from ICP to caudal MGB (D); this is supported by the work of Kudo and Niimi (1978). However, in the tree shrew, ICP would appear to project mainly to the deep dorsal nucleus of MGB (Oliver and Hall, 1978).

A synthesis of these results indicates that MGB (D) is supplied by axons originating from nuclei dorsal, caudal and lateral to ICC, and from the lateral tegmentum. Units in these regions have been shown to have similar properties in terms of lability and broad tuning to those we have described for MGB (D)(Aitkin et al., 1975; Jackson and Irvine, 1978).

It is clear that the afferent connections to the dorsal division of the MGB are complex. The picture is further complicated by the

knowledge emerging about the pathways between different parts of the
dorsal division and the auditory cortex (Raczkowski et al., 1976;
Andersen, 1979). We believe that further understanding of the connections - and, ultimately, function - of MGB (D) requires a careful
assessment of discharge characteristics in sectors of this nucleus,
followed by discrete injections of tracing substances to label the
pathways related to these physiologically-defined sectors. Until
these experiments are carried out, further speculations are unwar-
ranted.

Ventral Division of MGB

Our results have been restricted to pars lateralis (LV) of the
ventral division. The lateral-to-medial arrangement of low-to-high
BFs is now generally accepted in the literature (Aitkin and Webster,
1971, 1972; Gross et al., 1974; de Ribaupierre and Toros, 1976).
These studies, in one sense, simply confirm the classic observations
made over twenty years ago by Rose and Woolsey (1958) on the basis
of cochlear stimulation and field potential recordings. In all major
respects, this cochleotopic arrangement is consistent with the elegant Golgi maps of Morest (1964, 1965a) for MGB (LV).

However, it is apparent that a lateral-to-medial tonotopic organization forms only one dimension of the complete cochlear representation in MGB (LV). If discrete segments of the cochlear partition
are represented by isofrequency sheets of neurons disposed in the
above manner, it is likely from our results that the sheets concerned
with apical portions of the cochlear partition would have their
greatest representation caudally and ventrally within MGB (LV), as
well as laterally, while basal portions will occur rostrally and
dorsally. Thus, it would be expected that lateral-to-medial penetrations made in the rostral one-third of MGB (LV) would initially
encounter much higher BFs that those made in caudal MGB (LV) in the
same horizontal plane. Furthermore, penetrations made in a dorsal-
to-ventral direction through MGB (LV) should initially encounter
high best frequencies and the subsequent decrease in BF should be
minimal rostrally. This suggestion is compatible with published data
on dorso-ventral penetrations (Aitkin and Webster, 1972).

The connections between MGB (LV) and AI in the cat appear to
parallel the cochleotopic organization of the former structure from
lateral-to-medial (Andersen, 1979, Colwell and Merzenich, 1975).
The additional ventrocaudal-to-dorsorostral progression of low-to-
high BFs in MGB could also add weight to the suggestion by Rose and
Woolsey (1949), based upon small cortical lesions, that the anterior
portion of Mgp (presumably identical to Morest´s ventral division)
projects from its ventro-caudal to its dorsorostral limits to the
corresponding part of AI.

Continuing experiments in our laboratory seek to provide more details of the representation of high frequencies in MGB (LV) and their connections with ICC. Information is also being sought about pars ovoidea, since topographic arrangements here are still unclear, and we are proceeding with a similar analysis of the medial division of MGB.

ACKNOWLEDGMENTS

The authors wish to thank Karen Styles and Judy Sack for their help with illustrations and photography, and Despina Green who typed the manuscript. This research was supported by grants from the Australian Research Grants Committee.

REFERENCES

Aitkin, L. M., 1973, Medial geniculate body of cat: responses to tonal stimuli of neurons in medial division., J. Neurophysiol., 36: 275-283.

Aitkin, L. M. and Dunlop, C. W., 1969, Inhibition in the medial geniculate body of the cat, Exptl. Brain Res., 7: 68-83.

Aitkin, L. M. and Prain, S. M., 1974, Medial geniculate body: unit responses in awake cat, J. Neurophysiol., 37: 512-521.

Aitkin, L. M. and Webster, W. R., 1971, Tonotopic organization in the medial geniculate body of the cat, Brain Res., 26: 402-405.

Aitkin, L. M. and Webster, W. R., 1972, Medial geniculate body of the cat: organization and responses to tonal stimuli of neurons in ventral division, J. Neurophysiol., 35: 365-380.

Aitkin, L. M., Webster, W. R., Veale, J. L. and Crosby, D. C., 1975, Inferior colliculus. I. Comparison of response properties of neurons in central, pericentral and external nuclei of adult cat, J. Neurophysiol., 38: 1196-1207.

Altman, J. A., Syka, J. and Shmigidina, G. N., 1970, Neuronal activity in the medial geniculate body of the cat during monaural and binaural stimulation, Exptl. Brain Res., 10: 81-93.

Andersen, R. A., Roth, G. L., Aitkin, L. M. and Merzenich, M. M., 1980, The efferent projections of the central nucleus and the pericentral nucleus of the inferior colliculus in the cat, J. Comp. Neur., in press.

Andersen, R. A., 1979, Patterns of connectivity of the auditory forebrain of the cat. Ph. D. Thesis, University of California, San Francisco.

Calford, M. B. and Webster, W. R., 1979, Auditory properties of single units in dorsal and ventral divisions of the principal nucleus of cat medial geniculate, Proc. Aust. Physiol. Pharmacol. Soc., 10: 104P.

Colwell, S. and Merzenich, M. M., 1975, Organization of thalamocortical and corticothalamic projections to and from physiologically-defined loci within primary auditory cortex in the cat, Anat. Rec., 181: 336.

de Ribaupierre, F. and Toros, A., 1976, Single unit properties related to the laminar structure of the MGN, in: "Afferent and Intrinsic Organization of Laminated Structures in the Brain", O. Creutzfeld, ed., Springer, Berlin, 503-505.

de Ribaupierre, F., Toros, A. and de Ribaupierre, Y., 1975, Acoustical responses of single units in the medial geniculate body of the cat, Exptl. Brain Res., 23: 51.

Etholm, B., 1975, Inhibitory processes in the medial geniculate body, Acta Oto-Laryng. 80: 323-334.

Gross, N. B., Lifschitz, W. S. and Anderson, D. J., 1974, The tonotopic organization of the auditory thalamus of the squirrel monkey (Saimiri sciureus), Brain Res., 65: 323-332.

Jackson, G. and Irvine, D. R. F., 1978, Acoustic properties of neurones in cat mesencephalic reticular formation, Proc. Aust. Physiol. Pharmacol. Soc., 9: 179P.

Kallert, S., 1974, Telemetrische Mikroelektrodenuntersuchungen am Corpus geniculatum mediale der wachen Katze. Habilitationsschrift, Friedrich-Alexander-Universität,Erlangen-Nürnberg.

Kudo, M. and Niimi, K., 1978, Ascending projections of the inferior colliculus onto the medial geniculate body in the cat studied by anterograde and retrograde tracing techniques, Brain Res., 155: 113-117.

Lane, J. K., 1978, A protocol for horseradish peroxidase (HRP) histochemistry, in: "Neuroanatomical Techniques",Society for Neuroscience, Bethesda, Maryland, 59-61.

Moore, R. Y. and Goldberg, J. M., 1963, Ascending projections of the inferior colliculus in the cat, J. Comp. Neur., 121: 109-136.

Morest, D. K., 1964, The neuronal architecture of the medial geniculate body of the cat, J. Anat., Lond., 98: 611-630.

Morest, D. K., 1965a, The laminar structure of the medial geniculate body of the cat, J. Anat., Lond., 99: 143-160.

Morest, D. K., 1965b, The lateral tegmental system of the midbrain and the medial geniculate body: study with Golgi and Nauta methods in cat, J. Anat., Lond., 99: 611-634.

Niimi, K. and Naito, T., 1974, Cortical projections of the medial geniculate body in the cat, Exptl. Brain Res., 19: 326-342.

Oliver, D. L. and Hall, W. C., 1978, The medial geniculate body of the tree shrew, Tupaia glis. I. Cytoarchitecture and midbrain connections, J. Comp. Neur., 182: 423-458.

Powell, E. W. and Hatton, J. B., 1969, Projections of the inferior colliculus in cat, J. Comp. Neur., 136: 183-192.

Raczkowski, D., Diamond, I. T. and Winer, J., 1976, Organization of thalamocortical auditory system in the cat studied with horseradish peroxidase, Brain Res., 101: 345-354.

Rose, J. E. and Woolsey, C. N., 1949, The relations of thalamic connections, cellular structure and evocable electrical activity in the auditory region of the cat, J. Comp. Neur., 91: 441-466.

Rose, J. E. and Woolsey, C. N., 1958, Cortical connections and functional organization of the thalamic auditory system of the cat, in:"Biological and Biochemical Bases of Behavior",H. F. Harlow and C. N. Woolsey, eds.University of Wisconsin Press, Madison, 127-150.

Rouiller, E., de Ribaupierre, Y. and de Ribaupierre, F., 1979, Phase-locked responses to low frequency tones in the medial geniculate body, Hearing Res., 1: 213-226.

Semple, M. N. and Aitkin, L. M., 1979, Representation of sound frequency and laterality by units in central nucleus of cat inferior colliculus, J. Neurophysiol., 42: 1626-1639.

Sousa-Pinto, A., 1973, Cortical projections of the medial geniculate body in the cat, Adv. Anat. Embryol. and Cell Biol., 48: 1-42.

FUNCTIONAL ORGANIZATION OF THE MEDIAL GENICULATE BODY STUDIED BY

SIMULTANEOUS RECORDINGS OF SINGLE UNIT PAIRS

P. Heierli, F. de Ribaupierre, A. Toros and
Y. de Ribaupierre

Institute of Physiology
University of Lausanne
Bugnon 7, 1011 Lausanne, Switzerland

Information processing in the CNS involves the parallel activation of whole populations of neurons; as each unit often shares a complex synaptic connectivity with its neighbours, it seemed interesting to look for signs of functional interaction between several (usually pairs) of single units recorded with the same microelectrode.

The experimental procedure has been described elsewhere in detail (E. Rouiller et al., 1979). The concurrent activity of two and sometimes more single units was isolated from the MGB of 17 nitrous oxide anesthetized cats.

Off-line analysis of possible signs of functional correlation has been carried out by computation of Joint PST Histograms during spontaneous activity (Gerstein et al., 1972). Possible effects related to previous stimuli presentation could be annulated by using shifted and differential J PST H. When few data were at hand, histograms have been smoothed with a bell-shaped gaussian bin. Statistical methods allowed assessment of departure from the null hypothesis of independent firing of both cells (M. Abeles, in press).

The preliminary analysis presented here was intended to explore some of the possibilities as well as limits of this method, when applied to a complex structure. 856 pairs of cells have been studied by this technique. Shapes of histograms have been distributed in 7 broad interaction classes (Fig. 1). It appears that about 40 % of pairs have mutually independent activity. The rest is distributed in the main subnuclei of the MGB in more or less similar proportions (using strict criteria: CI = 16 %, E = 12 %).

   In a trial to isolate eventual subpopulations of pairs sharing
some specific interactive behaviour, histogram features have been
studied more quantitatively. It appeared that the degree of coupling
of the pair (evaluated by the ratio of interaction-dependent activity
on interaction-independent activity) as well as the duration of
correlated firing vary over a fairly wide range; these variables
follow an unimodal distribution.

   Some response characteristics of the cells were also studied in
relation to the 3 main classes of interactions (no correlation,
common input, excitatory interaction). The mean difference between
response latencies (DL) of paired cells is significantly shorter for
cells sharing a common input (DL to tone: 4.5 ms, to noise: 3.9 ms
and to clicks: 3.3 ms) than for cells firing independently (DL are
respectively 7.9, 6.7 and 6.5 ms). For excitatory interactions these
values are 5.7, 7.5 and 3.4 ms with mean latency of excitatory cells
comparable to that of follower cells.

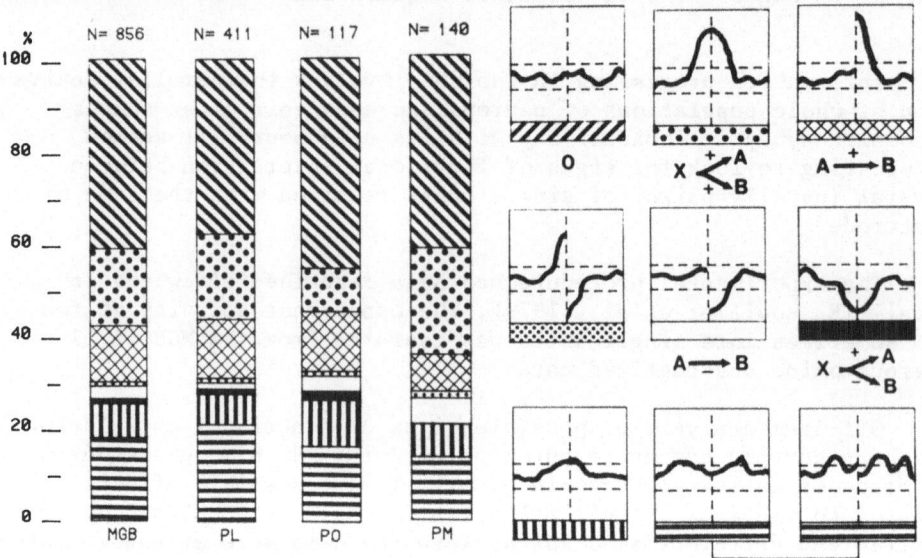

Fig. 1.   Distribution of interaction classes in the MGB. Right: main
          classes and representative J PST H; first row, from left
          to right: no correlation, common input (CI), excitatory
          synaptic interaction (E); second row: antisymmetric shape
          (?), inhibitory synaptic interaction, shared reciprocal
          synaptic input; third row: weak or doubtful interactions,
          and two examples of complex shapes that are difficult to
          interpret. Left: representation of each class in percents
          of the whole nucleus (MGB), pars lateralis (PL), pars
          ovoida (PO) and pars magnocellularis (PM).

The difference between the characteristic frequency of each member proved to be independent of the class of interaction to which the pair belonged.

We have been astonished that only 25 cells out of over 1700 showed evidence of inhibitory connections in a relay nucleus known to be rich in Golgi type II interneurons. This could reflect a sampling bias (larger cells being isolated presumably more often than smaller ones). On the other hand, it is possible that inhibitory activity in the MGB is mainly evoked in stimulus driven conditions. Indeed, when both spontaneous and stimulus driven activity were studied in the same pair only rarely did new or stronger correlation appear during stimulus presentation; on the contrary, histograms tended to flatten and sometimes displayed clear signs of inhibition (Fig. 2).

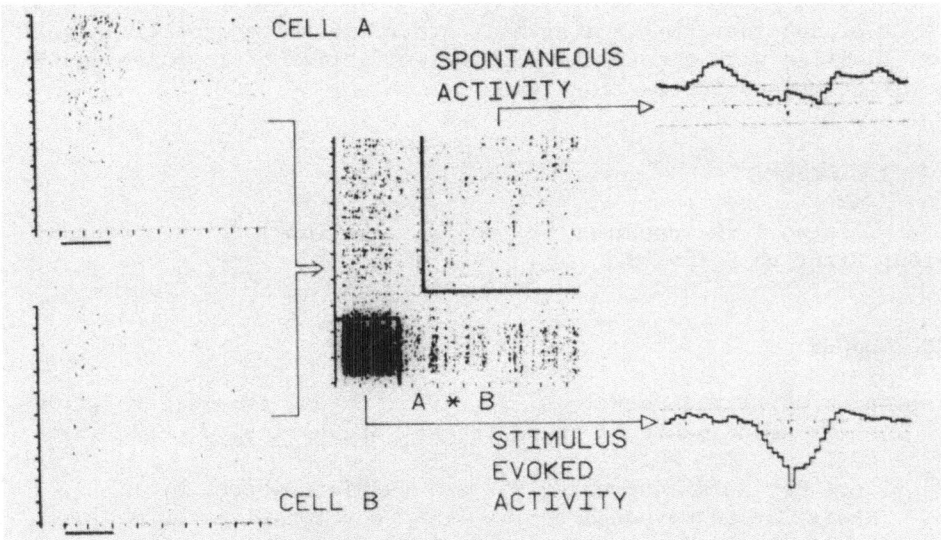

Fig. 2.   Stimulus driven versus spontaneous activity (cell pair B 14 C 11; stimulus 54.37 kHz, during 200 ms (presented every second), 80 to 0 dB, step: 10 stimulus-lines, contralateral ear). Left: dot displays of cells A and B. Middle: scatter diagram A.B computed over three stimuli sequences. Right top: differential J PST H of spontaneous period shows atypical signs of synchronization. Bottom: clear evidence of desynchronization during stimulus, possibly by shared reciprocal input.

It should be mentioned that sometimes a narrowing of the symmetrical peak indicating a common input was observed (sharpening of synchronization); in still other cases, histograms revealed no difference between both stimulus and spontaneous conditions. It could also be that very intense inhibitions cannot be detected at all, as the follower cell never fires.

Also striking is the similarity of the distribution of interaction classes between subnuclei, composed of distinct cell populations, with presumably specific connectivity between neighbour
cells.

The compared CF study of a pair suggests that the sensory
afferent pathway is distinct from an eventual common drive to both
cells. This is confirmed by several cases of pairs responding in
clearly distinct patterns to the same stimulus, despite evidence
of a strong common input in spontaneous activity.

A significantly shorter DL for cells sharing common inputs
would tend to support the alternate hypothesis.

We hope that this and several other unresolved questions will
be clasified with a more thorough investigation of response characteristics and interaction type.

ACKNOWLEDGMENT

This work is supported by the Swiss National Science Foundation, Grant no 3.239.69.

REFERENCES

Gerstein, G. L. and Perkel, D. H., 1972, Mutual temporal relation-
     ships among neuronal spike trains, Biophysical J., 12: 453-
     473.
Rouiller, E., de Ribaupierre, Y., and de Ribaupierre, F., 1979,
     Phase locked response to low frequency tones in the medial
     geniculate body, Hearing Res., 1: 213-226.

POSSIBILITIES OF RECORDING MULTIUNIT ACTIVITY IN THE AUDITORY

PATHWAY

E. David

Institute of Physiology and Biocybernetics

University of Erlangen-Nürnberg, F.R. G.

INTRODUCTION

It is of great interest to investigate the mechanism of neuronal processing of ecological signals that are of vital importance to individuals. However, the whole sound image of such signals is not depicted by the activity of individual or single cells. Several cells must be involved in creating such an image. Hence the multiunit activity of the cells must be observed in recognizing the pattern. This is possible either by using multichannel microelectrodes or microelectrodes having larger tip-diameter. In the latter case, it is necessary to discriminate different cell activities by computation.

METHODS

Cats in weak anaesthesia were used to observe the multiunit activity evoked by different auditory stimuli. The recording electrodes were made of tungsten which was sharpened by electrolytic etching and coated with an insulating material (Isonel). The tip was ultimately exposed by applying a high voltage (3 kV) for a few milliseconds. The tip-diameter was varied by inserting the electrode at different depths into the electrolyte during exposure. The electrodes were located in the brain by stereotactic device. The site of location was verified by controlling the acoustic activation and by histological findings.

The amplified multiunit activity was fed through an analog digital converter after pre-processing to a Linc 8 or PDP 11-40. Different programmes were used to compute amplitude histograms and

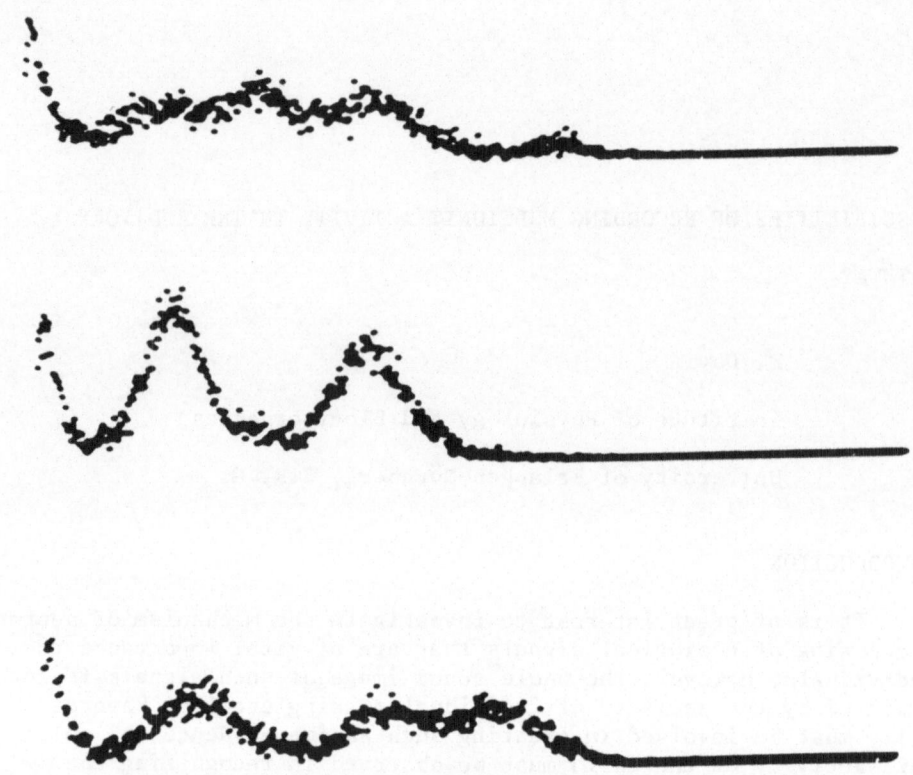

Fig. 1.   Amplitude histograms of multi unit activity in three dif-
          ferent functional stages. Recorded in the MGB, using a
          microelectrode with tip diameter of about 90 μm.

PST-histograms and also combined amplitude-PST-histograms. The
generation of the auditory stimuli was controlled by the computer.

RESULTS

     It was possible to differentiate three to six single units in
colliculus inferior and medial geniculate body by using amplitude
discrimination. The activity pattern derived from each amplitude
window showed a marked stimulus-dependent relation. The variation
of stimulus signal had significantly changed the amplitude-PST-
histogram as shown in Figs. 1 and 2. However, the cell reaction
to stimulus was different in each trace of the histogram. It is
possible to observe inhibition in one and an excitation in the other

trace of the same multiunit activity. The more complex the stimulus pattern the more pronounced is the variation.

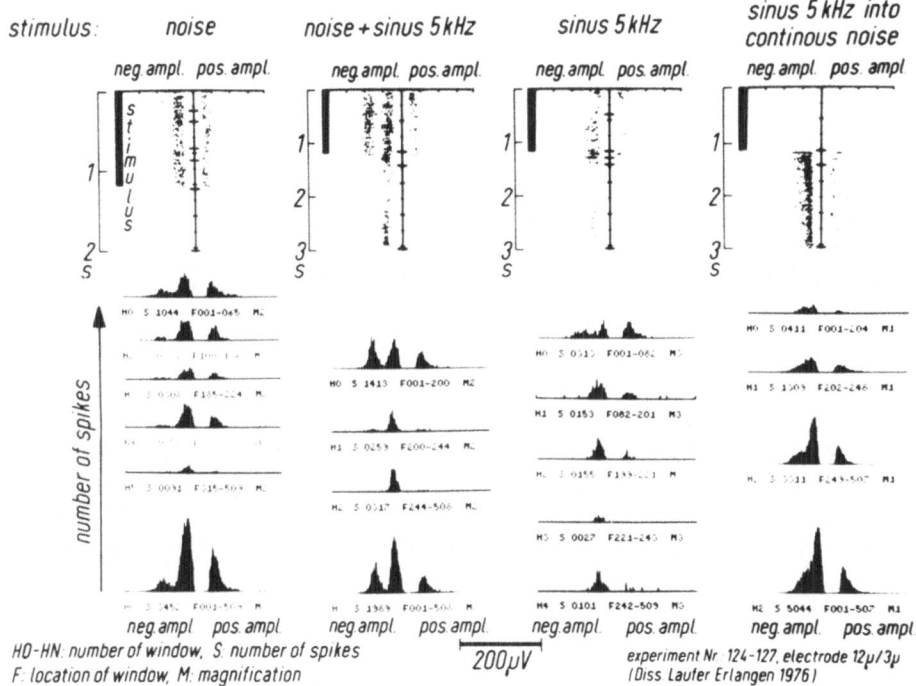

Fig. 2. Different combined amplitude stimulus time histograms and corresponding amplitude histograms at different times of the stimulus presentation.

In addition, it should be remarked that even a sharp discrimination of the spikes has not produced a better result. According to Lauffer (1980), it is useful to compute interval-histograms in addition to PST-histograms for every amplitude window. The interval histogram permits to evaluate the time pattern more exactly but only after carrying out the amplitude discrimination.

CONCLUSION

If we regard the spike amplitude as a function of the electrode distance from the spike generating cell, the data obtained by

computer processing (as described above) indicate that the coding
of information is comprehended not only as a temporal but also as
a local pattern. The specific information processing as described
by Keidel (1979) is also possible in an analogous manner. The results
are also in good agreement with similar recordings made by Bechtereva
(1969) in other regions of human brains. This type of coding information should be considered not only valid for the auditory system
(David et al., 1977) but also for the whole brain system.

REFERENCES

Lauffer, H., 1980, Über die Aussagekraft vom Amplitude-Zeitmustern
        bei multizellulärer Aktivität des Corpus gen. med. und lat.
        der narkotisierten Katze, Dissertation.
Keidel, W. D., 1979, "Handbook of Sensory Physiology", Springer,
        Berlin.
Bechtereva, N. P. 1969, "Physiologie und Pathophysiologie der
        tiefern Hirnstrukturen des Menschen", VEB Verlag, Berlin.
David, E. Keidel, W. D., Kallert, S., Bechtereva, N. P. and Bundzen,
        I. A., 1977, in: "Psychophysics and Physiology of Hearing",
        E. F. Evans and J. P. Wilson, eds., Academic Press, London.

POSTSTIMULATORY EFFECTS IN THE MEDIAL GENICULATE BODY OF GUINEA PIGS

Christoph Schreiner

Max-Planck-Institute of biophysical Chemistry
Dep. Neurobiology
Postbox 968, 3400 Göttingen, F.R.G.

INTRODUCTION

In recent years non-simultaneous masking methods were used in many psychoacoustical investigations. In particular, the forward-masking paradigm proves to be of great advantage concerning the temporal and spectral selectivity of the ear. Comparable results of neurophysiological investigations of forward-masking conditions were obtained at the level of auditory nerve (Smith 1977, Bauer 1978, Harris and Dallos 1979). Smith (1977) and Harris and Dallos (1979) concluded that there exists a reciprocal relationship between the poststimulatory recovery and the stimulatory adaptation; that both processes are directly related to the level of excitation; and that the two processes are independent of stimulus parameters. Bauer (1978), however, could demonstrate in some cases a dependence of poststimulatory "fibre-masking functions" on stimulus parameters, pointing to the poststimulatory influence of two-tone suppression.

Measurements of simultaneous and non-simultaneous two-tone suppression in the medial geniculate body (MGB) of guinea pigs yielded a close relation between the frequency characteristics of both kinds of tone-on-tone suppression (Schreiner 1979). This fact points to a poststimulatory suppression effect added to short-term adaptation. Therefore, the influence of excitatory and (simultaneous) inhibitory signals on the temporal characteristics of the poststimulatory suppression was studied.

METHODS

Experiments were done on unanaesthetized guinea pigs. The
responses of 20 single units of the MGB were measured. The conditioning signal (CS) was a tone of 100 ms duration. The frequency of
the CS was either situated at the characteristic frequency (CF) of
the unit or at that frequency which yielded the best inhibitory
effect on a CF response for simultaneous presentation. The averaged
ratio of inhibitory to excitatory stimulus frequency was 0.32 and
2.5, respectively. A probe tone (PT) of 30 ms duration was presented
simultaneously as well as non-simultaneously with the CS. The frequency of the PT was at the CF of the unit. The response to the uninfluenced PT was near the maximum firing rate of each unit. The level
of the CS was referred to the 50 % value of the averaged rate-intensity functions of the phasic CS-responses. The gap between CS
offset and PT onset, and the level of the CS served as parameters.

RESULTS

Simultaneous presentation of an excitatory CS and the PT
revealed a distinct adaptation process. At the beginning of the CS
a suppression of the PT response could not be observed. Shifting
of the PT to the end of the CS caused a suppression of the PT
activity of up to 55 % for high CS levels. Non-simultaneous presentation of the PT near the offset of the CS yielded a decrease of the
PT response, which has a good quantitative correspondence with the
amount of suppression during the tonic CS response. The temporal
characteristics of the poststimulatory recovery effect are shown in
Fig. 1. The recovery consists of two components. The first - fast -
part holds for about 20 ms. The time constant of the fast component
increases from 25 ms to about 70 ms for an increase of the CS level
from −30 dB to +30 dB. The second - slow - part has longer time
constants, ranging between 70 ms (−30 dB) and 390 ms (+30 dB). For
comparison, the results of Harris and Dallos (1979) are shown in
Fig. 1 (dashed lines) measured at the auditory nerve of chinchillas.
They also found two parts for the recovery process.

Several difference can be observed concerning an excitatory CS
when using an inhibitory tone as conditioning signal. A simultaneous
presentation of the probe tone with the onset of the inhibitory CS
results in a suppression of the PT activity of up to 40 % (see
Fig. 2). During the tonic part of the CS the PT response is diminished more than 70 % for high CS levels. Presenting the PT 1 ms
after the CS offset results in an instantaneous decrease of the
tone-on-tone suppression for high CS levels. However, the instantaneous recovery is not complete. The amount of suppression for a
CS level of +30 dB is reduced by a factor of 0.5: from more than
70 % for simultaneous suppression to about 36 % for poststimulatory
presentation of the PT. The remaining poststimulatory suppression

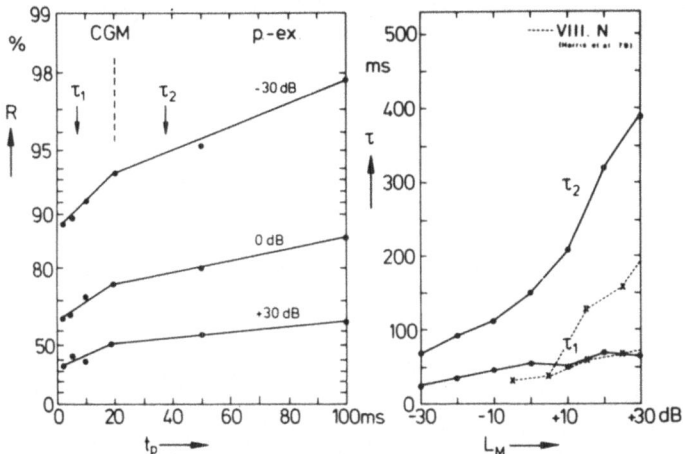

Fig. 1. Left: course of postexcitatory recovery in the MGB with CS
level as the parameter. Combined data of 18 units. Right:
time constants of the two parts of the recovery function
in the MGB (solid lines) and in the auditory nerve (dashed
lines, (Harris and Dallos 1979)).

effect is overlapped by the post inhibitory off-response, with its
maximum 20 ms after the end of the inhibitory CS. The recovery after
the off-response reveals rather long time constants of up to 500 ms
and more.

   Fig. 3 compares the amount of suppression (1 minus PT response)
at various time delays of the PT with the level of excitation during
the tonic part of the CS (the time delay is referred to the end of
the CS). In the left part of Fig. 3 the lower curve (solid line,
filled symbols) shows the normalized firing rate of the tonic part
of the CS (R/RO is referred to the maximum phasic response). It turns
out that both the amount of suppression during simultaneous presentation (-30 ms) and the poststimulatory suppression (2 ms, 20 ms) are
directly related to the level of the tonic excitation. The right
part of Fig. 3 shows that the amount of suppression for an inhibitory
CS is not related to the level of excitation (solid line, filled
symbols). For the simultaneous condition (-30 ms) this discrepancy
is well expressed. For the non-simultaneous case (2 ms) the amount
of suppression is about twice as much as the excitatory case would
suggest. Even at the maximum of the off-response (20 ms) the amount

of suppression does not seem to be linearly related with the excitatory activity during the inhibitory CS.

DISCUSSION

For an excitatory CS a close relation between the level of the
excitation and the amount of suppression could be established at the
MGB for simultaneous and non-simultaneous presentation of the PT.
This result agrees well with the findings at the auditory nerve
(Smith 1977, Harris and Dallos 1979). In both stations - auditory
nerve and MGB - the temporal recovery consists of two parts with

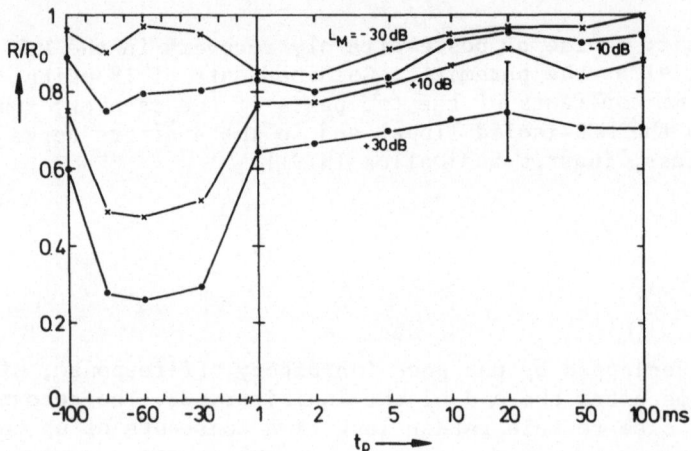

Fig. 2.   Simultaneous (left) and non-simultaneous (right) two-tone
          suppression of an excitatory probe tone by an inhibitory
          conditioning signal. The level of the CS is the parameter.

different time constants. The time constants of the first - fast -
part are found to be equal in MGB and auditory nerve (Harris and
Dallos 1979). In the MGB the time constants of the second - slow -
component of the recovery are twice the time constants found in the
auditory nerve (chinchilla). However, the differences between the
species (guinea pig and chinchilla) prevent one from concluding that
parts of the poststimulatory recovery remain uninfluenced between

the two stations. While at the auditory nerve the first part of the
recovery lasts for about 50 ms (Harris and Dallos 1979), in the MGB
the fast part terminated about 20 ms after CS offset. In spite of
some restrictions, these facts about the temporal characteristics
of postexcitatory recovery make it reasonable to postulate an
increase of its influence along the auditory pathway.

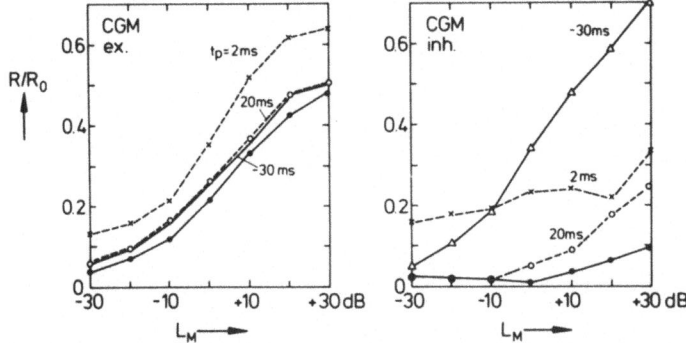

Fig. 3. Median rate-intensity function (solid line, filled symbols)
         of the tonic part of an excitatory (left) and an inhibitory
         (right) signal compared with the simultaneous (-30 ms)
         and the non-simultaneous (2ms , 20ms) reduction of the
         probe tone (1 minus probe response).

In the MGB the influence of inhibitory processes on the poststimulatory recovery is markedly stronger than at the auditory nerve
level. Thus, in the MGB forward masking tuning curves will not be
the same as frequency threshold curves. The "iso-forward masking
contours" will be a combination of the poststimulatory frequency
characteristics of inhibition and adaptation.

For comparing the psychoacoustics and the physiology of forward
masking the main difficulty is the contrary behavior of both kinds
of time constants. In psychoacoustics the time constants of the
recovery decrease with increasing CS level, whereas the recovery
time of single units increases. At least the psychoacoustical recovery exhibits two parts. The first - fast - component has a duration
of about 10 ms (Fastl 1979), which corresponds better with MGB
results 20 ms than with auditory nerve results (50 ms).

REFERENCES

Bauer, J. W., 1978, Tuning curves and masking functions of auditory-
    nerve fibres in cat. Sensory Processes, 2: 156-172.
Fastl, H., 1979, Temporal masking functions: III. Pure tone masker.
    Acustica, 43: 282-294.
Harris, D. M. and Dallos, P., 1979, Forward masking of auditory
    nerve fiber response, J. Neurophysiol., 42: 1083-1107.
Schreiner, Chr., 1979, Temporal suppression and speech processing,
    in: "Hearing mechanisms and speech", O. Creutzfeldt, H. Scheich,
    Chr. Schreiner, eds., Springer-Verlag, Berlin, Heidelberg,
    New York, 133-139.
Smith, R. L., 1977, Short-term adaptation in single auditory nerve
    fibers: some poststimulatory effects, J. Neurophysiol., 40:
    1098-1112.

HOW BIOSONAR INFORMATION IS REPRESENTED IN THE BAT CEREBRAL CORTEX

Nobuo Suga, Kazuro Kuzirai* and William E. O´Neill**

Department of Biology
Washington University
St. Louis, Missouri 63130 U.S.A.

INTRODUCTION

In mammals, an auditory signal sent to the brain by the cochlear nerve ascends from the cochlear nucleus to the auditory cortex
of the cerebrum through many intermediate nuclei. The auditory
signal is projected in parallel to each level of the central auditory system, because of multiple projections from the cochlea. This
multiple projection suggests that different types of auditory information or different attributes of the signals are processed both
hierarchically and in parallel. Except for the auditory cortex of
the mustached bat, Pteronotus parnellii rubiginosus***, however,
little is known about how this multiple representation is related
to functional organization beyond frequency representation.

The properties of an acoustic signal are commonly expressed
by the pattern of energy distribution in the coordinates of frequency, amplitude, and time. At the periphery of the auditory
system, there is an anatomical axis along the cochlear partitions
for expressing only the frequency of the signal. Amplitude and
duration are expressed by the magnitude and duration of excitation

---

*Present address: Department of Otolaryngology, Medical School,
Yokohama City University, Urafunecho, Minamiku, Yokohama, Japan.

**Present address: Center for Brain Research, University of
Rochester Medical School, Rochester, New York 14642 U.S.A.

***Pteronotus parnellii rubiginosus was previously called
Chilonycteris rubiginosa (Smith, 1972).

of sensory hair cells and primary auditory neurons. Recent progress
in auditory physiology has demonstrated that the central auditory
system has anatomical components of locations representing not only
frequency, but also amplitude, time and other important acoustic
parameters (Knudsen and Konishi, 1978, 1979; Suga, 1977; Suga and
O´Neill, 1979), that complex signals are processed by groups of
neurons examining different combinations of signal elements, and
that the individual groups are arranged at identifiable locations
in the brain (O´Neill and Suga, 1979; Suga and O´Neill, 1979; Suga
et al., 1978, 1979).

To explore the functional organization of the auditory system,
it is essential to (i) examine the properties of biologically
important sound, (ii) use these sounds as stimuli, (iii) measure
the "filter properties" of neurons by varying individual parameters
of these sounds, and (iv) examine how neurons with different filter
properties are arranged in the auditory system.

The Panamanian mustached bat emits orientation sounds (biosonar
signals) containing four harmonics, of which the second harmonic is
always predominant and the first is usually the weakest of the four.
Each harmonic consists of a constant-frequency (CF) and frequency-
modulated (FM) component. Therefore, there are eight components in
each emitted signal (CF_{1-4}, FM_{1-4}). Echoes eliciting behavioral
responses in the mustached bat always overlap temporally with the
emitted signal (Fig. 1A). As a result, biosonar information must be
extracted from a complex sound with up to 16 components. A long CF
sound is ideal for Doppler measurement, i. e., for obtaining information about target motion in the radial direction, because the
reflected sound energy is highly concentrated at a particular
frequency. The mustached bat uses the CF_2 at about 61 kHz for this
purpose and performs a very unique behavior called Doppler-shift
compensation*. The CF signal is also suited for target detection
for the same reason. A short FM sound is, on the other hand, more
appropriate for ranging, localizing, and characterizing a target
because of the wide distribution of sound energy over many frequencies.

---

*The mustached bat emits an orientation sound with the CF_2 at about
61 kHz when there are no Doppler-shifted echoes. When a Doppler-
shifted echo returns, say at 63 kHz, the bat reduces the frequency
by nearly 2 kHz, so that the Doppler-shifted echo is stabilized
at just above 61 kHz. Such behavior is called Doppler-shift compensation (Schnitzler, 1968, 1970). Because of the compensation, the
CF_2 of the Doppler-shifted echo is kept mainly within a range of
61–63 kHz. The frequency of the CF signal is different among sub-
species and among individuals of the same sub-species.

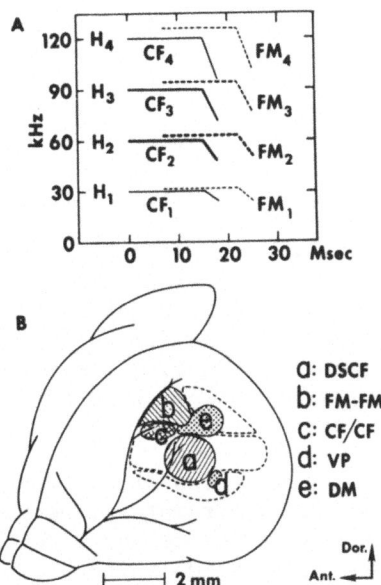

Fig. 1.   A. Schematized sonagrams of the synthesized mustached bat
orientation sound (solid lines) and the echo (dashed lines).
The four harmonics (H_{1-4}) of both the orientation sound and
the echo each contain a long CF component (horizontal bars:
CF_{1-4}) followed by a short FM component (oblique bars:
FM_{1-4}). Thickness of the lines indicates the relative
amplitudes of each harmonic in the orientation sound: H_2
is strongest, followed by H_3 (about -6 dB re H_2) and H_1 and
H_4 (-12 to -24 dB re H_2). Echo delay is measured as the time
interval between the onsets of corresponding components of
the orientation sound and the echo in a stimulus pair.
B. The dorsolateral view of the left cerebrum of the mustached bat. The areas within the dashed lines are the auditory cortex. Three areas specialized for the systematic
representation of biosonar information have been found:
DSCF, FM-FM, and CF/CF, VP, and DM areas indicated by a, b,
c, d, and e, respectively. The functional organization of
these areas are graphically summarized in Fig. 2. The
branches of the median cerebral artery are shown by the
branching lines. The longest branch is on the sulcus
(Suga, 1980b).

During the target-directed flight, both the duration (30-to-7 ms) and the emission rate (5-to-100/s) of the orientation sound as well as the echo amplitude and delay from the emitted sound change systematically. When the animal emits sounds, its ears are self-stimulated. To mimic such vocal self-stimulation, we delivered an electronically synthesized orientation sound from one loudspeaker (artificial vocal apparatus) placed anteroventrally to the bat stimulating both ears to the same extent. To mimic an echo, a similarly synthesized echo was delivered from the other loudspeaker (artifical target) mounted on an acoustic perimeter (hoop) of 160 cm in diameter. The individual parameters carrying different types of information important for echolocation (biosonar) were systematically varied. For instance, different target ranges were simulated by changing echo delay. Different target sizes were mimicked by modifying echo amplitude. Different target motions in the radial direction were introduced by shifting echo frequency. Different azimuthal target motions were introduced by moving the second loudspeaker horizontally. Beating wings of an insect were mimicked by modulating echoes in frequency and amplitude.

To study the response properties of cortical auditory neurons as a function of these stimulus parameters, we recorded action potentials from single neurons in the cerebral cortex of the mustached bats with microelectrodes. (The animals were lightly anesthetized by sodium pentobarbital. If not anesthetized, they were given neuroleptic analgesia, Fentanyl-Droperidol mixture, and local anesthetic was applied to the surgical wounds.) The auditory cortex is 0.9 - 1.0 mm thick and is about 14.2 mm^2, which is very large relative to the bat brain. The auditory cortex contains at least three major areas which are physiologically distinct, i.e., each processes or represents a different type of biosonar information. These areas are called DSCF, FM-FM, and CF/CF (Fig. 1A). In these areas, response properties of single neurons arranged orthogonally to the cortical surface are nearly identical. In this sense, there is columnar organization. Along the cortical surface, on the other hand, the response properties systematically vary and form an axis for systematic representation of a particular type of biosonar information. As will be described below, the DSCF area is organized along coordinates of frequency vs. amplitude and is devoted to processing information carried by the CF_2 of the Doppler-shifted echo (Suga. 1977). The FM-FM area examines the FM_{2-4} of the echo in reference to the FM_1 of the orientation sound and represents target range (O'Neill and Suga, 1979; Suga and O'Neill, 1979). The CF/CF area examines the $CF_{2,3}$ of the echo in relation to the CF_1 of the orientation sound (Suga et al., 1979) and represents Doppler shift (target velocity) in radial direction. There are two other areas, the VP and DM areas, the functional organization of which is still not clear, but which also appear to be important for processing biosonar information

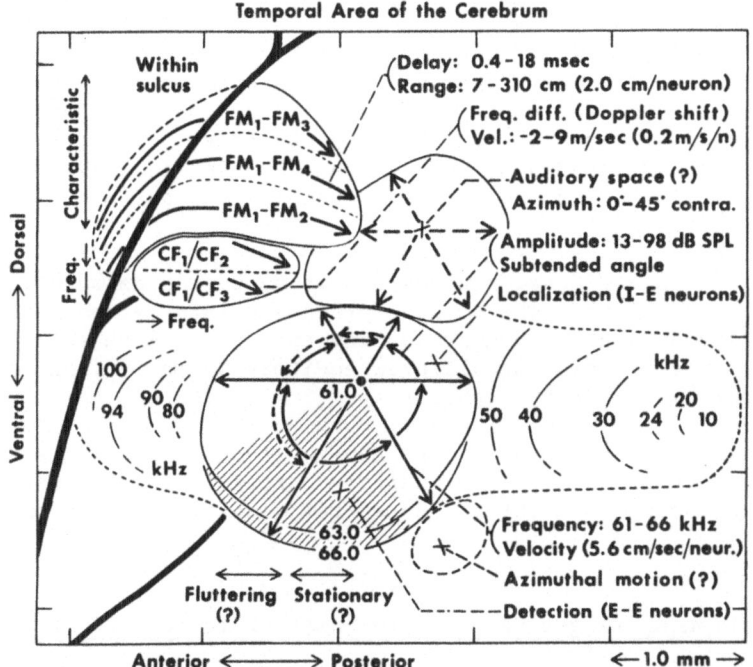

Fig. 2. Graphic summary of the functional organization of the
         auditory cortex. The tonotopic representation of the primary
         auditory cortex and the functional organization of the
         DSCF, FM-FM, and CF/CF areas are indicated by lines and
         arrows. The DSCF area has axes representing either target-
         velocity information (echo frequency: 61-66 kHz) or sub-
         tended target-angle information (echo amplitude: 13-98 dB
         SPL) and is divided into two divisions suited for either
         target detection (shaded) or localization. Its anterior and
         posterior halves are speculated as being suited for process-
         ing echoes from either fluttering or stationary targets.
         The FM-FM area consists of three major types of FM-FM range-
         sensitive neurons (FM_1-FM_2, FM_1-FM_3, and FM_1-FM_4) which
         form separate clusters. Each cluster has an axis represent-
         ing target range (echo delay: 0.4-18 ms), orthogonal to
         which target characteristics are probably represented. The
         CF/CF area consists of two major types of CF/CF facilita-
         tion neurons (CF_1/CF_2 and CF_1/CF_3) which aggregate in in-
         dependent clusters. Each cluster has two frequency axes and
         represents target velocity in the radial direction (cont'd)

(Fig. 2 cont'd) (-2 to 9 m/s). In the VP area, azimuthal-
motion-sensitive neurons have been found. The DM area
appears to have some sort of a neural map representing the
auditory space on the contralateral side in front of the
animal. The functional organization of the VP and DM areas
remains to be further studied. This summary figure is
based upon the data published by Suga and his coworkers
(Kujirai and Suga, unpubl.; Manabe et al., 1978; Suga,
1977; Suga and Jen, 1976; Suga and O'Neill, 1979; Suga et
al., 1979) and the data reported in this article (Suga,
1980b).

(Fig. 1B). In the VP area, neurons sensitive to the azimuthal motion
of a target were found. The DM area processes information carried
mostly by sounds higher than 70 kHz and appears to have some sort
of a neural map of the contralateral auditory space in front of the
animal (Kujirai and Suga, unpubl.).

The aim of this article is to review recent discoveries in the
functional organization of the auditory cortex of the mustached bat
and to describe the present status of our understanding of the auditory cortex.

Representation of Target Velocity Information in the DSCF Area

When the bat is flying toward a stationary object, its speed
is expressed by the frequency difference between the emitted sound
and the returning Doppler-shifted echo. The frequency information
of the emitted sound is available to the bat in the form of vocal
self-stimulation and perhaps efferent copy, about which little is
currently known. The frequency information of the Doppler-shifted
echo is available regardless of whether Doppler-shift compensation
is performed. With compensation, however, the measurement of echo
frequency becomes much more accurate because the echo frequency is
then analyzed by a group of filters in the cochlea which are un-
usually sharply tuned to sound between 60 and 62 kHz (Suga and Jen,
1977; Suga et al., 1975).

In the DSCF area, the majority of neurons are very sharply
tuned to particular preferred frequencies (best frequencies) between
61 and 63 kHz, the commonest frequencies of the CF_2 of Doppler-
shifted echoes. These neurons are systematically arranged according
to their best frequencies and form a radial frequency axis along the
cortical surface (Fig. 2). Along the frequency axis, best frequency
changes at a rate of 20 - 30 Hz/neuron, i.e., velocity information
is represented by increments of 5.6 - 8.4 cm/s/neuron. Therefore,
the representation of target-velocity information is very fine
(Suga, 1977; Suga and Jen, 1976). The representation of frequency
by the loci of activated neurons is called tonotopic representation.

Frequency resolution is directly related to the sharpness of the frequency-tuning curve. Extremely narrow frequency-tuning curves of peripheral neurons are further sharpened in the central auditory system by lateral inhibition. In the narrowest curve obtained so far, the best frequency is 61.51 kHz and the bandwidth remains at about 0.3 kHz over a broad amplitude range. The slope of the frequency-tuning curve is essentially infinite in a significant number of cortical neurons (Suga, 1977). These neurons act as narrow-band frequency detectors irrespective of stimulus amplitude. Since many neurons have an "excitatory" frequency-tuning curve (excitatory area) which is much narrower than that of peripheral neurons and which is bounded by "inhibitory" frequency-tuning curves (inhibitory areas), frequency representation in the DSCF area is discrete*.

When the target is a flying insect, its echo is modulated in frequency and amplitude by the beating wings. The Doppler shift in the echo consists of a "DC component" which is due to the change in distance between the bat and the insect and an "AC component" which is caused by the wings. Some neurons in the DSCF area vary their discharge rate synchronously with a frequency modulation as small as 0.05% (31-Hz frequency shift at 61 kHz) occurring at a rate of 100/s. Moths flying can produce a frequency shift as large as 800 Hz at 61 kHz carrier (Goldman and Henson, 1977). The wing beat information is thus expressed by rhythmic changes in discharge rate of neurons synchronous with the wing beat as well as by synchronous changes in excitation from one group of neurons to another with different best frequencies. Other neurons in the DSCF area are insensitive to frequency modulation, probably because of inhibitory areas that "sandwich" their narrow excitatory areas. These neurons may be considered to be suited for detection of the Doppler-shifted echoes from stationary targets.

Representation of the Subtended Target Angle in the DSCF Area

The size of a target is determined by its range and subtended angle. Target range is directly related to echo delay, while subtended target angle is largely related to echo amplitude, although several other factors are also involved. For simplicity, we may consider that the representation of the subtended target angle is equivalent to the representation of echo amplitude.

---

*Neural mechanism for the sharpening of a frequency-tuning curve starts to operate in the cochlear nucleus (Suga, et al., 1975). The mechanism was first demonstrated in cats (Katsuki et al., 1959). The same neural mechanism for sharpening of the cortical representation of a sensory signal was first demonstrated in the somatosensory system (Mountcastle, 1957).

At the periphery, auditory neurons increase the number of impulses per stimulus monotonically with stimulus amplitude. There is no anatomical basis for representation of echo amplitude. In the DSCF area, however, neurons are tuned not only to frequency, but also to amplitude. They respond maximally to particular preferred stimulus amplitude (best amplitudes) and are systematically arranged according to their best amplitudes, forming a circular amplitude axis (Fig. 2). The systematic representation of stimulus amplitude by the location of activated neurons is called amplitopic representation (Suga, 1977). The overlap of inhibitory areas with the excitatory area produces not only a sharper excitatory area, but also a particular amplitude-sensitivity curve (Suga, 1965; Suga, 1977). The neural mechanisms for amplitopic representation are systematic changes in the extent of lateral inhibition and the minimum threshold for excitation.

Functional Divisions in the DSCF Area

The frequency vs. amplitude coordinates found in the DSCF area are the first demonstration that the amplitude spectrum of a signal is represented by a spatial pattern of neural activity. The DSCF area is disproportionately large in the bat auditory cortex, and within it the representations of acoustic signals of 61.5 - 62.0 kHz and 30 - 50 dB SPL are disproportionately large. The frequency and amplitude of echoes from prey may be predominantly within this range from the moment of target detection through the early part of the target-directed flight.

Since the DSCF area is large, one might question whether there is a functional organization beyond the frequency vs. amplitude coordinates. In terms of binaural interaction, the DSCF area functionally consists of dorsal and ventral divisions which are dominated by I-E and E-E neurons, respectively. E-E neurons receive excitatory inputs from both ears. They have poor directional sensitivity, but are tuned to weaker echoes. They integrate (or even multiply) signals from both ears for effective target detection. On the other hand, I-E neurons receive inhibitory input from the ipsilateral ear and excitatory input from the contralateral ear. They are directionally sensitive and are tuned to stronger echoes. Therefore, these two divisions are suited for either target detection or localization (Fig. 2; Manabe et al., 1978).

The DSCF area contains neurons which respond to both AC and DC components of a Doppler shift and those which basically respond only to the DC component. It is not yet known whether they are aggregated into two functional divisions in the area. Such an arrangement would represent divisions for processing the echoes from either flying or stationary target.

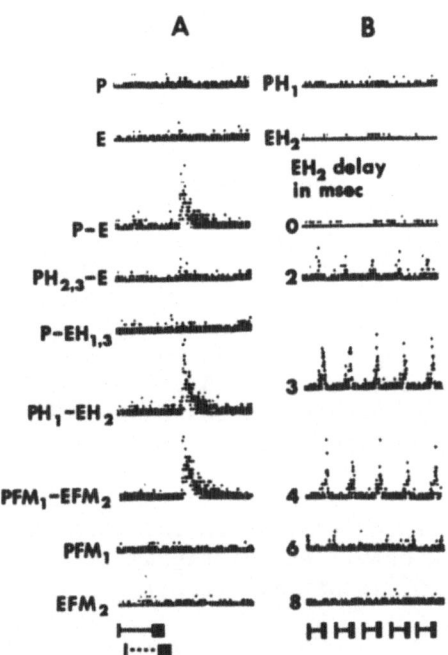

Fig. 3. The responses of single FM_1-FM_2 range-sensitive neurons
which are expressed by the number of impulses per unit time
against time after several milliseconds prior to and following the onset of a stimulus. Such a plot is called a peristimulus-time (PST) histogram.
A. An FM_1-FM_2 range-sensitive neuron tuned to a 7.0-ms echo
delay (120-cm target range). No response is seen to the
orientation sound alone (P) or the echo alone (E), but only
to the P-E pair. When either the first harmonic of P or the
second harmonic of E is eliminated from the P-E pair, there
is no clear response ($PH_{2,3}$-E and P-$EH_{1,3}$). The response to
the combination of the harmonic of P and the second harmonic of E (PH_1-EH_2) is the same as that to the P-E pair.
The response to the combination of the FM_1 of P and the
FM_2 of E (PFM_1-EFM_2) is also the same as that to the P-E
pair. Neither the elimination nor the continuous delivery
of the CF components of P and E has significant effect on
the response of this neuron. Therefore, it is called an
FM_1-FM_2 range-sensitive neuron. The neuron shows no response
to any single FM component (e.g., PFM_1 or EFM_2). PFM_1 sweeps
from 30.67 to 24.67 kHz at 66 dB SPL, and EFM_2 (cont'd)

(Fig. 3 cont'd) sweeps from 61.67 to 49.67 kHz at 52 db SPL.
These are the best FM sweeps for the facilitation response
of this neuron. The repetition rate of the delivery of the
paired or unpaired sound is 10/s. The echo delay is 7.0 ms.
The solid and dashed bars at the bottom are the stimulus
markers for the 34-ms-long orientation sound and echo,
respectively. The FM components in these sounds are 4-ms
long.
B. Another FM_1-FM_2 range-sensitive neuron tuned to a 3.2-ms
echo delay (55-cm target range). Response to neither P, E,
PH_1, nor EH_2 alone. No response to the P-E or the PH_1-EH_2
pair unless E or EH_2 delay is within the range between 2 and
6 ms. Note the remarkable response to each paired stimulus
with a 3 ms echo delay even at a high repetition rate, 100
pairs/s. FM_1 sweeps from 30.50 to 24.50 kHz at 61 db SPL,
and FM_2 sweeps from 63.60 to 51.60 kHz at 56 dB SPL. These
are the best FM sweeps and the best amplitudes for the
facilitation response of this neuron, indicating the importance of target range, size, and the relative velocity for
the excitation of the neuron (see also Fig. 4, A). The
stimulus markers at the bottom indicate the 7-ms-long orientation sound only. Each histogram is the sum of the
responses to 100 identical stimuli.

Representation of Target Range in the FM-FM Area

The primary cue for target ranging is the delay of an echo
from the emitted orientation sound. At the periphery, the echo
delay (target range) is coded by the interval between the responses of neurons to the orientation sound and the echo. There are
no anatomical components or locations representing range information. In the FM-FM area, however, most neurons respond poorly, or
not at all, to either the orientation sound or the echo alone,
but vigorously respond to orientation sound-echo pairs with
particular echo delays (Fig. 3B). Therefore, they are sensitive
to particular target ranges (O'Neill and Suga, 1979; Suga and
O'Neill, 1979; Suga et al., 1978).

Two classes of range-sensitive neurons were found: (i)
tracking neurons, whose preferred echo delay (best delay), i.e.,
preferred target range (best range), for response to an echo
following the orientation sound, becomes shorter and narrower as
the bat closes in on the target, and (ii) range-tuned neurons,
whose best range is constant, responding to the target only when
it is within a certain narrow fixed range (Fig. 4A) (O'Neill and
Suga, 1979; Suga et al., 1978). Range-tuned neurons are specialized for processing echoes from targets at particular ranges.
They are systematically arranged according to their best ranges
and form a neural axis representing target range, which runs from

7 cm to 310 cm (Fig. 2). This is called <u>odotopic representation</u>.
Target ranges between 50 and 140 cm are over-represented by the FM-FM
area, and best range varies at a rate of 2.0 cm/neuron along the
range axis (Suga and O´Neill, 1979).Neurons tuned to distances
further than 200 cm are small in population and appear to take a
limited role in ranging, because of their broad delay-tuning curves.
Several species of bats show the first sign of the approach phase
for a wire obstacle of 0.3 cm diameter at a distance of about 2.3 m
(Grinnell and Griffin, 1958) and can discriminate 1.2-2.5 cm distance
differences (Simmons, 1971). The neurophysiological data described
above thus have an interesting correlation with the behavioral data.

For the excitation of range-sensitive neurons, the essential
elements in the orientation sound-echo pairs are the first harmonic
FM component (FM_1) in the orientation sound and one or more higher
harmonic FM components (FM_{2-4}) in the echo (Fig. 3A).The frequency
of these FM components must sweep downward, as do the natural sounds,
otherwise some of the range-sensitive neurons do not respond. The CF
components have no significant effect on their excitation (Fig. 3A).
Neurons responding (or examining) the combination of FM_1-FM_3, FM_1-
FM_4 or FM_1-FM_2 form three major clusters, which are arranged dorsal
to ventral in this order in almost all brains studied (Fig. 2). In
each cluster, odotopic representation has been demonstrated (Suga
and O´Neill, 1979).

FM_1-FM_4 range-sensitive neurons are theoretically better suited
for fine characterization of small targets than FM_1-FM_2 range-sensitive neurons because FM_4 is much shorter than FM_2 in wavelength
and has a broader bandwidth. Most range-sensitive neurons are tuned
not only to a specific echo delay, but also to a particular echo
amplitude (Fig. 4A). Therefore, these neurons respond best to targets with particular cross-sectional areas at particular distances.
Furthermore, they respond best when echoes are Doppler-shifted by
approaching targets. Response properties of these neurons are, as
a result, quite complex (O´Neill and Suga, 1979; Suga and O´Neill,
1979). The functional organization of the FM-FM area and response
properties of the range-sensitive neurons suggest that the distribution of neural activity orthogonal to the range axis represents the
fine structure (amplitude spectra) of all FM components of an echo,
and therefore, target characteristics such as shape and size.

Representation of Target Velocity Information in the CF/CF Area

The DSCF area is primarily devoted to the fine and systematic
representation of the frequency and amplitude of the CF_2 in the
Doppler-shifted echo. The next obvious question is whether the
cerebrum has an area where neurons are sensitive to particular
differences in frequency between orientation sounds and Doppler-
shifted echoes, i.e., particular Doppler shifts. To represent a

Doppler shift systematically, two frequency axes are needed. There
are obviously no such coordinates in the cochlea. In the CF/CF area,
however, neurons are tuned to particular combinations of two CF
components (Figs. 4B and 5A), and form such frequency coordinates
(Fig. 6A).

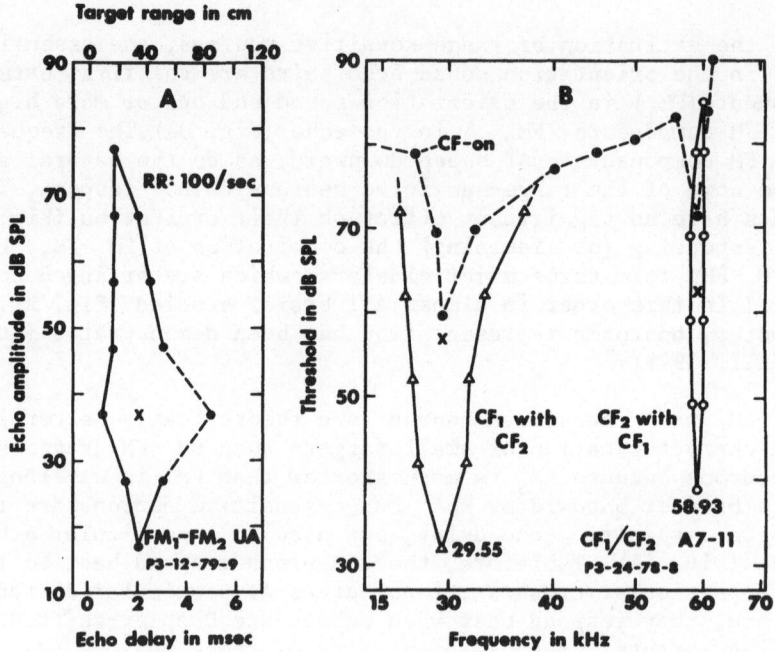

Fig. 4. A. Delay (range)-tuning curve of a single FM_1-FM_2 range-
         tuned neuron measured with orientation sound-echo pairs.
         Since the neuron was tuned to a 2.0-ms echo delay, the
         curve was measured with the paired stimulus delivered at
         a rate of 100 per second mimicking the sounds in the termi-
         nal phase of echolocation. The neuron shows the best
         response only when the echo delay is 2.0 ms and the echo
         amplitude is 37 db SPL ("x" marker). The duration of each
         sound is 7 ms (CF = 5 ms; FM = 2 ms) mimicking the orienta-
         tion sound in the terminal phase.
         B. Frequency-tuning curves of a single CF_1/CF_2 facilitation
         neuron. The dashed curve is the frequency-tuning curve
         (excitatory area) measured with a single CF tone. The solid
         curves are the "facilitation frequency-tuning (cont'd)

(Fig. 4 cont´d) curves" (facilitation areas) measured with a CF test tone which was delivered simultaneously with a conditioning CF tone. The conditioning tone used to measure the CF_1 facilitation area was 58.93 kHz and 63 dB SPL; for the CF_2 facilitation area, it was 29.55 kHz and 57 db SPL (x´s). These are the "best facilitation frequencies" of the CF_1 and CF_2 and are precisely harmonically related. The duration of these CF tone is 34 ms.

Neurons in the CF/CF area show poor responses to individual signal elements, but show remarkable facilitation of response when CF_1 is delivered together with CF_2 or CF_3, which are harmonically or quasi-harmonically related with CF_1 (Fig. 5A). Therefore, these neurons are named CF_1/CF_2 or CF_1/CF_3 facilitation neurons (Suga et al., 1978, 1979). The CF_1 best frequency for facilitation is 29.80 $\pm$ 0.8 kHz (mean $\pm$ standard deviation) for the 372 neurons studied. The CF_2 and the CF_3 best frequencies for facilitation are respectively 60.55 $\pm$ 0.97 for 204 neurons and 91.5. $\pm$ 1.04 kHz for 205 neurons*. Since these mean frequencies are not exact harmonics, most of the CF/CF facilitation neurons are maximally excited by echoes when the animal is compensating for a Doppler shift by reducing the frequency of its orientation sound, and not by the CF components of the emitted sound.

The CF_1/CF_2 and CF_1/CF_3 facilitation neurons form independent clusters (Fig. 2; Suga et al., 1979). In each cluster, the frequency of CF_1 is represented along the rostrocaudal axis, while that of CF_2 or CF_3 is represented along the dorso-ventral axis (Fig. 6A)**. These coordinates represent the deviation (D) of a CF_2 or CF_3 best facilitation-frequency from the exact harmonic relationship with the CF_1 best facilitation-frequency ($D = CF_2 - 2CF_2$ or $CF_3 - 3CF_1$). The primary function of the CF/CF facilitation neurons is to represent a Doppler shift, i.e., the relative velocity of a target in a radial direction. Iso-Doppler-shift (iso-velocity) contour lines are oblique to both frequency axes and the velocity representation stretches from 8.7 to -2.0 m/s (Fig. 6B). Since velocities from 0 to 2m/s are too slow for the flight speed of a hunting bat, and since the bat does not fly backward, the area representing velocities from 2 to -2 m/s

---

*These frequency axes are based upon the best frequencies of single neurons for facilitation. If the tonotopic representation in the CF/CF area were studied in the traditional way with a single CF tone, the poor responses of the neurons to such a stimulus would misrepresent the nature of this area.

**Out of the 372 neurons studied, 37 neurons showed facilitation for the combination of CF_1 with CF_2 and CF_3. There is a possibility that the responses of these neurons were due to simultaneous recordings of responses of CF_1/CF_2 and CF_1/CF_3 facilitation neurons.

probably responds to echoes from insects and conspecifics flying away from the echolocating bat. Velocities from 0 to 4m/s are somewhat over-represented. The relative speed of a target may be predominantly within this range in the approach and terminal phases of echolocation. There is also a possibility that the area tuned to 0 m/s is for the detection of orientation sounds emitted by conspecifics, because the frequencies of all CF components are in the exact harmonic relationship.

In contrast to FM-FM range-sensitive neurons, CF/CF facilitation neurons are equally sensitive to 0 - 10 ms echo delays. Their responses to orientation sound-echo pairs are similar, regardless of echo delay, when it is within this range. Hence, they are not suited for processing distance information. Their responses start to deteriorate when the delay becomes longer than 10 ms. At a delay longer than 20 ms, the facilitation becomes very poor and the facilitation threshold becomes high even though the two signal elements still overlap significantly (Fig. 5B). Such properties of the neurons act as a kind of time gate for echo processing. The greater horseshoe bat, Rhinolophus ferrumequinum, performs Doppler-shift compensation only when the echo delay is less than 15 ms (Schuller, 1974). The neurophysiological data described above show an interesting correlation with such behavior.

Protection of the Cortical Representation of Biosonar Information from Jamming

In spite of the fact that the second harmonic is always predominant in the orientation sound, neurons have not yet been found which examine the combinations of FM_2's or CF_2's in the orientation sound and a Doppler-shifted echo. Instead, neurons have been found which examine the nonfundamental FM or CF components in a Doppler-shifted echo with reference to the FM_1 or CF_1 in the orientation sound*. This unexpected result is perhaps a remarkable adaptation of the auditory system for the reduction of the jamming effect of sounds produced by conspecifics in processing biosonar information.

Many species of microchiropterans are colonial. Hundreds, or even thousands of bats roost in a single cave. They are frequently found in narrow elongated caves and culverts, where they fly in opposite directions without colliding. One of the important problems in echolocation is how the bats reduce the jamming effect of biosonar sounds produced by conspecifics or how the neural representation of biosonar information is protected from jamming. We can enumerate

---

*Combination-sensitive neurons which are not described in this article (e.g., CF_2-FM_2, CF_2/CF_3, etc.) have also been found, although the samples are small.

the following six possible mechanisms responsible for the reduction
of jamming: (i) the sharp directionality of the orientation sound,
(ii) the sharp directional sensitivity of the ear, (iii) binaural
hearing, (iv) the sequential processing of echoes, (v) the signature
(subtle difference) of orientation sounds used by individual bats,
and (vi) efferent copy originating from the vocalization system.
All these mechanisms would work together for successful echolocation.

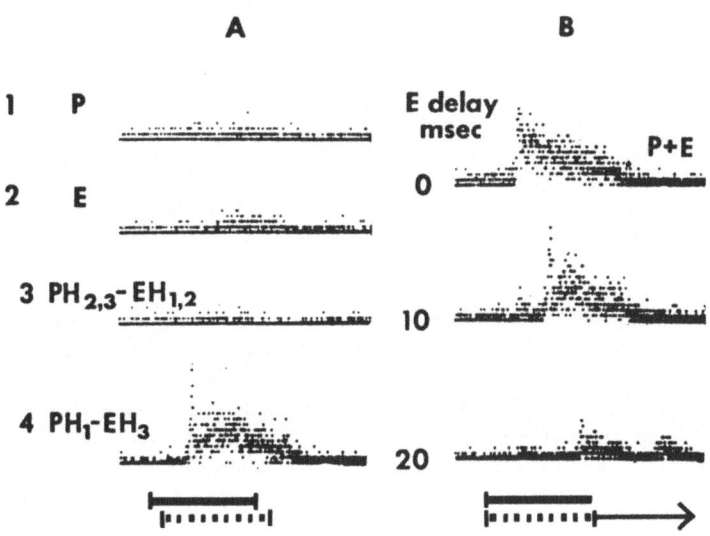

Fig. 5.   The response of a single CF_1/CF_3 facilitation neuron ex-
          pressed by PST histograms.
          A. Poor or no response to either orientation sound (P)
          alone or echo (E) alone or $PH_{2,3}$-$EH_{1,2}$ pair, i.e., P-E pair
          without PH_1 and EH_3. Vigorous response to either P-E (see
          B) or PH_1-EH_3 pair (the echo delay is 5 ms). Elimination of
          the FM components from the P-E has no effect on the
          response. The essential elements for the facilitation
          response are the CF_1 of P and the CF_3 of E.
          B. The response to the P-E pair becomes poor when the echo
          delay increases beyond 10 ms. In A and B, CF_1 is 29.05 kHz
          and 67 dB SPL, while CF_3 is 91.53 kHz and 54 dB SPL. These
          are the best frequencies and the best amplitudes for the
          maximum response of the neuron. This means that the CF_3
          should be Doppler-shifted by 4.38 kHz by a target approach-
          ing at a speed of 8.6 m/s. The duration of the CF and FM
          components in the sounds are 30 and 4 ms, respectively.
          Each histogram is the sum of the responses to 100 identical
          stimuli.

The response properties of neurons in the FM-FM and CF/CF areas indicate two other possible mechanisms which should be added to the above six. One of them is the adjustable time gate for sampling echoes and for excluding sounds produced by conspecifics. Neurons in the FM-FM area act as a kind of time gate, because they are tuned to particular echo delays between 0.4 and 18 ms. Likewise, neurons in the CF/CF area show facilitation of response only when the orientation sound and the Doppler-shifted echo overlap each other and when the echo delay is often not larger than 20 ms. Thus, the bat has a time gate which opens at the beginning of each emitted orientation sound and closes after a short period of time. The duration of this time gate probably changes according to the rate and duration of the sound emission.

The other mechanism may be called hetero-harmonic combination. In the past, we thought that biosonar information was obtained by processing the stronger harmonics in the echo in relation to the corresponding harmonics in the emitted sound. It was, therefore, totally unexpected that one of the critical components for the maximum excitation of neurons in the FM-FM and CF/CF area was always the weakest first harmonic in the orientation sound, while the other was one of the stronger, higher harmonics in Doppler-shift echoes. The functional role of this hetero-harmonic combination is to reduce jamming by sounds produced by the conspecifics in echolocation. For better understanding of this mechanism, one may consider the following two extreme hypothetical cases. When the energy of the first harmonic is not radiated at all, but is dissipated by bone conduction, the bone-conducted first harmonic stimulates the cochleae and conditions neurons in the FM-FM and CF/CF areas to be excited by echoes returning with a short time delay. In this case, it is essential for their excitation that the animal itself produces an orientation sound. Combinations of sounds produced by conspecifics would not excite these neurons. As a result, jamming of echolocation by conspecifics is reduced. When the first and higher harmonics are equally intense in the emitted sound, on the other hand, there is no such reduction of jamming by the hetero-harmonic combinations. The mustached bat falls between these two extreme situations, because the first harmonic is broadcast to some extent. It is, however, likely that the hetero-harmonic combination contributes to the reduction of jamming in this species.

The suppression of the emission of the first harmonic in the orientation sound introduces another important advantage to the mustached bat foraging for moths. The ears of many different species of moths are most sensitive to sounds between 15 and 40 kHz (minimum threshold: about 40 dB SPL), but their sensitivity becomes low when sounds are higher than 40 kHz (Fenton and Fullard, 1979; Roeder and Treat, 1957; Suga, 1961). The mustached bat emitting the orientation sound with the suppressed first harmonic can thus approach close to moths without being detected.

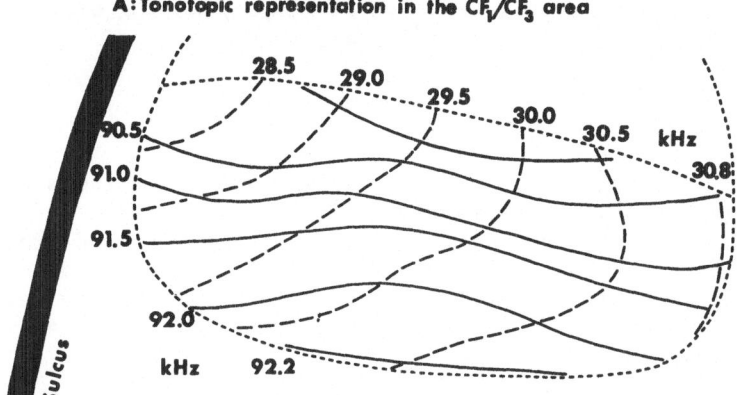

A: Tonotopic representation in the CF_1/CF_3 area

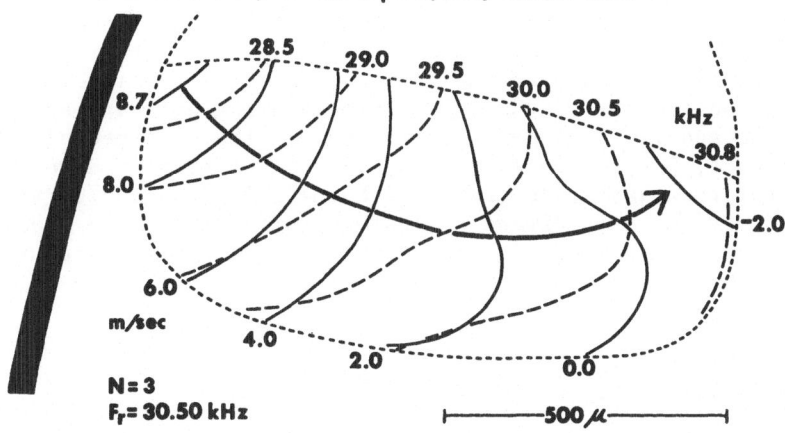

B: Iso-velocity and iso-CF_1 frequency contour lines

Fig. 6.   The functional organization of the CF_1/CF_3 area.
A. Iso-best-facilitation frequency contour lines for CF_1
(long dashed lines) and CF_3 (solid lines). These contour
lines are based on the data obtained from three unanesthetized mustached bats.
B. Iso-velocity contour lines (solid) are shown together
with the iso-best-facilitation frequency contour lines for
CF_1 (long dashed). The long arrow is the axis representing
Doppler shift, i.e., target velocity (8.7 to -2.0 m/s) in
the radial direction. The figure indicates that when the
bat emits an orientation sound with the CF_1 of 30.5 kHz
(resting frequency), neurons in the CF_1/CF_3 area are best
activated by targets moving with relative velocities from
-1.2 to 2.0 m/s, and when the frequency of the CF_1 is
reduced to 29.5 kHz, for example, they are stimu-(cont´d)

(Fig. 6 cont'd) lated best by targets moving with relative velocities from 2.4 to 6.1 m/s.

Representation of Auditory Space

Sound localization is based on interaural comparisons and cues provided by the pinna. Interaural amplitude (intensity) and time (phase) differences systematically vary with azimuth and are cues for sound localization in the horizontal plane. For sound localization in the vertical plane, the pinna, which modifies the waveform of an acoustic signal containing higher frequencies, is essential (Batteau, 1967).

The auditory system has no sensory epithelium upon which the auditory space in front of an animal is projected. For systematic representation of auditory space, the central auditory system should contain neurons which are sensitive to the binaural cues and to the cues provided by the pinna and systematically arrange them according to their tuning curves for these cues. Owls are superb in localizing a sound source and have an auditory space map in the midbrain (Knudsen and Konishi, 1978, 1979). Echolocating bats are specialized in hearing and may also have such a map. In the dorsomedial (DM) area (Fig. 1, Be), neurons are sensitive to particular directions. Their preferred azimuthes change with their loci in the cortical plane. However, their directional sensitivity curves are broad (Kujirai and Suga, unpubl.). The neural map of auditory space, as found in the barn owl, has not yet been demonstrated in bats.

Representation of Target Characteristics

One of the important target characteristics for insectivorous bats is whether a target is fluttering (e.g. like a flying insect) or is stationary. A fluttering target can be represented by neural activity in two possible ways: (1) the spatiotemporal change in neural activity synchronized with wing fluttering and (2) the responses of neurons which are insensitive to the carrier itself, but sensitive only to frequency modulation caused by wing fluttering. In the DSCF area, the spatiotemporal pattern of neural activity changes synchronously with flutter. We have not yet examined the second possibility.

The other target characteristics such as shape and size are related to the fine structure of the echo FM, so that it is theoretically best represented by the area where the FM signals are processed. As described above, neurons in the FM-FM area are tuned not only to particular echo delays, but to particular echo amplitudes, most probably particular amplitude-spectra of the FM compo-

nents of an echo. The distribution in neural activity along the axis
normal to the range axis perhaps represents target characteristics.

Theoretical Considerations

There are several different types of biosonar information related to the different attributes of a target. Some of these cannot
be processed in sequence, but in parallel. As summarized above,
different types of biosonar information are represented in different
areas of the cerebral cortex. This suggests that these are actually
processed in parallel in the auditory system. Parallel processing
is, however, undoubtedly incorporated with hierarchical processing,
because, as in the case of range-sensitive neurons, response properties are complex and are not readily explained by interaction
among primary of primary-like auditory neurons.

Our future research will be on the origin and destination of
biosonar information which is represented in the individual areas
described in this review. The problem of destination is directly
related to the questions of what the upper limit in specialization
(complexity of response properties) of single neurons is and how
overall target image is recognized by the brain. For recognition
of acoustic signals, information about the signals should be stored
in the brain and should be compared with incoming signals, i.e., the
acoustic signals should theoretically be cross-correlated with the
stored information. In auditory neurophysiology, the statement that
neurons respond to "sound X," but not to others is not quantitative,
so that we usually study the filter properties of the neurons by
measuring tuning curves for sound X by changing its individual
parameters (e.g. Fig. 4) (O'Neill and Suga, 1979; Suga and O'Neill,
1979; Suga et al., 1978, 1979). Therefore, we can treat the neurons
as filters. This is a theoretical advantage, since a filter acts as
a kind of a cross-correlator. Auditory neurons are filters which
correlate acoustic signals with their filter properties, i.e.,
stored information, and the degree of correlation is expressed by
the magnitude of the output of the filters. In other words, neurons
are maximally excited only when the properties of acoustic signals
perfectly match their filter properties (Suga, 1979).

All neurons in the auditory system, including peripheral ones,
act as filters. Specialized neurons expressing the outputs of neural
circuits tuned to particular information-bearing parameters (IBP's)*

---

*To generalize what has recently been found in the auditory cortex
of the mustached bat (O'Neill and Suga, 1979; Suga, 1977; Suga and
O'Neill, 1979 ; Suga et al., 1978, 1979) and the midbrain auditory
nucleus of the barn owl (Knudsen and Konishi, 1978; Knudsen et al.,
1979), the term, information-bearing parameter (IBP), is introduced,
because the term, "information-bearing element", does not include

may be called "IBP filters". Information is processed hypothetically
by many IBP filters (Suga, 1979, 1980a). Among our findings, the
following are important in understanding neural processing of complex
acoustic signals in general: (1) Complex sound is processed by neurons specialized for "examining", i.e., IBP filters tuned to different combinations of signal elements, (2) Different types of IBP
filters are aggregated separately in identifiable areas of the
cerebral cortex, and (3) in each aggregate, IBP filters are arranged
along axes for the systematic representation of information-bearing
parameter(s), i.e., signal variation which has biological importance
(Suga, 1980b). As found in the DSCF, FM-FM, and CF/CF areas, more
IBP filters are devoted to finer processing of biologically more
important aspects of echoes. (In these areas, IBP filters arranged
orthogonally to the cortical surface are characterized with nearly
identical filter properties. In this sense, there is columnar organization, and each column may be considered to be a functional unit
for processing sensory information, as proposed for other sensory
cortical areas.)

        One of our working hypotheses is that the recognition of the
overall acoustic image is directly related to neural activity (the
outputs of IBP filters) in a particular area, i.e., the area integrates all the subdivisions specialized for processing different
types of auditory information. An alternate hypothesis is that it
is directly related to the spatiotemporal patterns of neural activity
occurring separately at these functional areas. The latter appears,
at present, more likely. At the present stage of our research, however, we do not know the upper limit of neural specialization in
response to complex acoustic signals. Further research must be
conducted regarding the problems of the neural representation of
auditory information and the destination of information through the
neurons in each specialized area. Fortunately, neuroanatomical tracing techniques are available to explore these problems. The combination of neurophysiology and neuroanatomy promises to further advance
our understanding of the bat brain (Suga, 1979; 1980a).

        In the mustached bat, the large part of the auditory cortex is
primarily devoted for processing biosonar information. There are,
however, still significantly large areas, the function of which has
not yet been elucidated. Since this species forms large colonies
and uses several different types of complex sounds for communication,

---

interaural time and amplitude differences, interval between signals
(e.g., echo delay), and other parameters characterizing combinations
of information-bearing elements which are important for communication
and/or echolocation. An IBP is that limited part of a continuum which
carries information important for the species in nature. An identical
IBP can be quite different in its biological significance for different species of animals.

the auditory system remains to be studied also in terms of processing of the communication sounds.

The sonogram of a vowel shows several formants (CF components)
which are called F_1, F_2, and so on from the lowest formants. Vowels
are recognized by combinations of F_1, F_2, and F_3. The formant frequency varies among speakers, and this variation is biologically
important. The vowels are thus expressed by the areas which are at
identifiable loci in the coordinates of F_1 vs. F_2 frequencies and
F_1 vs. F_3 frequencies. Does the human brain have such coordinates
to represent vowels by the loci of the activated neurons? We have
no answer to this question, but we can point out that the CF/CF area
of the auditory cortex of the mustached bat has such frequency vs.
frequency coordinates to express important biosonar information. For
the recognition of many non-vowel phonemes and combinations of phonemes (words), transitions (FM components) are very important. When
an FM component is added to the F_2 of "a", for instance, the sound
is perceived as either "pa", "ta", or "ka", depending on the properties of the FM component. When another FM component is added to
the F_1 of "a", it is recognized as either "ba", "da", or "ga". This
example demonstrates that combinations of FM components, more generally, combinations of information-bearing elements, are very
important for speech recognition. Does the human brain have many
aggregates of neurons which are specialized for examining different
combinations of information-bearing elements? We cannot answer this
question, but we can point out that the brain of the mustached bat
contains several aggregates of combination-sensitive neurons which
are specialized for extracting different types of biosonar information from complex acoustic signals. For the generalization of the
basic neural mechanisms for processing complex sound found in the
mustached bat, the data comparable to these should be obtained from
different species of animals. Unfortunately, these are not yet
available.

ACKNOWLEDGMENTS

We are grateful for the support of the U.S. National Science
Foundation (grant BNS 78-12987) and for the cooperation of R. W.
Coles, E. G. Jones, and P. Wasserbach. This article is the expanded
version of an article written by N. Suga (1980b).

REFERENCES

Batteau, D. W., 1967, The role of the pinna in human localization,
     Proc. Royal Soc. Lond., 168B: 158-180.
Fenton, M. B. and Fullard, J. H., 1979, The influence of moth hear-
     ing on bat echolocation strategies, J. Comp. Physiol., 132:
     77-86.

Goldman, L. J. and Henson, O. W. Jr., 1977, Prey recognition and
    selection by the constant frequency bat, Pteronotus p.
    parnellii, Behav. Ecol. Sociolbiol., 2: 411-419.
Grinell, A. D. and Griffin, D. R., 1958, The sensitivity of echolo-
    cation in bats, Biol. Bull., Woods Hole, 114: 10-22.
Katsuki, Y., Watanabe, T., and Suga, N., 1959, Interaction of audi-
    tory neurons in response to two sound stimuli in cat, J. Neuro-
    physiol., 22: 603-623.
Knudsen, E. I. and Konishi, M., 1978, A neural map of auditory space
    in the owl, Science, 200: 795-797.
Knudsen, E. I. and Konishe, M., 1979, Mechanisms of sound localiza-
    tion in the barn owl Tyto alba , J. Comp. Physiol., 133: 13-
    21.
Manabe, T., Suga, N., and Ostwald, J., 1978, Aural representation
    in the Doppler-shifted-CF processing area of the primary audi-
    tory cortex of the mustache bat, Science, 200: 339-342.
Mountcastle, V. B., 1957, Modality and topographic properties of
    single neurons of cat's somatic sensory cortex, J. Neuro-
    physiol., 20: 408-434.
O'Neill, W. E. and Suga, N., 1979, Target range-sensitive neurons
    in the auditory cortex of the mustache bat, Science, 203: 69-
    73.
Roeder, K. D. and Treat, A. E., 1957, Ultrasonic reception by the
    tympanic organ of noctuid moths, J. Exp. Zool., 134: 127-158.
Schnitzler, H.-U., 1968, Die Ultraschall-Ortungslaute der Hufeisen-
    Fledermause (Chiroptera-Rhinolophidae) in verschiedenen Orien-
    tierungs-situationen, Z. vergl. Physiol., 57: 376-408.
Schnitzler, H.-U., 1970, Echoortung bei der Fledermaus Chilonyc-
    teris rubiginosa, Z. vergl. Physiol., 68: 25-38.
Schuller, G., 1974, The role of overlap of echo with outgoing
    echolocation sound in the bat Rhinolophus ferrumequinum,
    Naturwissenschaften, 61: 171-172.
Simmons, J. A., 1971, The sonar receiver of the bat, Ann. N. Y.
    Acad. Sci., 188: 161-174.
Smith, J. D., 1972, Systematics of the chiropteran family
    Mormoopidae, Univ. Kansas, Museum Nat. Hist. Misc. Pub., 56.
Suga, N., 1961, Functional organization of two tympanic neurons
    in noctuid moths, Jap. J. Physiol., 11: 666-677.
Suga, N., 1965, Functional properties of auditory neurones in the
    cortex of echolocating bats, J. Physiol. (Lond.), 181: 671-700.
Suga, N., 1977, Amplitude-spectrum representation in the Doppler-
    shifted-CF processing area of the auditory cortex of the
    mustache bat, Science, 196: 64-67.
Suga, N., 1979, Representation of auditory information by the brain
    (II), Shizen, Chuokoronsha, Tokyo, Japan, 79: 6, 70-81. (in
    Japanese).
Suga, N., 1980a, Functional organization of the bat's auditory
    cortex beyond tonotopic representation, in: "Multiple Cortical
    Somatic Sensory-Motor, Visual, and Auditory Areas and Their
    Connectivities", C. N. Woolsey, ed., New Jersey, U.S.A. (in

press).

Suga, N., 1980b, Cortical representation of biosonar information in
    the mustached bat, Publ. House Hungarian Acad. Sci. (in press).

Suga, N. and Jen, P.H-S., 1976, Disproportionate tonotopic represen-
    tation for processing species-specific CF-FM sonar signals in
    the mustache bat auditory cortex, Science, 194: 542-544.

Suga, N. and Jen, P.H.-S., 1977, Further studies on the peripheral
    auditory system of "CF-FM" bats specialized for the fine
    frequency analysis of Doppler-Shifted echoes, J. Exp. Biol.,
    69: 207-232.

Suga, N. and O'Neill, W. E., 1979, Neural axis representing target
    range in the auditory cortex of the mustached bat, Science,
    206: 351-353.

Suga, N., O'Neill, W. E., and Manabe, T., 1978, Cortical neurons
    sensitive to particular combinations of information bearing
    elements of bio-sonar signals in the mustache bat, Science,
    200: 778-781.

Suga, N., O'Neill, W. E., and Manabe, T., 1979, Harmonic-sensitive
    neurons in the auditory cortex of the mustache bat, Science,
    203: 270-274.

Suga, N., Simmons, J. A., and Jen, P.H.-S., 1975, Peripheral
    specialization for fine analysis of Doppler-shifted echoes in
    "CF-FM" bat Pteronotus parnellii., J. Exp. Biol., 63: 161-192.

THREE-DIMENSIONAL STUDY OF EVOKED FIELD POTENTIALS IN THE AUDITORY

CORTEX OF THE CAT

Márk Molnár, George Karmos and Valéria Csépe

Institute of Psychology
Hungarian Academy of Sciences
Budapest, Hungary

INTRODUCTION

Both evoked potential studies (Woolsey, 1961) and single cell investigations (Brugge and Merzenich, 1971) demonstrated the tonotopic organization of the auditory cortex of the cat. Less attention was paid to the distribution and alterations of field potentials in different subregions of the auditory cortex in behavioral situations and to mechanisms underlying the observed changes of these evoked responses.

METHODS

15 adult cats of either sex were used in the experiments. Bilateral stereotaxic implantation of electrodes was carried out under Nembutal anesthesia into the acoustic cortex, dorsal hippocampus and medial geniculate body. Clicks were generated at 3/s rate by 4 kHz sine wave bursts of 1 ms duration and were administered through bone conduction to maintain the acoustic input constant. Middle ear muscles were cut prior to the experiments.

A square array of electrodes was used for epidural recordings. It consisted of 36 electrodes arranged as a 6 x 6 matrix embedded in dental cement and was implanted above the acoustic cortex. For intracortical recordings a new type of multielectrode was constructed. It was an array of six 62 μm thick nichrome wires entering the cortex independently, and by the end of the implanting procedure the recording tips were in intact tissue arranged vertically on each other. The interelectrode distance was about 300 - 400 μm. This electrode allowed the simultaneous recording of field potentials from five different depths of the cortex and from the surface.

Experiments were carried out in a sound attenuated chamber on freely moving cats. Electrical activity was stored on a multichannel tape recorder for further analysis.

RESULTS

Mapping the distribution of surface evoked potentials on the subareas of the acoustic cortex in our experiments confirmed the data of Karmos (1980). It must be emphasized that the intensity of the click applied in our experiments was well above threshold (80dB). Therefore it was not surprising that no sharp changes were found in the evoked potential wave recorded in the different areas. Evoked responses from the surface and from three different depths of the A I area are shown in Fig. 1.

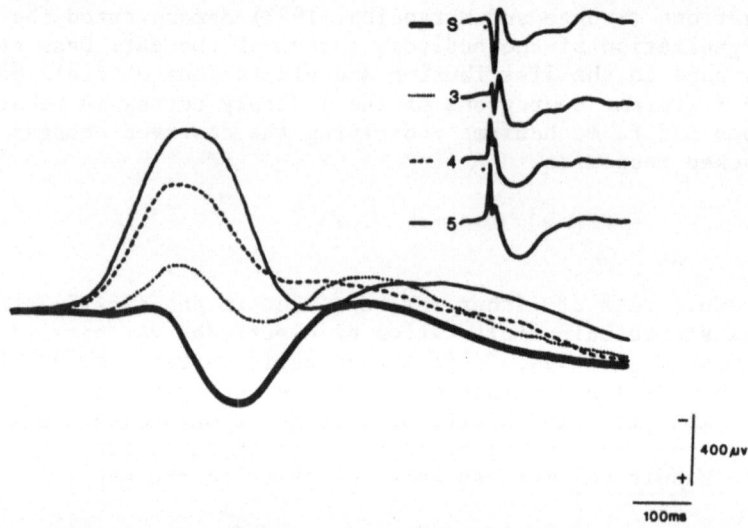

Fig. 1. Evoked responses from the surface (S), the 3rd, 4th and 5th intracortical electrodes. For better visualization responses from the 1st and 2nd intracortical electrodes are not shown. Each potential represents the average of 32 individual responses. On the right 256 ms of the response are shown, on the left the first components are super-imposed. For more detailed explanation see text.

Considering the whole 256 ms of the responses (right in Fig. 1.)
the first positive component on the surface seems to have a phase
reversal in the depth. Late components have the same polarity in the
whole depth of the cortex. If the first components are superimposed
on a larger time scale (left in Fig. 1) it becomes clear that none
of the components has a real phase reversal. The difference between
the peak latencies of the first components recorded from the surface
and from the deepest electrode is in this case about 2 ms. Stimulating the ventrolateral part of the medial geniculate body complex
wave could be recorded on the surface and in different depths of the
auditory cortex, depending upon the intensity of the stimulation
(Fig. 2).

CGM STIM. QUIET

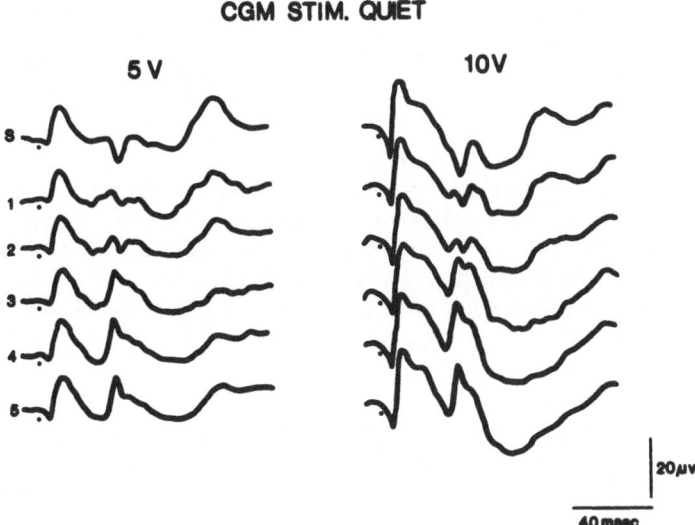

Fig. 2.   Averaged evoked responses (128 ms) recorded from the audi-
          tory cortex elicited by the stimulation of the medial
          geniculate body by pulses of 0.1 ms duration. Numbers refer
          to progressively deeper layers.

An early positive component was recorded in every depth with
no sign of phase reversal. This first early positive component was
followed by a late wave complex with the latency of about 40 ms.
The shape of these late waves changed as the intensity of the stimulation was increased. It is quite probable that this late wave
complex represents the activity of a subcortical loop. Further
evidence supporting this assumption is that the administration of
barbiturates entirelly suppresses its appearance.

DISCUSSION

In spite of a cumulating body of data, our knowledge of the genesis of cortical evoked potentials is rather poor. Classical studies were carried out in anesthetized animals where late components do not appear. The dipole theory maintains that the potential recorded on the surface is the volume conducted mirror image of the potential originated in the depth of the cortex (Bishop and Clare, 1952). Our results rule out the existence of a simple dipole in the cortex. The early components recorded in different depths might represent the activity of different cellular processes. In late components no phase reversal could be observed indicating homogeneous polarization of cortical tissue during this period. Stimulation of the specific afferent pathway to the auditory cortex produced complex waves, probably representing the activity of different generators.

REFERENCES

Bishop, G. H. and Clare, M. H., 1952, Sites of origin of electrical potentials in striate cortex, J. Neurophysiol., 15: 201-210.
Brugge, J. F. and Merzenich, M. M., 1971, Representation of frequency in auditory cortex in the macaque monkey, in: "The Physiology of the Auditory System", M. B. Sachs, ed., National Educational Consultants, Baltimore.
Karmos, G., 1980, Auditory cortical correlates of motivation, in: "Recent Progress in Hungarian Neurobiology". vol. 10, Lissák, K., ed., Akadémiai Kiadó, Budapest.
Woolsey, C. N., 1961, Organization of cortical auditory system, in: "Sensory Communication", W. Rosenblith, ed., M. I. T. Press, Cambridge, Mass.

DIFFERENTIAL DIAGNOSIS OF HEARING DISORDERS - CLINICAL FINDINGS

CONTRIBUTING TO INFORMATION PROCESSING IN THE AUDITORY PATHWAY

H. von Specht

HNO-Klinik, Medizinische Akademie

Magdeburg, GDR

The far-field recording of auditory evoked potentials (AEP) has been gaining importance to neurophysiological, audiological and recently also to otoneurological approaches. Both electrocochleography (ECochG) and recording of early AEPs (brainstem potentials, latency 2 - 10 ms) are highly reliable and independent of subjects´ vigilance; they are important to the diagnosis of cochlear and retrocochlear hearing disorders as far as the brainstem region. Late AEPs (latency more than 50 ms) are reliable only with recording conditions of constant vigilance, preferably in wakefulness. Late AEPs are advantageous in that they may be used to determine hearing thresholds for lower frequencies, while for early AEPs the lower frequency limit is around 1 kHz.

The differences of the frequency bands of early and late AEPs of 100 Hz - 3 kHz and 0.1 - 30 Hz, respectively, are of relevance with regard to noise superimposed on the signal. The high-frequency fractions of the EEG that are superimposed on early AEPs, and potentials of myogenic origin are of a largely random character, being little correlated with the signal. In contrast, often the EEG superimposed on late AEPs cannot be considered as a stochastic process and is also correlated with the AEP (Freigang and von Specht, 1977). Accordingly, in recording the potentials by the averaging technique, the recording of early AEPs may be regarded as more reliable since the requirements of a random noise (that is superimposed on the signal without correlating with it) are met in better terms.

While early AEPs are independent of vigilance, the effect of vigilance on late AEPs is substantial. Thus, for clinical applications problems are encountered in the event of vigilance altera-

225

tions, but also for examinations in natural and drug-induced sleep.
In neurophysiological respects, however, investigations into the
variation of late AEPs with the vigilance may be of great value
(Kevanishvili and von Specht, 1979).

Even though, because of the above-mentioned influences, the
late AEPs are less reliable than the early components, the good correlation between these potentials (e.g. the amplitude N1P2) and
sensation offers decisive advantages for comparisons with psychophysical measurements. However, the correlation between electrophysiological and psychophysical examinations should in each case
be interpreted with utmost care and more in qualitative terms
(Pratt and Sohmer, 1977).

Yet, it is possible, e.g. by means of late AEPs, to prove the
recruitment, i.e. a sensation of loudness increasing with the
intensity (dB, SL) more vehemently than in the normal case. The
gradients of intensity-amplitude characteristics and also of
intensity-latency characteristics in patients with recruiting ears
were higher-sloped when compared with those of normal ears, which
finding was in agreement with their subjective sensations. This is
demonstrated, as an example, by Fig. 1 for patients with Menière´s

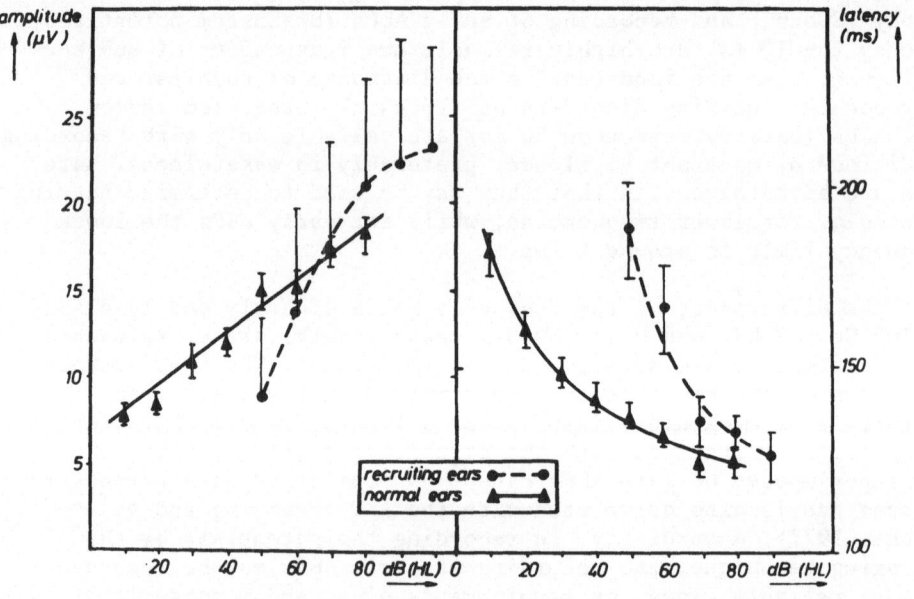

Fig. 1.   N1P2-amplitude vs. intensity and N1-latency vs. intensity
          characteristics of normal and recruiting ears (mean values
          and standard deviation of 20 normal and 14 recruiting ears,
          stimulation rate 0,4/s, 1 kHz tonebursts of 0.5 s),
          according to Freigang and von Specht (1977).

syndrome. Such comparative studies require consistent stimulation and recording conditions.

A similar increase in slope of the characteristics has been even observed for the early AEPs (brain stem potentials, BERA) and compound action potentials (AP) of the auditory nerve (ECochG) because of the peripheral localization of the hearing disorder in Menière's disease. The ECochG is especially advantageous to differential diagnosis of Menière's disease. In addition to a differentiation between various potential modes Eggermont (1978) also described diagnostic possibilities with the aid of AP tuning curves. Such tuning curves may be also obtained subjectively and by means of early AEPs. Since the considerable time required for determining AP tuning curves is further increasing when tuning curves are to be obtained on the basis of early AEPs, routine application in audiological diagnostics will hardly be possible. A subjective determination is not suited either, owing to the high demands on the patient's ability to cooperate. Yet, investigations of this kind when performed in individual cases, may contribute to the better understanding of disorders in the processing of auditory information and associated abnormal hearing sensations (e.g. recruitment and diplacusis).

The recording of early AEPs has been found to be very efficient in the diagnosing of processes in the brain stem region. When sufficiently suprathreshold stimuli (in excess of 75 dB, HL) are applied, delays of the latencies of the first five prominent peaks (PI - PV) of early AEPs provide highly meaningful diagnostic information. In this context it is important that the method is noninvasive. (The range of normal values for peak PV latencies is 5.1 - 5.9 ms, with the limit for pathogical values being deliberately chosen as low as 6 ms. It is in this way that even slight pathological changes in latencies can be evidenced, though at the risk of some false positive findings).The latency differences between individual peaks and interaural latency differences may provide further information about the localization of the hearing disorder (topodiagnostics).

It is, however, of great importance to the diagnostics in the brain stem region that the recording of early AEPs is accomplished within the framework of a test battery (various subjective audiological tests, vestibular diagnosis, X-ray examination). Pathological events in the brain stem region are often accompanied by auditory missensations, partly restricted to certain frequency bands, which are not detected by the common audiological tests. Determining such phenomena through interviewing of patients in the course of electrophysiological examinations is governed by intelligence, musicality, and the compliance to cooperate. We believe, however, that the acquisition of subjective audiological and psycho-auditory data in conjunction with the results of electrophysiological, X-ray and

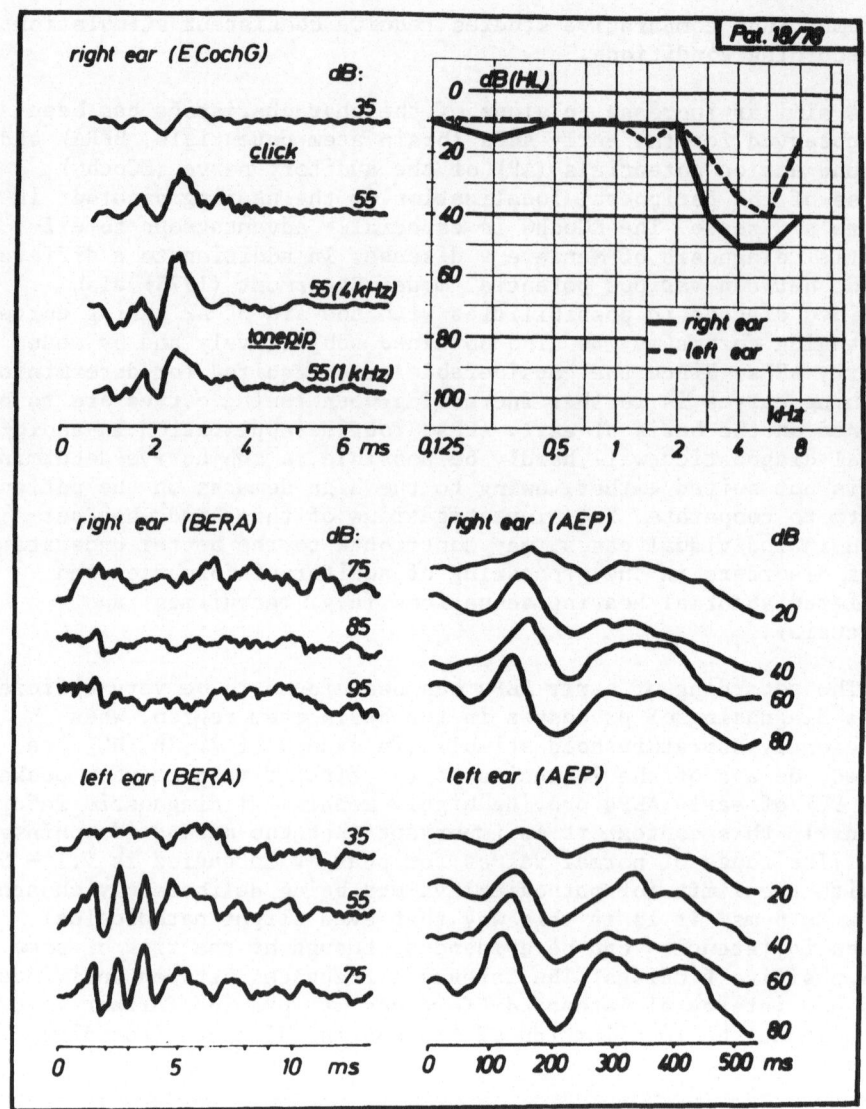

Fig. 2. Electrophysiological data (AP, early and late AEPs) and
         subjective audiogram (noise-induced hearing loss due to
         occupational noise exposure) of a patient with a cere-
         bello-pontine angle tumor (approx. 3 cm dia.) on the
         right side. Note the absence of peaks PII - PV for early
         AEPs (BERA) of the right ear; parameter: dB (SL),stimu-
         lation rate for BERA and ECochG: 20/s, for late AEPs:
         0.5/s (1 kHz tonebursts of 0.3 s duration).

and computer-tomographic examinations along with neurosurgical
findings can contribute in the long run to a better understanding
of the processing of acoustic information, at least in the lower
parts of the auditory pathway.

Finally, an example is used to demonstrate that even with the
very reliable method of recording early AEPs, contradictions may
occur between electrophysiological findings and subjective sensations, even though the efficiency of the method with respect to
diagnosing pathological events is not disputed. Fig. 2 shows, in
addition to the thresholds, the results of ECochG and recording of
early and late AEPs obtained for a patient with cere-bello-pontine
angle tumor, neurosurgically evidenced in the right brain stem
region. APs and late AEPs could be recorded in agreement with the
perception of acoustic stimuli in the affected ear, with its hearing
being of disturbed quality. Even for quite suprathreshold intensities
(dB, SL) the recording of early AEPs did not yield any noticeable
responses, except for peak I corresponding to the AP. This finding
was also obtained for a frequency-specific stimulation with tone
pips and a stimulation rate reduced to 5/s. The assumed cause was
an insufficient synchronization in the damaged brain stem region,
in which case brain stem potentials cannot be recorded through
application of the averaging technique. Moreover, the responses
originating in different parts of the auditory pathway and recorded
by means of averaging may be considered anyhow as electrical by-
products of hearing.

REFERENCES

Eggermont, J. J., 1979, Compound action potentials: tuning curves
    and delay times, Scand. Audiol., Supp. 9: 129-139.
Freigang, B. and von Specht, H., 1977, Möglichkeiten und Grenzen der
    Objektivierung psychoakustischer Messungen mittels langsamer
    AEPs. Prom. B, Magdeburg.
Kevanishvili, Z. Sh. and von Specht, H., 1979, Human slow auditory
    evoked potentials during natural and drug-induced sleep,
    Electroenceph. clin. Neurophysiol., 47: 280-288.
Pratt, H. and Sohmer, H., 1977, Correlations between psychophysical
    magnitude estimates and simultaneously obtained auditory nerve,
    brain stem and cortical responses to click stimuli in man,
    Electroenceph. clin. Neurophysiol., 43: 802-812.

SESSION V
CENTRAL AUDITORY MECHANISMS B
Chairman: F. de Ribaupierre

BINAURAL INTERACTION IN THE CAT INFERIOR COLLICULUS: COMPARISON

OF THE PHYSIOLOGICAL DATA WITH A COMPUTER SIMULATED MODEL

Y. Sujaku*, S. Kuwada, and T.C.T. Yin

Dept. of Neurophysiology
University of Wisconsin Medical School
Madison, Wisconsin 53706 U.S.A.

INTRODUCTION

Many low frequency cells in the inferior colliculus (IC) have
been shown to be sensitive to differences in the time of arrival
of sound to the two ears, thereby suggesting that they are involved
in sound localization. Most studies of this phenomenon have employed
dichotically presented sinusoids with variable delays in the onset
of the tone to one ear with respect to the other (Rose, Gross,
Geisler, and Hind, 1966: Moushegian, Rupert, and Stillman, 1971).
For many IC neurons the discharge rate as a function of interaural
delay is cyclic with a period equal to that of the stimulating
frequency. The repetitive cycling nature of these interaural delay
curves indicates that these neurons are sensitive to changes in
interaural phase created by the delays. That this is so is indicated
by the finding that a stimulus which creates an orderly and continuously changing interaural phase difference independent of the
stimulus onset, i.e. a binaural beat stimulus, yields results similar
to those obtained by the interaural delay method (Kuwada, Yin, and
Wickesberg, 1979). The response of these phase sensitive cells often
modulates above and below the level of firing for monaural stimulation alone. This suggests that a simple temporal summation mechanism
is inadequate to explain "cycling" and indicates that additional
inhibitory mechanisms must be present. Most of these IC neurons show
an equal amplitude modulation as a function of interaural phase,
with no particular sensitivity to the dynamics of phase change, e.g.
the rate or direction of change. In addition to this typical
response characteristic, we have also studied a sizable number of

---

*On research leave from Oita University, Japan.

cells that have more unusual response properties. For example, there
are cells whose cyclic delay curve varies as a function of stimulus
off time, interaural intensity differences, or the ear which initially received the signal. Furthermore, there are neurons that respond
selectively to particular rates and/or direction of interaural
phase change. This paper describes a computer simulated model of
a binaural cell which can be made to behave like the typical interaurally phase sensitive IC neuron as well as to mimic some of these
more unusual properties.

THE MODEL

Structure

The model consists of two inputs, one from each ear, which
make excitatory synaptic connections onto the binaural cell (B).
In addition there is a crossed collateral presynaptic inhibition
from each input. In order to account for possible conduction,
synaptic, and mechanical delays, four discrete delays D1 - D4 are
introduced on the input axons.

Input

The inputs to the model are sequences of impulses from each
ear. The pulse trains are assumed to be time-varying Poisson
processes with dead time equal to 1/10 of the period of the stimulus.
The effect of this dead time is to create a different distribution
of impulses than would be seen in a classic Poisson distribution.
The rate of occurrence is proportional to the amplitude of the
half-wave rectified acoustic stimulus.

Synaptic Interaction

Presynaptic region. Impulses arriving at each of the four
presynaptic regions cause instantaneous release of a neurotrans-
mitter. The amount of transmitter released, Δq, for each impulse
is a certain fraction kA of the available transmitter. Let the
arrival time of an impulse be t_o and the quantity of available
transmitter be q, then we assume that

$$q(t_o^+) = (1-kA)q(t_o^-), \quad 0 \leq kA \leq 1$$

and

$$\Delta q = q(t_o^-) - q(t_o^+) = kAq(t_o^-)$$

where k is a constant and A is the amplitude of the presynaptic
impulse.

After each impulse, we assume that the quantity of the available transmitter increases up to a certain amount, which in this case equals 1. Then we have

$$dq/dt = \tau_q^{-1} q^n (1-q), \qquad n = 0 \text{ or } 1$$

where τ_q is a time constant. This curve is exponential when $n = 0$ and logistic when $n = 1$ (Oono and Sujaku, 1975).

Postsynaptic region. At each of the four postsynaptic sites, the amplitude of the postsynaptic potential increases instantaneously in Δp increments proportional to the amount of transmitter released. Thus

$$\Delta p = m \Delta q$$

where m is a constant.

Following this increment, the postsynaptic potential p decreases exponentially with a time constant τ_p.

$$dp/dt = -\tau_p^{-1} p$$

Presynaptic inhibitory region. At the two presynaptic inhibitory synapses, the postsynaptic potential p decreases the amplitude A1 (=1) of the impulse traveling along the postsynaptic axon by p, so that in response to a presynaptic impulse the amplitude of the postsynaptic impulse becomes A1−p; if $p > A1$, then the postsynaptic impulse is totally blocked.

Discharges

The binaural cell (B) discharges when the sum of the two postsynaptic potentials at the excitatory synapses exceeds the threshold, $\Theta(=1)$, of the cell. After each discharge the postsynaptic potential of B drops to zero. All the simulated discharge patterns shown are the output of cell B.

RESULTS AND DISCUSSION

Space limitations do not permit a full description of the comparisons between the results obtained from the neural data and from the model. However, Fig. 2 illustrates representative comparisons of these results. Data obtained from single cell recordings in the IC of cats are shown for two different cells in the left hand column while the results from the model are shown in the right hand column. Figs. 2A and 2B illustrate the responses of a typical neuron to the binaural beat and interaural delay stimuli, respectively. Fig. 2A (left) shows a peristimulus time

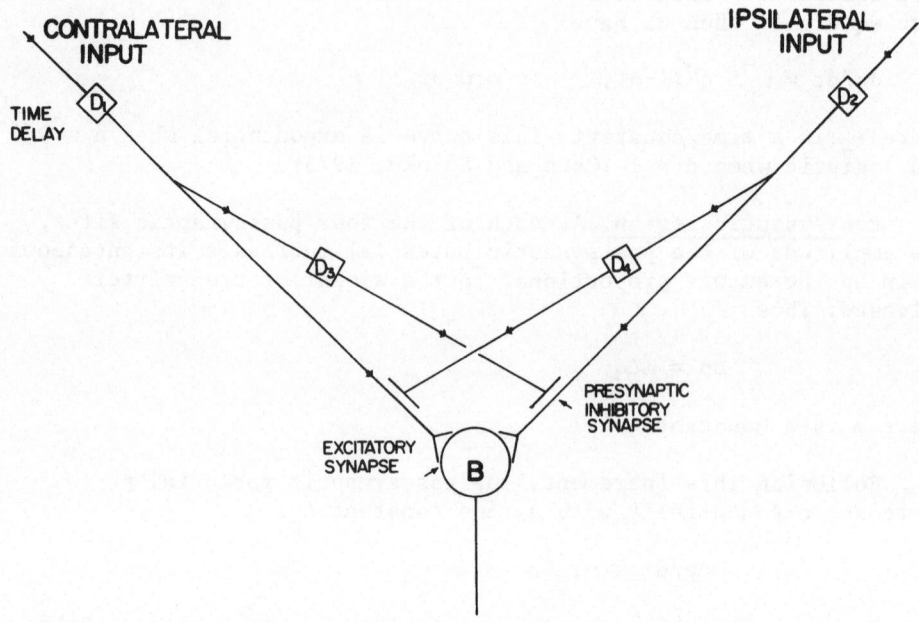

Fig. 1. Structure of the model.

histogram of the response to a 3 second duration binaural beat
stimulus with a 1 Hz beat frequency. The neuron discharges with 3
bursts during this period, each burst occurring at the same interaural phase difference. Fig. 2B (left) is the delay curve obtained
from the same cell at the same stimulating frequency and intensity.
As expected, the delay curve is cyclic at the period of the stimulating tone. The estimates of peak interaural phase sensitivities
obtained from these two methods are very similar (discrepancy of
.02 phase). On the right of Figs. 2A and 2B are shown the results
from the model for the same binaural beat and interaural delay
stimuli used to obtain the neural data. The arrows in Fig. 2B show
the level of discharge for monaural stimulation alone. The model
is able to simulate the neural data even to the point of modulating
the response below the level of monaural stimulation. The silencing
of the response seen in both the model and the neural data at un-
favorable interaural phase differences must be due to an inhibitory
mechanism which we have modeled as a crossed collateral inhibition.

Fig. 2C shows beat frequency period histograms for a cell
with an unusual response property, namely a sensitivity to the
direction of interaural phase change. The two directions are
produced by reversing the frequencies of the tones to the two ears.
This cell responds more vigorously when the tone to the contralateral

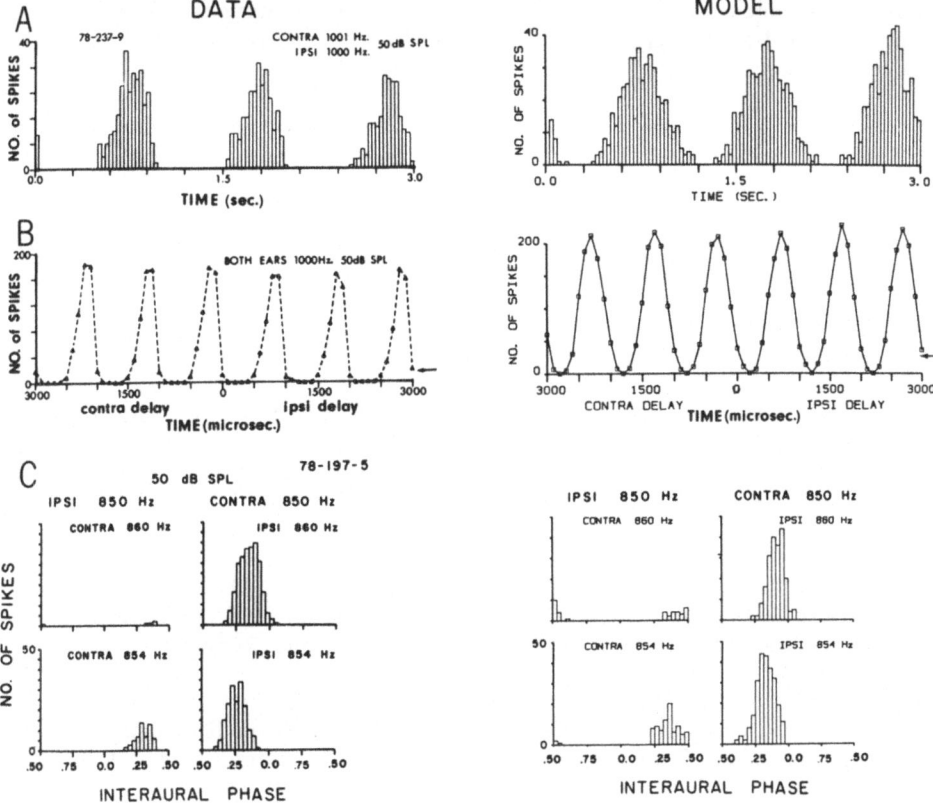

Fig. 2.   Comparisons of the neural data with the computer-simulated
           model. A. Peristimulus time histograms of the response to
           a binaural beat stimulus, 3 seconds in duration presented
           5 times at an intensity of 50 dB SPL (re 2x10^{-5}Pa).
           Frequency to the contralateral ear and ipsilateral ear
           was 1001 and 1000 Hz, respectively. B. Interaural delay
           curves derived from delaying the signal to each ear from
           3000 μs contralateral delay to 3000 μs ipsilateral delay
           in steps of 100 μs. At each interaural delay a 1000 Hz
           tone of 1 second duration, at 50 dB SPL, was presented
           once. Arrows indicate the level of response to monaural
           stimulation. C. Beat frequency period histograms derived
           from the response to binaural beat stimuli of 3 seconds
           duration, repeated 10 times. The left and right pairs
           of each set reflect opposite directions of phase change
           while the upper pair involved higher rates of phase
           change (i.e. higher beat frequency).

ear is lower in frequency. This response suggests that the neuron
is sensitive to the direction of movement of a sound source. By

a judicious choice of model parameters, such an asymmetrical response can be obtained from the model by eliminating one of the crossed collateral inhibitory connections.

We have also been able to simulate results from other IC cells that appear to show a sensitivity to the speed but not the direction of interaural phase change (thereby suggesting a rate sensitivity) as well as ones whose interaural delay curves can be altered by variations in the stimulus off time or the ear which first received the signal. A full description of these results will be forthcoming.

ACKNOWLEDGMENTS

Supported by N.I.H. grants NS12732 and EYO2606 and the Japan Ministry of Education.

REFERENCES

Kuwada, S., Yin, T.C.T., and Wickesberg, R.E., 1979, Response of cat inferior colliculus neurons to binaural beat stimuli: possible mechanisms for sound localization, Science 206: 586-588.

Moushegian, G., Stillman, R.D., and Rupert, A.L., 1971, Characteristic delays in superior olive and inferior colliculus, in: "Physiology of the Auditory System", M.B. Sachs, ed., National Educational Consultants, Baltimore, Md.

Oono, Y. and Sujaku, Y., 1975, A model for automatic gain control observed in the firings of primary auditory neurons, Trans. Inst. Elect. Comm. Engin. Japan, 58: 352-359.

Rose, J.E., Gross, N.B., Geisler, C.D., and Hind, J.E., 1966, Some neural mechanisms in the inferior colliculus of the cat which may be relevant to the localization of a sound source, J. Neurophysiol., 29: 288-314.

CODING PROPERTIES OF THE DIFFERENT NUCLEI OF THE CAT´S MEDIAL

GENICULATE BODY

A. Toros-Morel, F. de Ribaupierre and E. Rouiller

Institute of Physiology
University of Lausanne
Bugnon 7, 1011 Lausanne, Switzerland

Morest´s subdivisions of the cat´s medial geniculate body (MGB) differ from one another in neuronal morphology, ascending connections and cortical projections (Morest, 1964, 1965; Sousa-Pinto, 1973). Differences in auditory coding, particularly frequency tuning properties, have been observed between the pars magnocellularis and pars lateralis, whereas only few cells in the pars dorsalis respond to acoustic stimulation (Aitkin, 1966). The aim of this study was the analysis of coding properties of the different nuclei of the MGB to simple acoustic stimuli, with emphasis on response latencies, temporal response patterns, frequency tuning and binaural properties.

Experiments were performed on 29 nitrous oxide anaesthetized cats. Methods of surgery, stimulation, recording and analysis are described in a previous paper (Rouiller et al., 1979). Responses to tone bursts, noise bursts and clicks were characterized for over 3000 units distributed in the 6 nuclei of the MGB that were histologically delimited on Nissl stained sections.

Comparison of the proportion of cells driven by the different stimuli used (Fig. 1) reveals that responsiveness to both tones and broad band stimuli (class "CL+NO+TO") is higher in the pars magnocellularis (PM) and the suprageniculate nucleus (SG) than in the pars lateralis (PL), pars ovoida (PO) and pars ventrolateralis (PV). In these latter nuclei more cells responded selectively to tones and not to noise and clicks (class "TO"). In the pars dorsalis (PD) few cells were driven by simple acoustic stimulation.

Quantitative distribution of the different response character-istics in the 6 nuclei was evaluated. The same types of temporal

response patterns were found in all nuclei, in more or less similar
proportions: The majority of units responded to tone or noise
bursts with an "ON" transient pattern of discharge. Sustained excitatory "THROUGH" responses to tones were evoked in 7-15% of the
units whereas noise bursts evoked this type of response in 24-39%
of the units. Mean and median latency values of the first spike of
the response are indicated in Table 1, for tones, noise and clicks:
for the three stimuli mean latencies were significantly shorter in
PM and SG than in PL, whereas they tended to be longer in PO and
PV.

     Frequency tuning properties were evaluated for over 300 units
by reconstruction of their tuning curve. Four types were observed:
narrow primary-like, narrow "U" shaped, broad and multipeak. All
types were found in PL, PO and PM but in different proportions:
primary-like and "U" shaped, regrouped in the narrow category,
represented more than half of the tuning curves exhibited by units
in PL (54 %) whereas they were less frequently encountered in PM
and PO (31-32% of the units). In the same way, the proportion of
broad tuning curves observed for 36% of the units in PL, was higher
in PM and PO (47-54%). Few cells in PL were of the multipeak type
(10%) whereas it represented 14% of the units in PO and 22% in PM.
In SG and PV the few tuning curves that were observed, were distributed in only two categories, broad and multipeak, with usually a
predominancy of the broad type (50-63%). When considering the
distribution of the width of the frequency range of response, in
octaves, at an average intensity of 60 dB SPL, a similar picture
was obtained except that then PO resembles PL more than the other
nuclei. Moreover very broad ranges, exceeding 6 octaves, were more
frequently encountered in PM, SG and PV.

     Units in the 6 nuclei could not be differentiated on the basis
of their binaural properties: for all three stimuli they show mostly
binaural inputs, with a contralateral dominance. Of the various
binaural interactions (summation, inhibition and occlusion, as
defined by Hall et al., 1968) the most usual are summation and
inhibition, very few being occlusion. Their distribution is quite
similar in the 6 nuclei.

     In comparing the frequency distribution of auditory response
characteristics in the 6 nuclei of the MGB, some could be associated
for sharing comparables coding properties: higher selectivity to
tones, longer mean latencies and narrower frequency tuning in PL and
PO suggest that these two nuclei are mainly concerned with a precise
frequency analysis. Less selectivity to the acoustic stimulus,
shorter mean latencies and broad tuning in PM and SG are comparable
with a more direct and global analysis of the acoustic stimulus.
Units in PV present coding properties intermediate between the two
precedent groups of nuclei whereas few cells in PD responded to a
simple acoustic stimulation.

Table 1. Mean and Medium Values of Responses Latencies to Tone, Noise and Clicks in the 6 Subdivisions of the Medial Geniculate Body

	TONE			NOISE			CLICKS		
	N	Mean ± SD (ms)	Median (ms)	N	Mean ± SD (ms)	Median (ms)	N	Mean ± SD (ms)	Median (ms)
PV	55	22.1* ± 12.5	18	37	21.2** ± 9.2	20	20	14.6* ± 6.8	12
PO	225	18.8 ± 7.9	17	110	21.1** ± 11.0	19	82	12.8* ± 7.5	9
PD	10	24 ± 16.9	15	7	18.4 ± 9.2	15	4	12.8 ± 4.2	10
PL	965	18.4 ± 16.9	15	439	17.1 ± 9.4	14	412	10.7 ± 6.7	9
PM	335	15.4** ± 9.1	12	268	15.6* ± 9.2	12	216	8.9** ± 4.7	8
SG	98	14.9* ± 14.6	11	80	13.5** ± 9.3	10	68	7.9** ± 3.7	7

N = number of cells.

* = mean latency significantly different at the 0.05 level from the PL mean latency.

** = mean latency significantly different at the 0.01 level from the PL mean latency.

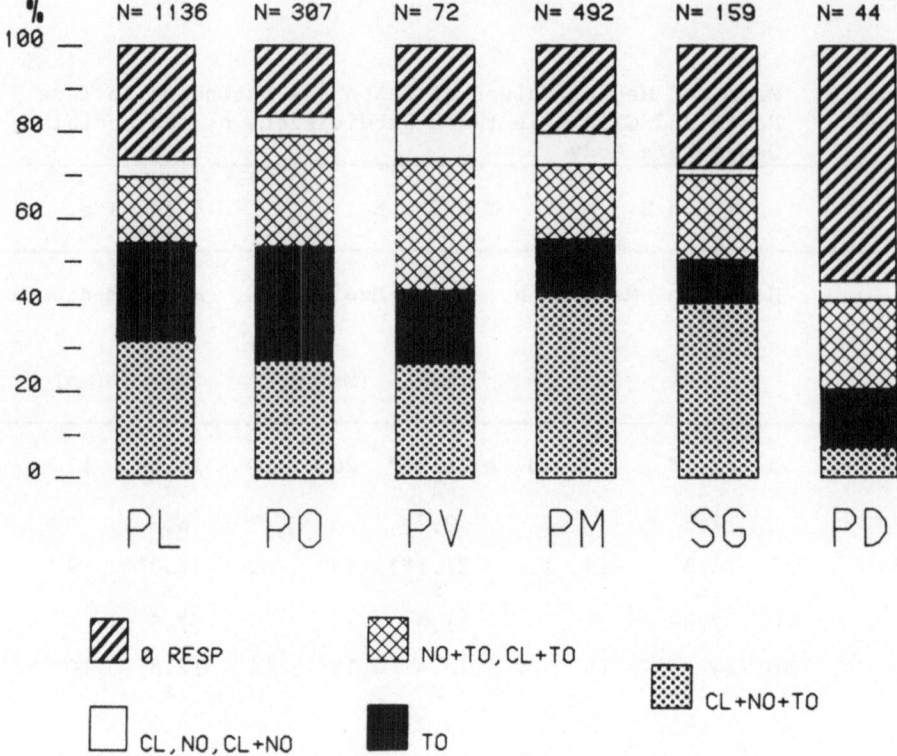

Fig. 1. Distribution of units responsiveness to clicks, noise and
         tone, in five classes, for the 6 nuclei of the MGB (abbre-
         viations as mentioned in the text).

         N = sample size for each nucleus of units tested to the
         three stimuli.

    It is interesting to note that some nuclei present a functional
heterogeneity: for example, the anterior part of PL is characterized
by shorter mean latencies, narrower tuning properties and a higher
proportion of ipsilateral dominance than in the posterior PL (Toros-
Morel et al., 1979).

ACKNOWLEDGMENT

    This work is supported by the Swiss National Science Founda-
tion, Grant no 3.239.69.

REFERENCES

Aitkin, L. M., 1966, Medial geniculate body of the cat: responses
    to tonal stimuli of neurons in medial division, J. Neuro-
    physiol., 36: 275–283.
Hall, J. L. II and Goldstein, M. H. Jr., 1968, Representation of
    binaural stimuli by single units in primary auditory cortex
    of unanesthetized cats, J. Acoust. Soc. Am., 43: 456–461.
Morest, D. K., 1964, The neuronal architecture of the medial
    geniculate body of the cat, J. Anat. (Lond.), 98: 143–160.
Morest, D. K., 1965, The lateral tegmental system of the midbrain
    and the medial geniculate body: study with Golgi and Nauta
    methods in cat, J. Anat. (Lond.), 99: 611–634.
Rouiller, E., de Ribaupierre, Y. and de Ribaupierre, F., 1979,
    Phase-locked responses to low frequency tones in the medial
    geniculate body. Hearing Res., 1: 213–226.
Sousa-Pinto, A., 1973, Cortical projections of the medial geniculate
    body in the cat, Adv. Anat. Embryol. Cell Biol., 48.
Toros, A., Rouiller, E., de Ribaupierre, Y., Ivarsson, C., Holden,
    M. and de Ribaupierre, F., 1979, Changes of functional
    properties of medial geniculate neurons along the rostro-
    caudal axis, Neurosci. Lett. Suppl., 3: S5.

INTERAURAL DELAY SENSITIVE UNITS IN THE MGB OF THE CAT

C. Ivarsson, Y. de Ribaupierre, A. Baroffio and
F. de Ribaupierre

Institute of Physiology
University of Lausanne
Bugnon 7, 1011 Lausanne, Switzerland

The simplest physical clue allowing partial localization of a sound source is the delay in its time of arrival at the two ears. Interaural delay sensitive units have been described at different levels of the central nervous system (Altman, 1968; Brugge et al., 1969; Goldberg and Brown, 1969). The aim of this study is to describe the effect of this interaural delay DT on the extracellularly recorded activity of MGB neurons.

Cells affected by DT could expected to present two types of activity depending on whether they are phase detectors or delay detectors. Delay detectors should present a maximal response at a fixed delay domain. When tested with tones their response to that same delay will be repeated in a periodic manner.

Phase detectors require a given phase relation between the signals at the two ears to have a maximum of activity. These detectors typically respond periodically to interaurally delayed tones. The periodicity is the inverse of the frequency of the tone and the portion of the period to which the unit responds is independent of the frequency, therefore the range of the response to DT widens as the frequency of the stimulating tone decreases. These two types of detectors can be distinguished by the evolution of their response with frequency.

Out of 908 neurons (36 electrode tracks located in the MGB of 12 cats), 203 were studied for the influence of different interaural delays. Recording and stimulating techniques have been described elsewhere (Rouiller et al., 1979). Broadband noise and tone bursts were passed through two channels each having a delay line (Philips

TDA 1022) and the difference between these two delays is the interaural delay DT. Using these delay lines requires filtering that cuts
off high frequencies (3 dB at 12 kHz), therefore the initially broad-
band noise will lack high frequency components. When units were
tested with delayed tone bursts, the frequency used was below 2.5
kHz. Each test consisted of 100 stimuli presented at a rate of 1/s,
with a stimulus duration of 200 ms. Each stimulus presentation was
at a different interaural delay, spanning linearly in 100 steps one
of the following ranges: from -.5 to +.5 ms, 0 to 1 ms or -.5 to
+3.4 ms.

The response pattern of the tested units were unaffected (43%)
or the rate of discharge was quantitatively modified (27%) or it was
qualitatively modified (30%) and turned into another type of pattern.
Changes in pattern were for example from a transient ON response to
a sustained THROUGH response.

From the 66 cells responding to tones below 2.5 kHz, 40 were
sensitive to DT. Most of these cells were tested for large interaural delays, covering several times the period of the stimulating
tone, and present a periodic response. For 10% of them sufficient
exploration enables us to decide that they are phase detectors, none
can be proven to be delay detectors and 23% could be either delay or
phase detectors but were not tested at several frequencies as well
as to noise and therefore cannot be categorized. Unit Z 12 C 14
(Fig. 1, left) is typically a phase detector. One can note the similarity between the maximum of response of the unit (on the dot
display) and the modulated signal (left of the dot display) that
represents the sum of the signal at the two ears. The response of
the unit to two tones of different frequencies are shown. Below each
dot display one finds superposed to a schematic representation of
the cells response the function COS ($2\pi f \ast DT$) that is there to indicate the expected periodicity of the detector. The remaining cells
have responses that cannot be simply interpreted. A schematic
representation of the responses of 6 units are shown on the right
side of Fig. 1 with the function COS ($2 \pi f \ast DT$).

Among 140 cells tested to delayed noise 77 present a change
in their response when DT is changing. They are ordered in Fig. 2
according to their domain of response by right (A), respectively left
(B), limit of the strong part of the response. 26% of them present
more than one strong response domain. For these units the response
domain represented in the figure is the one closest to zero delay.
62% of the units present a change in their response within $\pm$ 200 µs,
which is the physiological range of DT for the cat. If within the
200 µs the change in the response of these units codes the position
of the sound source, it is interesting to note that the number of
transition coding the left side (16) is comparable to the number of
transitions coding the right side (21) and the importance given to
the left and right field is proportional to these numbers. The simi-

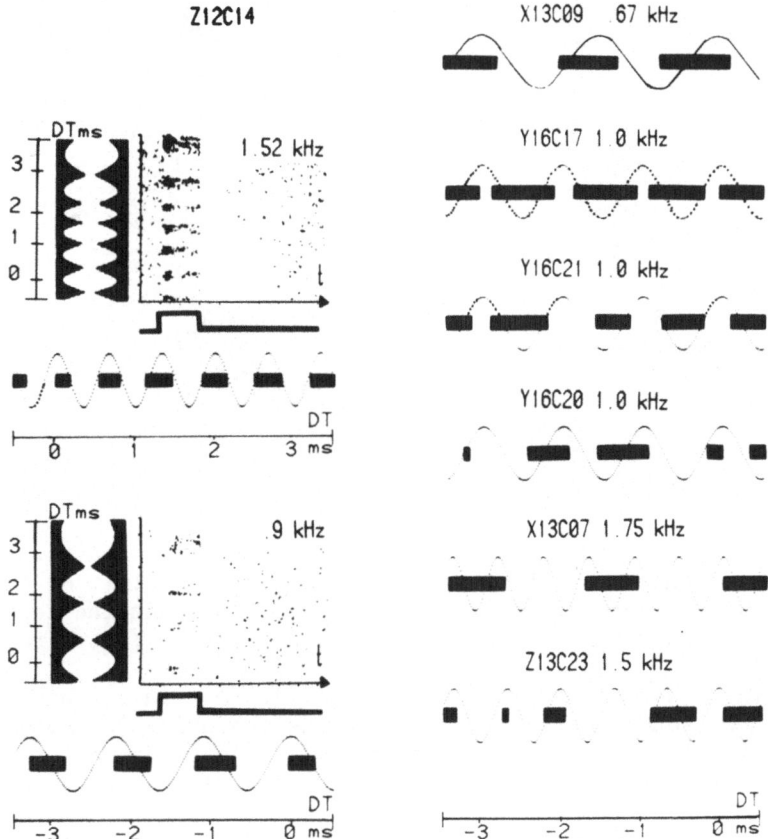

Fig. 1. Right: schematic representation of the response of 6 dif-
         ferent units at the indicated frequencies to which the
         function COS (2πf*DT) is superposed. Left: dot displays
         with below them indicated the stimulus duration (200 ms)
         and the schematic representation of the responses of unit
         Z 12 C 14 for the two indicated frequencies. The modulated
         signal is the envelope of the sum of the signal at the two
         ears.
         DT: Interaural delay.

larity of the two grafics suggests that the two boundaries play an
equivalent role. 21 units present a transition around zero delay.

    A map of auditory space built with narrow channels, of which
a great number is necessary to allow for the fine spatial resolution, responding in a sector of space would require that most of
the units studied present some sensitivity to DT. This is not the
case and we found no units responding in a narrow enough range to
code by themself for a precise orientation.

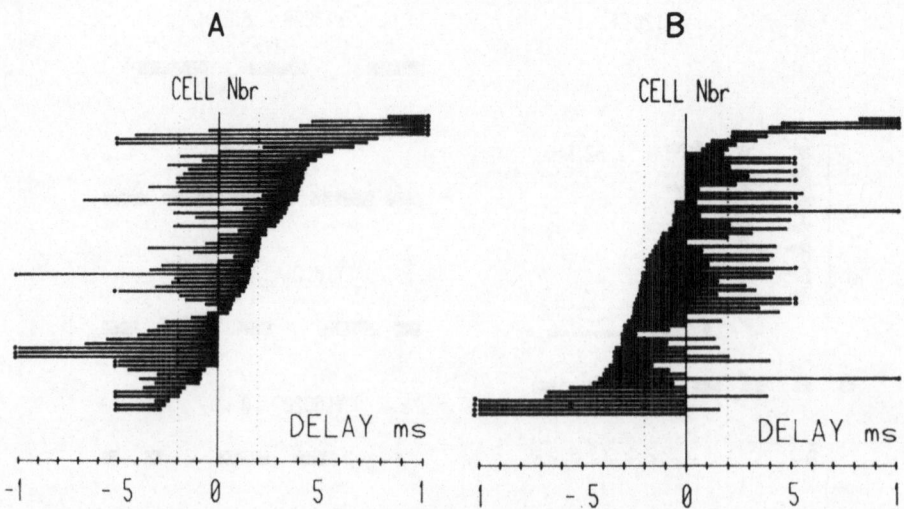

Fig. 2.   Response to interaurally delayed noise of 70 units. They
          are ordered by the right (A) and left (B) limit of the
          strong part of their response. The black dot indicates that
          the unit still responded at the end of the tested range.

On the other hand one could hypothesize that an orientation can
be obtained even with the type of response the units in the MGB
present. Indeed this could be obtained with the combined activity
of units responding over distinct wide ranges of interaural delays.

Since phase detectors have response ranges to DT changing with
the frequency, they are not good candidates for mapping space. Being
relatively frequency selective they could participate in a mechanism
of sound source segregation. The comparatively large delays associated to this phase sensitivity (up to 2 ms) could reflect the
particular case of echo processing.

Irregularities in the periodicity of the response to interaural
delay could be due to the modulation of the neural delay by other
centers.

ACKNOWLEDGMENT

This work is supported by the Swiss National Science Foundation,
Grant no 3.329.69.

Altman, Ya. A., 1968, Are there neurons detecting directional sound
    source motion? Exptl. Neurol., 22: 13-25.
Brugge, J. F., Dubrovsky, N. A., Aitkin, L. M., and Anderson, D. J.,
    1969, Sensitivity of single neurons in auditory cortex of cat
    to binaural tonal stimulation; effects of varying interaural
    time and intensity, J. Neurophysiol., 32: 386-401.
Goldberg, J. M. and Brown, P. B., 1969, The response of binaural
    neurons of the dog superior olivary complex to dichotic tonal
    stimuli: some physiological mechanisms of sound localization,
    J. Neurophysiol., 32: 613-636.
Rouiller, E., de Ribaupierre, Y., and de Ribaupierre, F., 1979,
    Phase locked response to low frequency tones in the medial
    geniculate body, Hearing Res. 1: 213-226.

TEMPORAL INFORMATION IN THE MEDIAL GENICULATE BODY

E. Rouiller, Y. de Ribaupierre, A. Toros and
F. de Ribaupierre

Institute of Physiology
University of Lausanne
Bugnon 7, 1011 Lausanne, Switzerland

If some temporal information of acoustic signals is to be used
and processed in the auditory cortex, it should be transfered
through the medial geniculate body (MGB) with sufficient precision.

Response characteristics to repetitive clicks were analysed for
418 single units recorded in the MGB of 35 nitrous oxide anaesthetized cats. The methological procedures have been described elsewhere (Rouiller et al. 1979). Trains of repetitive clicks of frequency going from 10 to 2000 Hz, had a duration of 500 ms, and were
delivered once per second. Generally, the stimulus was presented
simultaneously in both ears, except when a monaural stimulation was
more efficient, at an intensity of 34 to 44 dB SPL. The duration of
the individual clicks was 50 microseconds.

The responses of MGB units to trains of clicks can be divided
in the same categories defined by de Ribaupierre et al. (1972) in
the primary auditory cortex. The majority of MGB units were classified as "locker". This type of unit shows discharges precisely
time-locked to the individual clicks of the trains, as illustrated
in Fig. 1. This synchronization, which is obvious on the dot display
at 25 or 50 Hz, is still present at 100 Hz, as shown by the interspike interval histogram; the intervals between successive spikes
during the response are distributed in peaks corresponding to inte-
gral multiples of the stimulus period. It appears from the period
histograms that the discharges are localized in a restricted portion
of the stimulus period. By this approach, an upper limiting rate of
locking can be defined for each "locker", which, in the MGB, varies
between 10 and 800 Hz, but is rarely higher than 300 Hz. The precision of locking can be estimated by the standard deviation (SD)

251

measured on the spikes distribution in the period histogram (assimilated to a normal circular distribution). For the unit of Fig. 1,
the SD is equal to 4.6 ms, 1.6 ms and 0.7 ms at respectively 25, 50
and 100 Hz. The SD was measured in this way for 85 MGB "lockers" to
different rates of repetitive clicks. The extreme values and the mean
of the SD observed are plotted in Fig. 2A as a function of the
corresponding click rate. The SD's variability is greatest for low
click rates, and the average SD progressivelly decreases for increasing rates, showing a significant improvement of the precision
of locking with frequency. To compute the period histograms shown
in Fig. 1, the time was reset to zero at each click, so that these
spikes distributions can be assimilated to post-click histograms.
The mean of the discharges distribution is thus an estimation of
the mean response latency. Most "lockers" (70 %) show a decrease of
that mean latency for rates going from 25 to 100 Hz; in the pars
lateralis, the mean response latency averaged for 33 cells is 10.9
ms at 25 Hz, 9.4 ms at 50 Hz and 7.8 ms at 100 Hz. On the other
hand, if one considers the mode of the period histogram instead of
the mean of the discharges as a measure of the latency, it is generally independent of the click rate; this strongly suggests that
the decrease of the mean response latency for click rates going from
25 to 100 Hz is mainly due to the increase of the precision of
locking demonstrated in Fig. 2A.

In a previous report (Rouiller et al., 1979), we have demonstrated that phase -locked activity is present in a small population
of MGB neurons, giving excitatory sustained responses to tone
bursts. A significant synchronization of the discharges with the
phase of the tonal stimulus was found in response to low-frequency
tones up to about 1000 Hz. A simple model taking into account the
jitter of the successive synapses along the auditory pathways was
proposed to describe the progressive decrease of this upper frequency limit of synchronization observed as one moves up in the auditory
pathways. The synaptic transmission delay is characterized by its
mean value D and its standard deviation J or jitter. For N synapses,
the total delay is

$$D = \sum_{i=1}^{N} D_i \quad \text{and the total jitter } J = \sqrt{\sum_{i=1}^{N} J_i^2}$$

The synchronization of the discharges with the phase of the tonal
stimulus was appreciated in period histograms, and quantitatively
estimated by the computation of a degree of synchronization R, as
proposed by Goldberg and Brown (1968). The degree of synchronization R and the jitter J are linked by the following relation
proposed by Anderson (1973):

$$R = \pi/4 \{ \exp [- (2 \pi . F . J)^2/2]\}$$

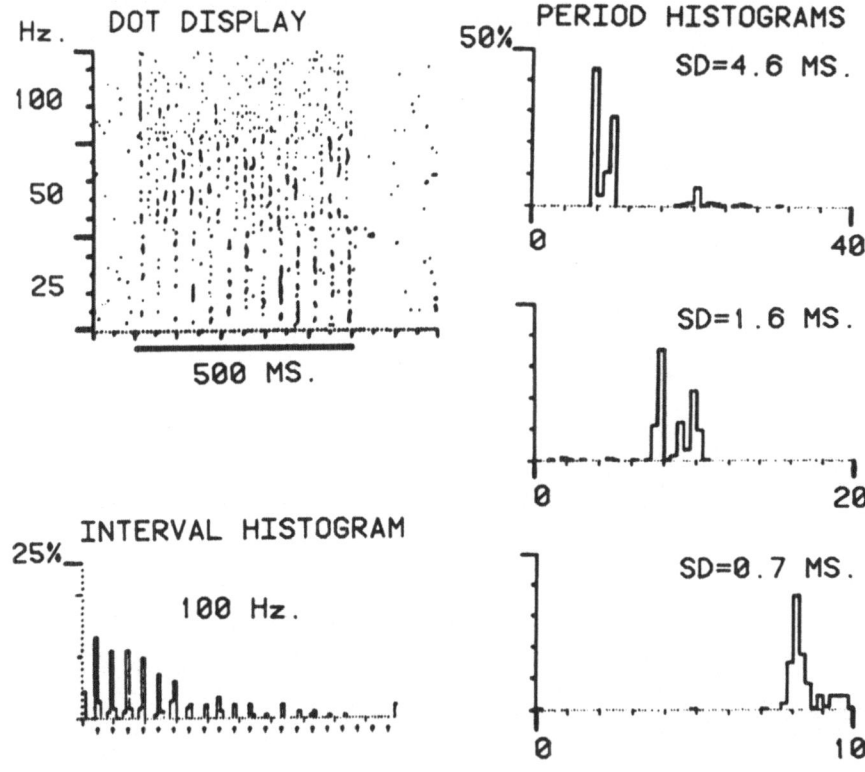

Fig. 1.   Upper left: dot display showing the typical response of
          a "locker" to repetitive clicks. Ordinate: number of
          stímuli delivered (25 at each indicated click rate).
          Abscissa: time, the duration of the stimulus (500 ms) is
          indicated by the horizontal bar. Lower left: interspike
          interval histogram computed during 25 presentations of
          100 Hz click train. Ordinate: relative number of intervals
          in % in each bin. Abscissa: intervals duration, the arrows
          indicate the integral multiples of the stimulus period.
          Bin size = 2 ms. Right column: period histograms. Firing
          probability within a period of the stimulus. The length
          of the abscissa is equal to the period (40, 20, 10 ms for
          respectively 25, 50, 100 Hz). Ordinate: relative number
          of spikes in each bin. Binwidth = 1/50 of the period. SD
          = standard deviation of the spikes distribution (see text).
          Unit 49-28-1.

This allows to compute the total synaptic jitter J from the degree
of synchronization R of the discharges evoked by a pure tone of
frequency F. The total synaptic jitter up to the MGB is plotted in
Fig. 2B against the frequency of the sinusoidal stimulus for a
population of 16 MGB phase-locked cells. The synaptic jitter is

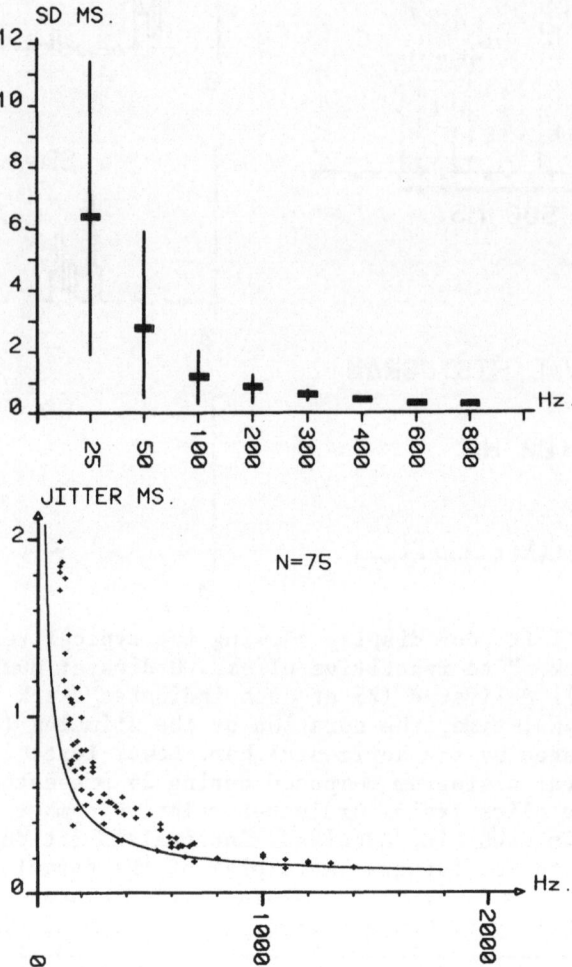

Fig. 2. A: standard deviation of the spikes distribution (SD) as
a function of the rate of repetitive clicks. For each click
rate, the average SD is represented by the black rectangle,
and the extremities of the vertical lines are the maximal
and minimal SD observed. The number of values (cont′d)

(Fig. 2 cont'd) measured was 62, 68, 51, 29, 7, 5, 1, 1 at respectively 25, 50, 100, 200, 300, 400, 600, 800 Hz. B: The jitter of the synaptic transmission delay up to the MGB is plotted against the frequency of the tonal stimulus for 16 phase-locked MGB units. The synaptic jitter is computed from period histograms to sinusoidal stimuli, as explained in the text. N = number of frequencies tested.

a function of the tonal stimulus frequency: it is close to 2 ms near 100 Hz and progressively decreases to reach a constant value of 0.2 ms for frequencies above 700 Hz. For phase-locked cells, an overall transmission delay can be measured from the phase shift observed on period histograms obtained at different frequencies (de Ribaupierre et al. 1980). If the cochlear delay is subtracted from this overall delay, a neural delay from the VIIIth nerve to the MGB is obtained and appears to be significantly greater by 2 to 3 ms for neurons having a CF below 300 Hz. This longer delay can be due to anatomical reasons (larger number of synapses and/or slower conduction velocity pathway for very low frequencies) or to functional properties: the synaptic facilitation could become optimal for spike intervals obtained at frequencies above 300 Hz. This second explanation would also be consistant with the decrease of the synaptic jitter observed for higher frequencies, as well as the decrease of the synchronization variability to repetitive clicks. A minimal value observed for these two variables is close to 0.2 ms. This value is compatible with a precise conservation of temporal information up to frequencies of 1000 Hz, which is sufficient to deal with the fundamental frequency of most species vocalizations and their fine temporal structure, including human speech.

ACKNOWLEDGMENT

This work was supported by the Swiss National Fund for Scientific Research, Grant no 3.239.69.

REFERENCES

Anderson, D. J., 1973, Quantitative model for the effects of stimulus frequency upon synchronization of auditory nerve discharges, J. Acoust. Soc. Am., 54: 361-364.

Goldberg, J. M. and Brown, P. B., 1969, Response of binaural neurons of dog superior olivary complex to dichotic tonal stimuli: some physiological mechanisms of sound localization. J. Neurophysiol. 32: 613-636.

Ribaupierre F. de, Goldstein M. H. Jr and Yeni-Komshian G., 1972, Cortical coding of repetitive acoustic pulses, Brain Res., 48: 205-225.

Ribaupierre F. de, Rouiller E., Toros A. and Ribaupierre Y. de,

1980, Transmission delay of phase-locked cells in the medial
geniculate body, Hearing Res. (in press).
Rouiller E., Ribaupierre F. de and Ribaupierre Y. de, 1979, Phase-
locked responses to low frequency tones in the medial geniculate
body, Hearing Res., 1: 213-226.

SOME INVESTIGATIONS OF ACOUSTICAL EVOKED POTENTIALS FROM PERIPHERAL AND CENTRAL STRUCTURES OF THE AUDITORY PATHWAY IN RABBITS

M. Biedermann, E. Emmerich and H. Kaschowitz

Institute of Physiology
Friedrich-Schiller University
Jena, G.D.R.

For our experiments with sound-injured animals we wanted to examine simultaneously responses of peripheral and central structures of the auditory system. Some further conditions should be realized: Measurements should be possible and comparable over a longer period and the animal should be awake. Here we want to report some preliminary results of cochlear microphonics (CM), brain stem potentials, and acoustically evoked responses of the cortex in normal rabbits.

A silver-silver chloride electrode was implanted into the cavity of the flocculus cerebelli in its frontal bony wall with the help of local anaesthesia. With this position of the electrode CM and brain stem responses could be received. Furthermore electrodes were fixed on the surface of the acoustic cortex. In chronic experiments only the margin could be reached because the acoustical cortex of the rabitt is covered by the great mandibular joint. All electrodes were attached to the skull with the help of "Duracryl". During the experiments the rabbits were in a sound isolated chamber with their legs fixed on a board. After some sittings to habituate the rabbits to the experimental situation, a relaxed awake condition was reached. The animals were binaurally stimulated in a free soundfield by clicks (a half sinus wave 1.6 kHz) produced by a special generator (home production). CM and brain stem potentials were amplified by an audio frequency spectrometer (Brüel and Kjaer) and averaged by a NTA 512B averager. The cortical responses were amplified by an EEG-amplifier (Zwönitz) and were averaged, too, by the NTA. The EEG of the rabbits were watched and evaluated simultaneously.

As shown in Fig. 1(a) there is a linear dependency between the intensity of clicks and the amplitude of CM, as might be expected.

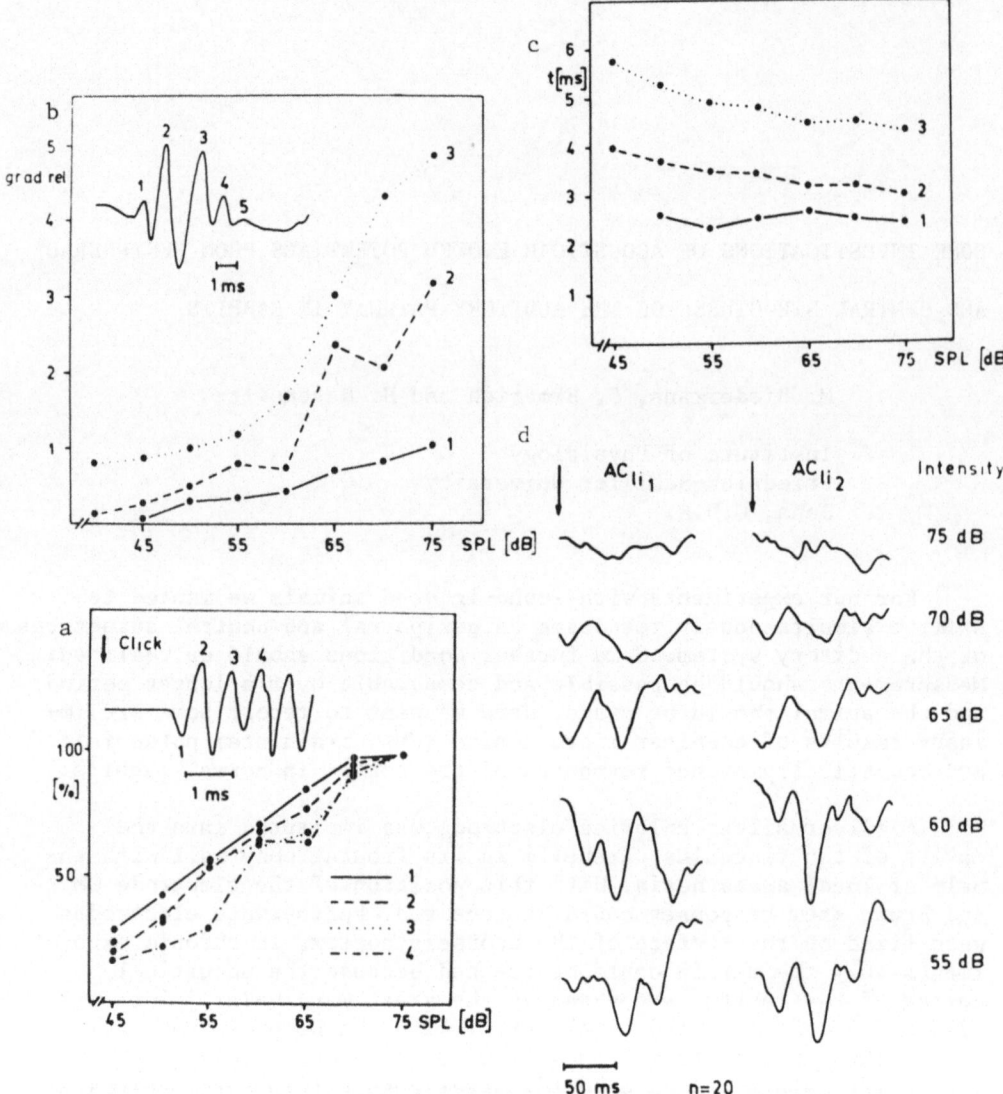

Fig. 1.   (a) Cochlear microphonics (waves 1–4) as a function of
          click intensity. The amplitudes of the largest potentials
          (click intensity 75 dB SPL) are set equal 100%. Inset: click
          response. n = 100. (b): Amplitudes of brain stem potentials
          (waves 1–3) as a function of click intensity. The CM were
          eliminated by changing the click polarity. Ordinates:
          Amplitudes in a relative graduation. Inset: A typical brain
          stem response of a rabbit. n = 100 pos. and 100 neg. clicks.
          (c): Latencies of brain stem potentials (waves 1–3) as a
          function of stimulus intensity. There is no shift in wave
          1(the summating action potential of acoustic (cont'd)

(Fig. 1 cont'd) nerve) but a considerable shift in wave 2
and 3. (d) Cortical evoked potentials from two points of
the acoustical cortex as a function of click intensity. Most
distinct response at 60 dB SPL. Thresholds for the cortical
evoked potentials were at 40-45 dB SPL. n = 20. Click in
(a) - (d) : A half sinus wave (1.6 kHz).

The waves of the brain stem potentials obtained from the rabbit
(to be seen in Fig. 1(b) as an inset) are similar to those of other
mammalian species. The amplitudes become larger with rising stimulus
intensity too, visible in the input-output curves. Wave 1 is the
summating action potential of the acoustical nerve and the linear
ascent is particularly distinct. The latencies shown in Fig. 1(c)
behave in a different way. In wave 1 the latency is constant for
all intensities whereas the latencies become shorter with rising
click intensities for the later waves. Under constant experimental
conditions all potentials are consistently obtainable from the same
rabbit for months. Click-evoked responses from two points on the
acoustic cortex are shown in Fig. 1(d). In comparison with other
species it is relatively difficult to record responses from the
acoustical cortex of the awake rabbit. The most important require-
ment is a good habituation of the animal to the experimental
situation. With increasing click intensity above the threshold, in
our experiments at 40 - 45 dB SPL (not shown in Fig. 1(d)), a quick
increase of the cortical responses in time range of 40 - 60 ms is
to be seen. Above a value of 60 dB SPL, at which the maximal cortical
response occurs, the amplitudes begin to decrease again and even at
a click intensity of 75 dB SPL a cortical evoked potential could
hardly been perceived. The latencies at 60 dB SPL possibly seem to
be at a minimum. Presumably clicks of a low intensity may have a
high level of information for rabbits.

It may be that more precise quantitative studies of potential
parameters will give further insight into some aspects dealing with
processing of information in the rabbits auditory pathways, since
there only exists a small knowledge of the auditory system of
rabbits.

SESSION VI
AUDITORY LOCALIZATION
Chairmen: L. Aitkin and J. Syka

ANATOMICAL-BEHAVIORAL ANALYSES OF HINDBRAIN SOUND LOCALIZATION

MECHANISMS

R. B. Masterton, K. K. Glendenning and R. J. Nudo

Psychology Department
Florida State University
Tallahassee, Florida 32306, U.S.A.

The representation of visual and somesthetic sensory hemifields on the contralateral side of the nervous system is an axiom of neurology. In sharp contrast to these rather strict contralateral representations, however, it is usually held that the representation of space in the central auditory system is bilateral. That is, stimulation of one ear alone, or stimulation by a sound source located in one hemifield, evokes activity on both sides of the brain. The bilaterality of this evoked activity is easily deducible anatomically and clearly demonstrable electrophysiologically (e.g., Held, 1893; Stotler, 1953; Rosenzweig, 1954; and see review by Brugge and Geisler, 1978). Recently, it has also become possible to visualize its distribution by means of the radioactively-labeled 2-deoxyglucose (2-DG) technique for marking neural tissue of high metabolic activity (Sokoloff et al., 1977).

Fig. 1 shows (14_c)-2-DG autoradiographs of the auditory mid-brain of two 6-week old kittens, each exposed to pulsing white noise, one after the right ear was destroyed, the other from a sound source 60° to the left of its midline. At the midbrain and still higher levels of the mammalian auditory system, the metabolic activity evoked by the stimulation is greater on the side contra-lateral to the sound source (i.e., on the right side in Fig. 1). However, the auditory structures ipsilateral to the sound source (i.e., on the left side in Fig. 1) also show heightened metabolic activity and its presence there constitutes evidence of the pre-sumed anomaly of the auditory system which is the topic of this review.

The clarity and ease of demonstrations of bilateral activity evoked either by monaural sound or by a lateral sound source have

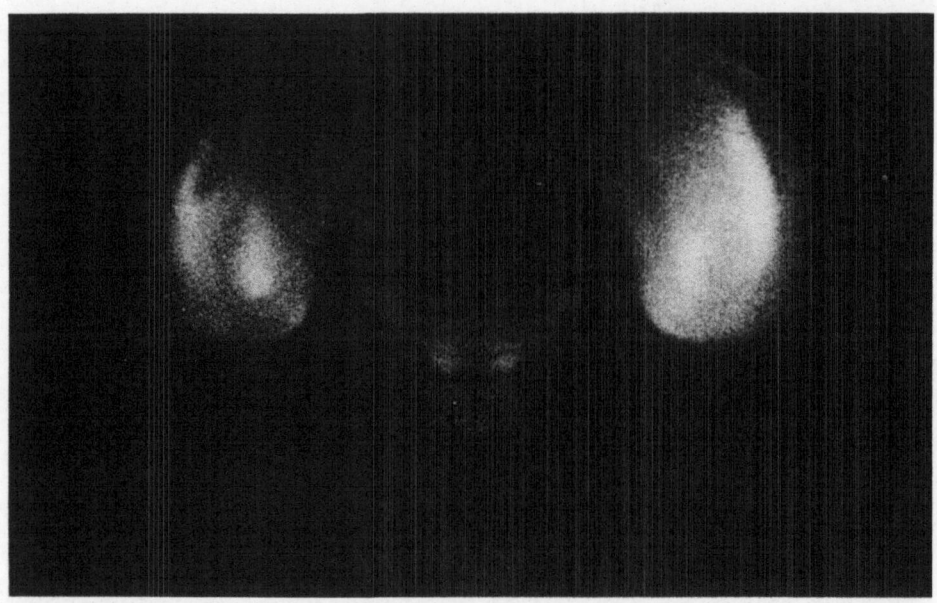

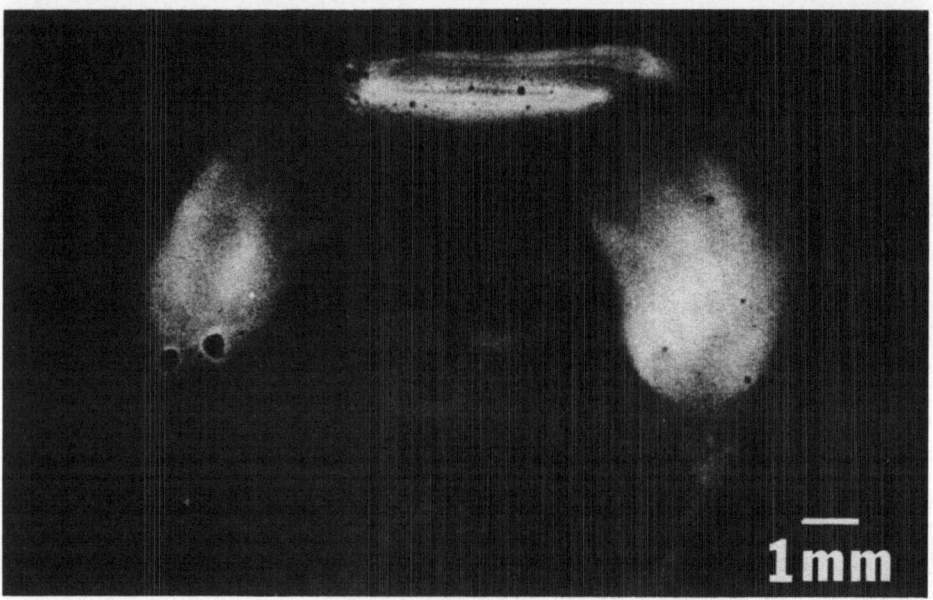

Fig. 1. (14$_C$)-2-Deoxyglucose autoradiographs through inferior colliculi of two 6-week old kittens stimulated with pulsing white
noise. Upper: after right ear was destroyed; lower: with
sound source 60° to left of midline. Note that activity
level is higher on contralateral (right) side but that
ipsilateral (left) side also shows heightened activity.

encouraged us to presume that the converse of this fact is also true—that both sides of the auditory system are involved in the neural representation of the direction of a sound source (e.g., see Rosenzweig, 1954; Starr, 1974; Benson and Teas, 1976). However, despite these demonstrations, and despite the wide acceptance of the converse idea that these demonstrations are tantamount to demonstrating a bilaterality of the neural representation of the direction of a sound source, direct ablation-behavior data have now begun to accumulate that suggest that the central representation of each hemifield of auditory space—that is, a representation sufficient for the accurate localization of a sound in the horizontal plane—is wholly contained on only one side (the contralateral side) of all but the very lowest levels of the auditory system.

The first suggestion that this might be the case came from the studies of Neff and his colleagues and students (e.g., Moore et al., 1974; Casseday and Neff, 1975). They showed that among the several central commissures which the auditory system might use to compare the activity of left and right sides (for example, corpus callosum, commissure of inferior colliculus, commissure of Probst, etc.) only the trapezoid body, the lowest auditory commissure, is indispensable for the ability to localize the source of a sound. Other commissures can be sectioned without measurable deficit. Although the significance of this discovery seems to have remained unappreciated, it now can be seen that it is exactly what would be expected if whatever comparisons between left and right sides that might be necessary for sound localization were accomplished entirely in the hindbrain, no further nor higher-level comparison between left and right being necessary.

A second suggestion that the two halves of auditory space might each be represented contralaterally came from the work of Strominger (1969), and Strominger and Oesterreich (1970), who showed that cats with unilateral lesions of auditory cortex or of the brachium of the inferior colliculus suffered a deficit in localizing the source of a sound in the contralateral hemifield. Of course, this is exactly the result one would expect if the representation of the auditory hemifields in the forebrain were separate and contralateral. However, the behavioral deficits seemed to be either relatively minor elevations in threshold or they disappeared after postoperative training. Therefore, the significance of this discovery also remained largely unappreciated—the deficits were usually explained as the secondary result of a passive amnesia, learning deficit, or more general discumfiture rather than a true sensory or perceptual loss.

Still a third suggestion that the auditory hemifields might be represented contralaterally arose from examination of human patients with unilateral lesions of cerebrum or brainstem (Wortis and Pfeffer, 1948; Sanchez-Longo et al., 1957, 1958). In particular, Sanchez-Longo et al. (1957) showed that in five carefully studied patients with temporal lobe lesions, each revealed an impairment in sound localization in their contralateral auditory field. Thus, evidence both from clinical neurology and from direct ablation-behavior experimentation converged toward the same conclusion--after a unilateral lesion, something changed in the contralateral auditory field. This change was apparently restricted to the contralateral hemifield but it was not clear whether the change was permanent or temporary or whether it should be characterized as essentially sensory or perceptual in nature or as the secondary result of an ill-defined dementia.

Though these early experiments clearly suggested that each auditory hemifield might be represented contralaterally at levels above the hindbrain, none of the reports nor their combination proved to be sufficient to challenge the less direct but more easily verifiable evidence of apparent bilaterality provided by electro-physiological recordings. In either the experimental or clinical reports, the lack of clearly permanent or profound behavioral deficits following unilateral lesions allowed other interpretations that were more easily reconciled with the obvious bilaterality of monaural evoked activity. For these reasons, the idea that the auditory system was an exception to the general rule of contralateral representation remained without serious challenge until very recently.

Deficits in Reflexive Orientation to Sounds in Contralateral Hemifield

The first clear demonstrations of profound and possibly permanent contralateral deficits resulting from unilateral lesions in the auditory system were provided by Thompson (see Thompson and Masterton, 1978). In these experiments cats with unilateral lesions of either the lateral lemniscus, brachium of the inferior colliculus, or auditory cortex were tested for their ability to make reflexive orientations of their head in the direction of an unexpected sound. Fig. 2 shows the outcome of such an experiment for a cat with a section of the right lateral lemniscus. As can be seen in the figure, the cat suffered a profound deficit in the accuracy with which it oriented to a sound emanating from the left hemifield--that is, the hemifield contralateral to the lesion. At the same time, it suffered little or no measurable deficit in the accuracy of the response to a sound in the right hemifield--the one ipsilateral to the lesion.

This same result was obtained for cases with unilateral lesions at either the level of the brachium of the inferior colliculus or auditory cortex. Although the deficits following lesions at these higher levels were different in nature from those seen in lemniscal cases (for example, bilateral deficits in the latency of the response were sometimes seen), a profound deficit in the accuracy of the reflexive response was confined to the sound field contralateral

to the lesion while the accuracy in responding to the sound field
ipsilateral to the lesion remained within normal limits (Thompson
and Masterton, 1978).

The consistency of these contralateral deficits regardless of
the level of the lesion, from lateral lemniscus to auditory cortex,
reemphasizes the importance of the prior discovery that only the
lowest level auditory commissure (the trapezoid body) is necessary
for the localization of sounds (Moore et al., 1974; Casseday and

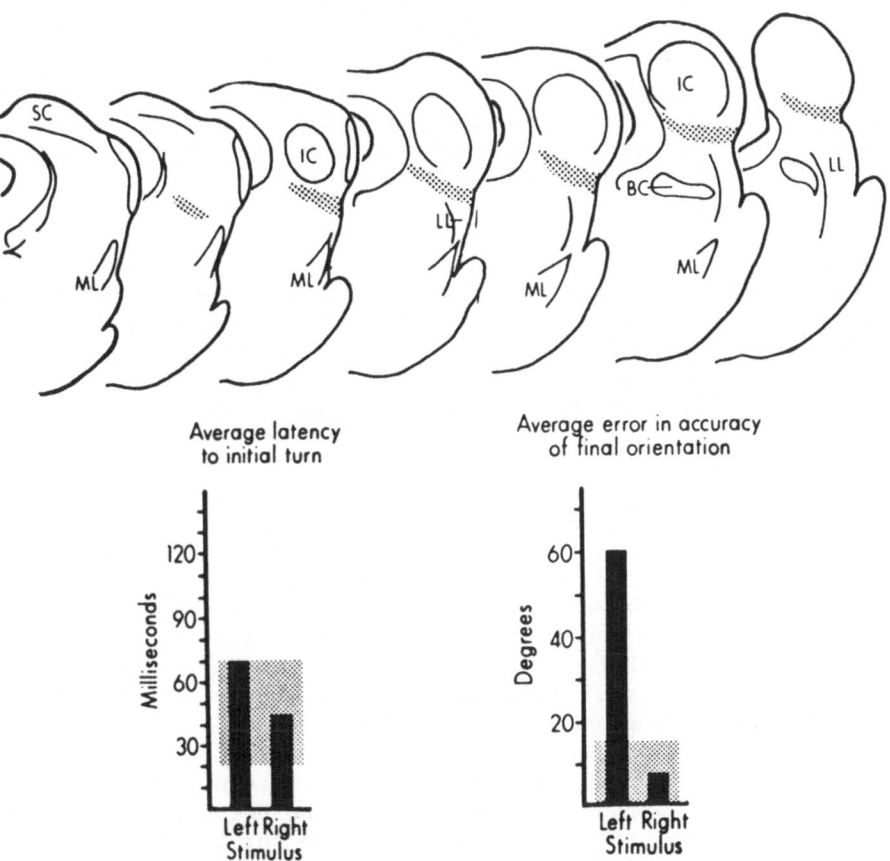

Fig. 2. Reconstruction and behavioral results in a case with section of the right lateral lemniscus. Histograms allow comparison of latency and accuracy of responses to left or to
right with normal cats (shaded areas). Note that deficit
in accuracy of reflexive orientation to an unexpected sound
is confined to contralateral (left) hemifield following
section of the right lateral lemniscus.

Neff, 1975). For if bilateral comparison of activity at higher
levels were indeed necessary, unilateral lesions would result in
bilateral, not unilateral, deficits.

Deficits in Instrumental Responses to Sounds in Contralateral Hemi-field

On the basis of these results, (Jenkins 1980; Jenkins and
Masterton, 1979a) undertook a series of experiments designed to
test explicitly the possibility of a strict contralateral representation of the auditory hemifields. In these experiments cats with
unilateral lesions were trained to discriminate the azimuth of
sounds emanating from any of seven directions from -90° (opposite
the left ear) through 0° (opposite the nose) to +90° (opposite the
right ear). Instead of relying on reflexive orienting responses,
however, these cats were given extensive training, testing, remedial
training, and retesting for true instrumental sound-localization.
That is, they were put in a position where they had to localize
accurately in order to obtain their daily ration of water. The
training and testing procedures were designed to bring out any
residual ability for sound localization that an animal might possess
and thus, allow an assessment of the permanency and profundity of
whatever deficit they might have.

Fig. 3 illustrates the training and testing apparatus that was
used: a semicircular enclosure with seven carefully matched speakers
arranged along its circumference. At the center of the semicircle
was the "starting point" and above this point, looking down on the
cat, was a television camera by which the alignment of the cat's
head at the start of a trial could be observed and videotaped. When
the cat aligned itself at the start point with its head pointed
towards the center speaker (i.e., 0°), a single, brief click was
triggered from one of the seven speakers. From trial to trial, this
sound was randomized both in direction among the seven speakers
and in intensity (in 5-dB steps over a 20-dB range). If the thirsty
cat proceeded to the speaker which had produced the click, it was
rewarded with a measured amount of water. If it responded to any
other speaker, it went unrewarded and had to return to the starting
point and trigger another trial to receive another opportunity for
reward. In order to provide remedial training for the directions
where it was most needed, the locus of the sound source was not
changed after an incorrect response--the same locus was repeated
again and again until a correct response was eventually made.
Although these repeated trials did not enter into the scoring of
the cat's performance, this procedure allowed the cat to receive
hundreds of trials distributed over months of daily remedial training at exactly the directions where it might be suffering a deficit.

Cats with unilateral lesions either of the lateral lemniscus, inferior colliculus, brachium of the inferior colliculus, medial geniculate, or auditory cortex were each tested in this way. The results of a case with a unilateral section of the left lateral lemniscus are shown in Fig. 4. It can be seen that the cat suffered a profound deficit that was confined to the right (that is, the contralateral) hemifield, while localization of sounds in the left (or ipsilateral) hemifield was unimpaired. The deficit in this case continued without appreciable change for over 9 months and several hundreds of remedial training trials--until it eventually became necessary to sacrifice it for histological verification of the lesion. It should be noticed that this case, with a lesion quite similar to the case shown in Fig. 2, also had a closely similar deficit. That is, despite the differing nature of the required responses--a reflexive response in the first case, a conditioned instrumental response in the second, the deficits were the same-- gross inaccuracies in sound localization confined to the hemifield contralateral to the lesion. This similarity in the deficits despite the differences in the response requirements suggests that it was

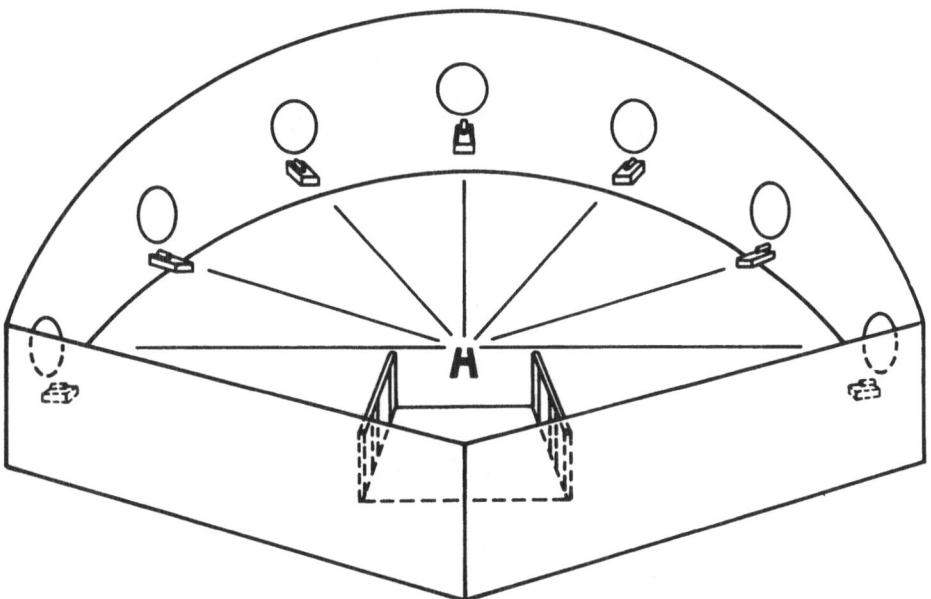

Fig. 3.   Seven-choice test apparatus for sound localization ability. The cat triggered a single brief click of random intensity from one of the seven speakers by aligning head over "H" on floor with its nose pointing toward centerspeaker. Water reward for correct responses was delivered through troughs below speakers.

the common feature, the mechanism for sound localization, that was disrupted by the lesions. Further, and more to the present topic, it seems obvious that whatever the mechanism disrupted by the lesions, it is somehow tuned to sounds originating in the contralateral hemifield.

With other cases, it was shown that unilateral lesions either of the inferior colliculus, the brachium of the inferior colliculus, the medial geniculate, or auditory cortex each resulted in a deficit with one common element: if the lesion was complete enough to result in a measurable deficit, the deficit--regardless of its nature--was found in the sound field contralateral to the lesion. At the same time, responses to sounds originating in the sound field ipsilateral to the lesion were essentially unaffected.

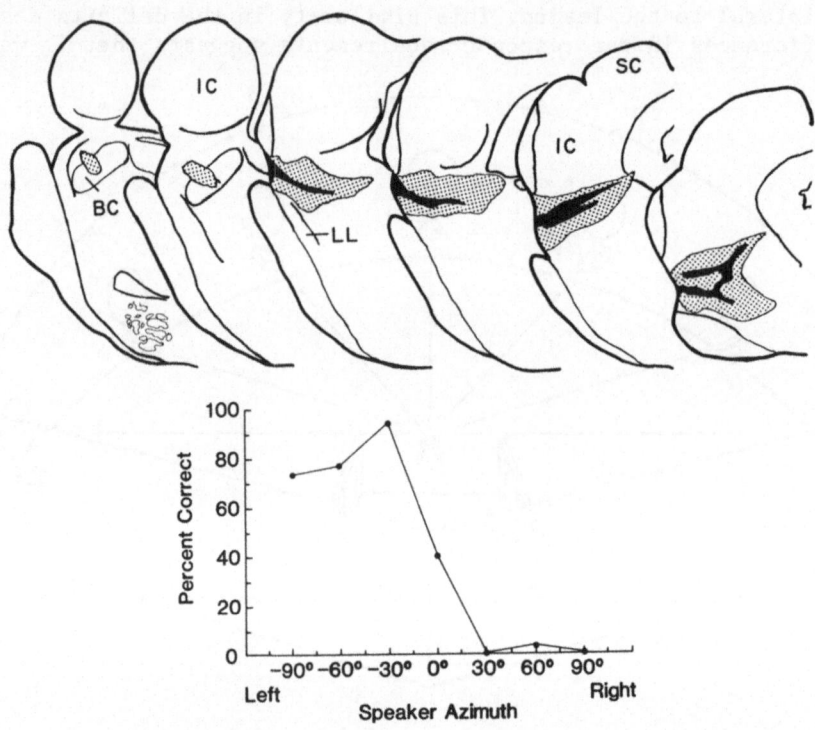

Fig. 4. Reconstruction of lesion and behavioral results in a case with section of left lateral lemniscus. Graph shows percentage correct in each of seven directions--three to the left and three to the right of midline. Note that deficit in localization was confined to right hemifield in the field contralateral to the (left) lesion.

Differences in Laterality of Sound Localization Deficits Resulting from Lesions above or below the Superior Olivary Complex

Having shown that unilateral lesions anywhere along the auditory pathway from lateral lemniscus to cerebral cortex resulted in contralateral deficits, Jenkins (1980; Jenkins and Masterton, 1979a) also showed that lesions at the lowest levels of the auditory system result in entirely different deficits. Cats either with one ear destroyed or with lesions in the trapezoid body-superior olivary complex did not have contralateral deficits in sound localization ability. Instead, they suffered deficits that were either ipsilateral or bilateral. These results would seem to indicate that a marked rearrangement or transformation of neural activity takes place in the hindbrain somewhere between the trapezoid body and the lateral lemniscus--almost certainly in the superior olivary complex.

Fig. 5 summarizes the sound localization abilities of the 15 cases with lesions at one or another level of the auditory system just described and allows their comparison with the range of performance of 5 normal cats.

To begin with, since the data used for the histograms were obtained over several months of daily postoperative testing, remedial training and retesting, they show that almost every case had a long-lasting, probably permanent deficit.

Proceeding through the upper histogram from left to right, the first three cases illustrate that lesions at or below the level of the superior olivary complex result in bilateral or ipsilateral deficits in sound localization. These results are essentially similar to those obtained by Casseday (see p. this volume) in tree shrews.

The remaining 12 cases in the upper histogram show that any lesion above the superior olivary complex large enough to produce a measurable deficit produced it in the contralateral hemifield. Since in each of these 12 cases the localization of sounds in the field ipsilateral to the lesion remained in the normal range, it can be concluded that one side of the auditory system alone is sufficient to maintain normal sound localization ability in the contralateral sound field.

As in the previous experiments using reflexive responses to sound, the nature of these sound localization deficits varied depending on the size and level of the lesion. These differences in the deficits will be discussed elsewhere. The important point for the present purpose is that in no case with a lesion above the level of the superior olives did a significant deficit occur in the ipsilateral sound field.

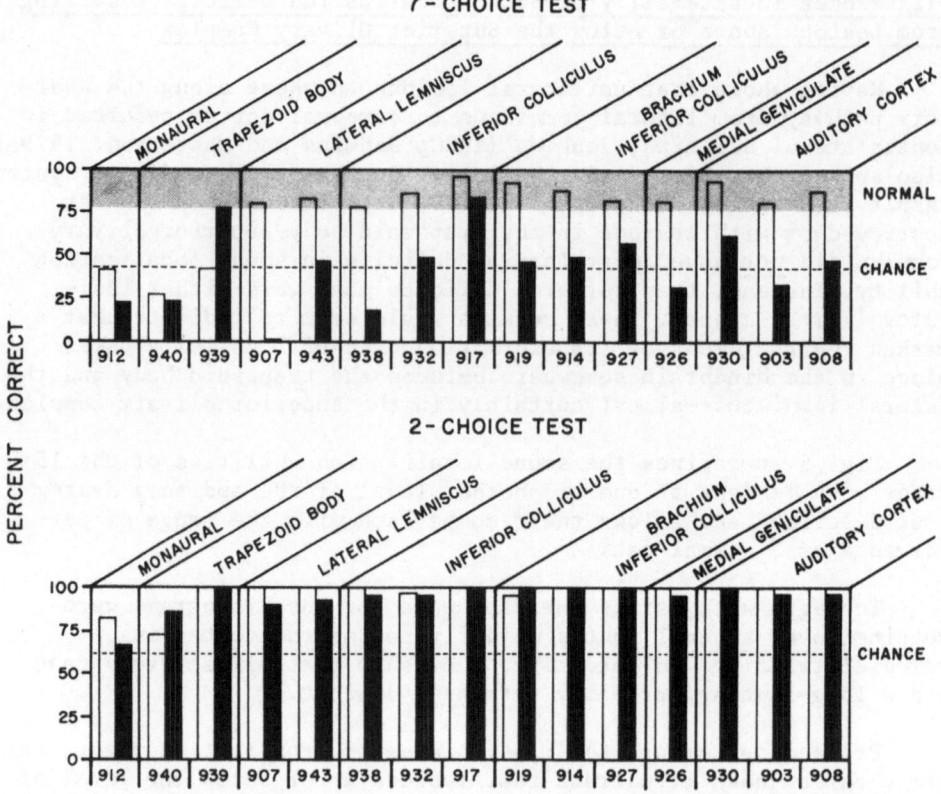

Fig. 5. Summary of 15 cases tested for sound localization ability
in each of two ways. Top histogram shows performance in
the seven-choice test apparatus. Bottom histogram shows
performance of same cases after apparatus was reduced to
only two choices--one to left and one to right of midline.
In both histograms, the range of normal performance is
indicated by shading. Open bars: performance to hemifield
ipsilateral to lesion; filled bars: performance in hemifield contralateral to lesion. Note that for cases with
lesions above superior olivary complex in upper histogram,
open bars fall within range of normal performance. This
result shows that one side alone is sufficient to maintain
normal localization ability in the contralateral hemifield.
The lower histogram shows performance of the same animals
in the same apparatus but with only two speakers. This
less-refined procedure is not usually sensitive to unilateral deficits.

It is important to note that because with extensive remedial training a remnant of localizing ability often could be elicited in the hemifield contralateral to a lesion, it can not be concluded that each side of the central auditory system is necessary for sound localization in the contralateral hemifield—only that it is sufficient. In short: if there is a unilateral lesion there may not be an obvious deficit; but if there is a unilateral deficit, the lesion causing it is contralateral.

Given the consistency in the outcome of these experiments, it is only natural that the question arises as to why the contralateral representation of the auditory hemifields was not discovered previously. The probable answer to that question can also be seen in Fig. 5 in the lower histogram. The results depicted there were obtained in the same cases and in the same apparatus as the ones shown in the upper histogram. However, instead of having seven speaker locations and therefore seven possible sound directions, the apparatus was reduced to only two speakers, symmetrically placed one to the left and one to the right of midline.

This simpler version of the apparatus was virtually identical to the apparatus used in almost every previous experiment on sound localization (e.g., see reviews by Neff et al., 1975; or Masterton and Diamond, 1973; Heffner and Masterton, 1975; Jenkins and Masterton, 1979b; Masterton et al., 1975). As the lower histogram in Fig. 5 shows, not one of the animals with unilateral lesions in the upper levels of the auditory system revealed any permanent deficit whatever. That is, the same animals that suffered markedly on the seven-speaker test performed the two-speaker test within normal limits admitting, at most, only a transient difficulty. These results replicate almost exactly the results obtained by previous experimenters using the same apparatus to test animals with the same lesions. Clearly, if we had relied on this less refined version of the sound localization task, the deficits might have been missed entirely.

In summary, several lines of ablation-behavior and clinical evidence indicate that at the level of the lateral lemniscus and anywhere higher, each side of the auditory pathway contains a neural representation of its contralateral hemifield sufficient to support normal sound localization within it. The evidence also indicates that the integration of binaural activity into this unilateral representation of the contralateral sound field begins in the medulla, almost certainly in the trapezoid body-superior olivary complex.

Returning to the more general question of whether or not the auditory system is essentially different than the visual and somatosensory systems in its central representation of sensory hemifields, we would now suggest that despite the large amount of suggesting the contrary, the auditory system is not appreciably dif-

ferent from these other modalities. Furthermore, the evidence seems
to suggest that the trapezoid body-superior olivary complex is
essentially analogous to the optic chiasm for vision or the spino-
bulbar decussations for somesthesis, at least in the sense that
lesions below the respective decussations result in ipsilateral or
bilateral deficits while lesions above them result in contralateral
deficits. In these respects, at least, it now appears that the auditory system is no exception to the general rule of contralateral
representation of sensory hemifields.

REFERENCES

Benson, D. A. and Teas, D. C., 1976, Single unit study of binaural
    interaction in the auditory cortex of the chinchilla, Brain
    Res., 103: 313-338.
Brugge, J. F. and Geisler, C. D., 1978, Auditory mechanisms of the
    lower brainstem, Ann. Rev. Neurosci., 1: 363-394.
Casseday, J. H. and Neff, W. D., 1975, Auditory localization: Role
    of auditory pathways in brainstem of the cat, J. Neurophysiol.,
    38: 842-858.
Heffner, H. E. and Masterton, R. B., 1975, Contribution of auditory
    cortex to sound localization in the monkey (Macaca mulatta),
    J. Neurophysiol., 38: 1340-1358.
Held, H., 1893, Die centrale Gehörleitung, Arch. Anat. Physiol.
    Anat.-Abt., 201-248.
Jenkins, W. M., 1980, "Sound localization: Effects of unilateral
    lesions in the central auditory system," Doct. diss., Florida
    State University.
Jenkins, W. M. and Masterton, R. B., 1979a, Contralateral represen-
    tation of auditory hemifield in upper levels of brainstem
    auditory system, Anat. Rec., 193: 575-576.
Jenkins, W. M. and Masterton, R. B., 1979b, Sound localization in
    pigeon (Columba livia), J. Comp. Physiol. Psychol., 93: 403-
    413.
Masterton, R. B. and Diamond, I. T., 1973, Hearing: Central neural
    mechanisms, in: "Handbook of Perception, Vol. III," Academic
    Press, New York.
Masterton, R. B., Thompson, G. C., Brunso-Bechtold, J. K. and
    RoBards, M. J., 1975, Neuroanatomical basis of binaural phase-
    difference analysis for sound localization: A comparative study,
    J. Comp. Physiol. Psychol., 89: 379-386.
Moore, C. N., Casseday, J. H. and Neff, W. D., 1974, Sound localiza-
    tion: The role of the commissural pathways of the auditory
    system of the cat, Brain Res., 83: 13-26.
Neff, W. D., Diamond, I. T. and Casseday, J. H., 1975, Behavioral
    studies of auditory discrimination: Central nervous system,
    in: "Handbook of Sensory Physiology, Vol. V/2," Springer-
    Verlag, New York.
Rosenzweig, M. R., 1954, Cortical correlates of auditory localiza-

tion and of related perceptual phenomena, J. Comp. Physiol.
Psychol., 47: 269-276.

Sanchez-Longo, L. P. and Forster, F. M., 1958, Clinical significance of impairment of sound localization, Neurol., 8: 119-
125.

Sanchez-Longo, L. P., Forster, F. M. and Auth, T. R., 1957, A clinical test for sound localization and its applications,
Neurol., 7: 655-663.

Sokoloff, L., Reivich, M., Kennedy, C. D., Rosiers, M. H., Patlak,
C. S., Pettigrew, K. D., Sakurada, O. and Shinohara, M.,
1977, The (14_C)deoxyglucose method for the measurement of
local cerebral glucose utilization: Theory, procedure and
normal values in the conscious and anesthetized albino rat,
J. Neurochem., 28: 897-916.

Starr, A., 1974, Neurophysiological mechanisms of sound localization,
Fed. Proc., 33: 1911-1914.

Stotler, W. S., 1953, An experimental study of the cells and connections of the superior olivary complex of the cat, J. Comp.
Neurol., 98: 401-432.

Strominger, N. L., 1969, Localization of sound in space after unilateral and bilateral ablation of auditory cortex, Exp. Neurol.,
25: 521-533.

Strominger, N. L. and Oesterreich, R. E., 1970, Localization of
sound after section of the brachium of the inferior colliculus,
J. Comp. Neurol., 138: 1-18.

Thompson, G. C. and Masterton, R. B., 1978, Brain stem auditory
pathways involved in reflexive head orientation to sound,
J. Neurophysiol., 41: 1183-1202.

Wortis, S. B. and Pfeffer, A. Z., 1948, Unilateral auditory-spatial
agnosia, J. Nerv. Ment. Dis., 108: 181-186.

EFFECTS OF UNILATERAL ABLATION OF ANTEROVENTRAL COCHLEAR NUCLEUS ON LOCALIZATION OF SOUND IN SPACE

J. H. Casseday and H. A. Smoak

Departments of Surgery (Otolaryngology) and Psychology
Duke University
Durham, North Carolina 27710, U. S. A.

INTRODUCTION

The anteroventral cochlear nucleus is the main source of projections to the medial and lateral superior olives (Stotler, 1953; Warr, 1966; Strominger and Strominger, 1971; Jones, 1979), structures which process binaural information (Brugge and Geisler, 1978) and which most likely play a role in the ability of animals to localize sounds in space (Masterton et al., 1967; Casseday and Neff, 1975). The anteroventral cochlear nucleus also projects to the central nucleus of the inferior colliculus, as do the medial and lateral superior olives. In our laboratory, experiments on the tree shrew (Tupaia glis) show that the projections from the anteroventral nucleus converge in the lateral part of the central nucleus of the inferior colliculus with projections from medial and lateral superior olives (Jones, 1979). Thus, the anteroventral cochlear nucleus, the medial and lateral superior olives and the lateral part of the inferior colliculus seem to be part of a single pathway, the function of which may involve sound localization. To test this idea we trained tree shrews (Tupaia glis) to localize sound, and then tested their ability to localize after unilateral lesions in the anteroventral cochlear nucleus. In this report we will give preliminary evidence that such lesions affect localizing behavior in a unique way. To account for the possibility that the deficit seen after this lesion is merely due to unilateral damage to the auditory system, we also tested animals with unilateral ablation of the cochlea or of the auditory cortex.

METHODS

Tree shrews were trained to localize sound in an apparatus which consisted of a semi-circular cage; around the curved perimeter of the cage, loudspeakers can be placed at as many as 7 different positions. A starting platform is located at the middle of the flat edge of the cage. In the first experiments, the positions of the loudspeakers were 0° or directly in front of the starting platform, and 30°, 60° and 90° to the right and left of 0°. In later experiments we found it unnecessary to use the locations at 60°. Drinking spouts are located at the loudspeakers and at the starting platform. The apparatus is automated so that when the animal licks the spout at the starting platform a sound occurs at one of the loudspeakers; the animal is rewarded with a few drops of water if it licks the spout in front of the loudspeaker that sounded. The next trial does not occur until the animal returns to the starting platform and licks the spout there. More than 100 trials a day can be obtained this way, but we have found that the best performance occurs if we limit the number of trials to 70-80 per day. The stimuli consist of broadband noise bursts, 50 ms in duration, or clicks. The interval between noise bursts is 450 ms and between clicks is 200 ms. During the initial training the stimuli remain on until the animal makes a response. After the animal is sufficiently trained only one, two or three stimuli are presented on any one trial. The number of stimulus presentations is changed randomly from trial to trial. We have found that this procedure results in more consistent performance than if only one stimulus is presented on each trial.

RESULTS AND DISCUSSION

The results of a normal animal in localizing noise bursts are shown as response histograms in Fig. 1A. In this figure and in those that follow, there is a response histogram for each stimulus location. The filled bars represent correct responses. For example the histogram for 90° left in Fig. 1A shows that about 65 % of the animals´ responses were to the location of the sound (90° left) and about 35 % to the adjacent location (60°)left . The other histograms in Fig. 1A show that for each location, except 90° left and 60° right, the animal´s performance was between 80 % and 90 % correct.

Following this preoperative testing, a lesion was placed in the left anteroventral cochlear nucleus. A reconstruction of the cochlear nucleus in parasagittal view is shown in Fig. 1B to indicate the position of the lesion. Although the anterior part of the anteroventral cochlear nucleus remained intact, fibers to this part, in the ascending branch of the auditory nerve, were transected. Fibers of the descending branch, to posteroventral and dorsal cochlear nucleus, remained intact.

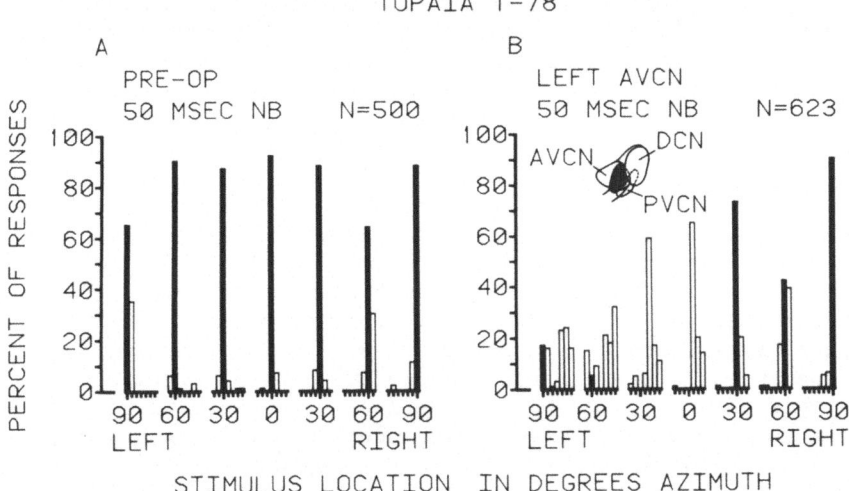

Fig. 1. Histograms to show percent of responses to each location
in the localization apparatus. The number under each histogram indicates the stimulus location in degrees. A: preoperative performance. B: performance following a lesion
in the anteroventral cochlear nucleus. The location of the
lesion is shown in black in the reconstruction at upper
left. Abbreviations: AVCN, anteroventral cochlear nucleus;
DCN, dorsal cochlear nucleus; N, number of trials; NB,
noise burst; PVCN, posteroventral cochlear nucleus.

Postoperative testing revealed a striking deficit: the animal
failed to localize sounds at the left, that is, at the side of the
lesion. Moreover, the errors are directional; they are almost always
to the right of the actual location of the stimulus. The most
striking aspect of the deficit is that the further to the right the
stimulus is presented, the less the margin of error. When the sound
was 90° left, responses occurred at all locations, including 90°
left and 90° right. But when the sound occurred at the center location (0°), over 60 % of the errors occurred at the adjacent location to the right (30° right); no responses were recorded at the
actual location of the sound. In short, it is as if the animal does
not identify the locus of a sound presented at the left but does
identify the locus, although incorrectly, of a sound presented
straight ahead.

Of course these results are preliminary, and to interpret them
we need additional data after lesions in the cochlear nucleus as
well as after unilateral lesions in other parts of the auditory

pathway. We have tested other animals after lesions intended for the anteroventral cochlear nucleus, and the deficit is not always as striking as that shown in Fig. 1B. We have not yet examined the lesions histologically in these cases, so we cannot yet say whether the deficit is related to the position or size of the lesion.

We have, however, tested animals with lesions in other parts of the auditory pathway, at the cochlea and at the auditory cortex, and the behavior after either lesion is different from behavior after lesions of the anteroventral cochlear nucleus. Fig. 2A shows that an animal with its left cochlea ablated fails completely to distinguish between right and left in that it nearly always goes to the right, regardless of where the stimulus occurs. However, at the right the animal discriminates between 30° and 90° significantly better than chance (p <.05). If this last result can be replicated, it may mean that we will have to include, in our ideas on processing of the cues to localization, a monaural component, possibly related to direct pathways from cochlear nucleus to inferior colliculus.

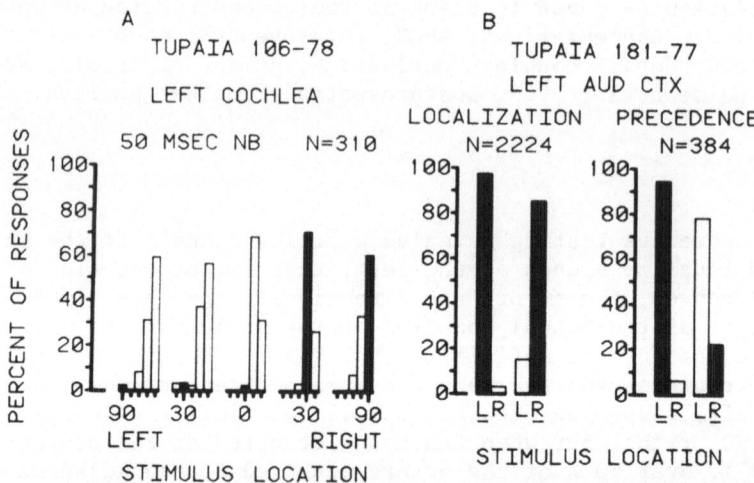

Fig. 2.   A. Histograms to show proportion of responses to each location of the sound source after ablation of the left cochlea. Note that the animal seldom approaches locations to the left. B. Histograms to compare performance in localization (left) and precedence (right) after ablation of left auditory cortex. In the precedence tests clicks are presented at both left (L) and right (R) locations with a 5 ms lag between locations. The underlined location (L or R) (cont'd)

(Fig. 2 cont'd) indicates the leading stimulus. In localization tests the underlined location indicates the stimulus
location.

In contrast to the above results, animals with unilateral ablation of auditory cortex show little or no deficit in localizing
single sound sources. Fig. 2B shows this result in the graph labeled "Localization". For convenience in comparing localization with
precedence performance, we have collapsed responses to all left
stimuli as well as responses to all right stimuli. The graph labeled
"Precedence" in Fig. 2B shows a marked deficit when the preceding
sound is in the auditory field contralateral to the ablated cortex
but not when the preceding sound is in the ipsilateral field. This
observation replicates, in the tree shrew, earlier results (Whitfield et al., 1972) on the cat and serves to emphasize a difference
between auditory cortex and lower auditory centers in the processing
of spatial stimuli.

CONCLUSIONS

The results show that these behavioral measures are sensitive
enough to reveal subtle deficits after unilateral lesions at different levels in the auditory system, e. g., cochlear nucleus vs
auditory cortex. The sensitivity of these tests should enable us to
determine whether or not different parts of the auditory pathways
in the lower brain stem, such as subdivisions of the cochlear
nucleus or superior olivary complex, play different roles in sound
localization.

ACKNOWLEDGMENT

This research was supported by NIH grant NS 12322.

REFERENCES

Brugge, J. F. and Geisler, C. D., 1978, Auditory mechanisms of the
    lower brain stem, Ann. Rev. Neurosci., 1: 363-394.
Casseday, J. H. and Neff, W. D., 1975, Auditory localization: role
    of auditory pathways in brain stem of the cat, J. Neurophysiol.,
    38: 842-858.
Jones, D. R., 1979, Auditory pathways in the brain stem of the tree
    shrew, Tupaia glis. Ph. D. Thesis, Duke University.
Masterton, B., Jane, J. A. and Diamond, I. T., 1967, Role of brain
    stem auditory structures in sound localization. I. Trapezoid
    body, superior olive, and lateral lemniscus, J. Neurophysiol.,
    30: 341-359.
Stotler, W. A., 1953, An experimental study of cells and connections

of the superior olivary complex of the cat, J. Comp. Neurol.,
    98: 401-432.
Strominger, N. L. and Strominger, A. I., 1971, Ascending brain stem
    projections of the anteroventral cochlear nucleus in the rhe-
    sus monkey, J. Comp. Neurol., 143: 217-242.
Warr, W. B., 1966, Fiber degeneration following lesions in the an-
    terior ventral cochlear nucleus of the cat, Exp. Neurol.,
    14: 453-474.
Whitfield, I. C., Cranford, J., Ravizza, R. and Diamond, I. T.,
    1972, Effects of unilateral ablation of auditory cortex in cat
    on complex sound localization, J. Neurophysiol., 35: 718-731.

BINAURAL INTERACTION MODELS AND MECHANISMS

H. Steven Colburn and Peter J. Moss

Research Laboratory of Electronics
Massachusetts Institute of Technology
Cambridge, Massachusetts 02139 U.S.A.

Binaural interaction models can be developed at the psycho-acoustic level or at the level of a single nerve cell with inputs from both cochleae. Our goal is to encompass both levels of modeling within a single framework, or at least to develop models on both levels that are compatible.

Almost all binaural psychoacoustic models are constructed around internal measurements of interaural time delay, interaural intensity difference, or interaural cross-correlation function for each frequency band (Colburn and Durlach, 1978). Models based on the internal measurement of interaural time delay or of interaural cross-correlation have been successful in describing many binaural phenomena, including binaural unmasking in detection, binaural creation-of-pitch effects, and interaural time sensitivity; how-ever, since the cross-correlation function does not change when the amplitudes of the signals are interchanged, it is not surprising that these models are inadequate for the processing of interaural intensity differences (Colburn and Latimer, 1978). On the single cell level, these models can be realized by a simple coincidence network (Jeffress, 1948) in which fibers with a common character-istic frequency, one from each cochlea, innervate a cell that fires only when both input fibers have fired in the recent past (i.e., only when input firings are "coincident"). This type of model cell is consistent with most of the physiological data from the typical EE cell in the MSO (Goldberg and Brown, 1968, 1969). Not only does the behavior of the model cell exhibit the observed dependence on the interaural phase of low-frequency sinusoidal stimuli, but it also predicts the observed relations between monaural and binaural stimulation. For these predictions it is very important that the properties of the input patterns be taken into account. For example,

the observation that the average rate of response of the cell is‘
less than the monaural rate for some phases of the stimulus is a
consequence of the nature of the input firing patterns (specifically,
the fact that the instantaneous rate of firing approaches zero
during half of each stimulus cycle) and does not require an inhibitory process at the level of the MSO cell. We conclude that, except
for interaural intensity effects, the coincidence network model
is successful in describing both psychoacoustic behavior and single-
cell response patterns from a population of binaural neurons.

The successes of the coincidence network modeling encourages
the development of a network model for the processing of interaural
intensity differences. Moreover, there is evidence from a variety
of areas that interaural time and interaural intensity are processed
separately in the peripheral nuclei (Masterton et al., 1967; Hausler
et al., 1979) and it has been argued that interaural intensity
processing is one of the functions of the LSO (Boudreau and Tsuchitani, 1970; Goldberg, 1977). This paper is an exploration of this
hypothesis.

We consider a population of model cells with a deterministic
rule that relates the firing times of the cell to the firing times
of the input fibers. Each model cell is assumed to be innervated
by two types of input fibers, excitatory inputs from the ipsilateral
cochlea and inhibitory inputs from the contralateral cochlea. Each
input firing gives rise to a unit step change in the membrane potential that decays to zero exponentially. The potential changes
are positive for excitatory inputs and negative for inhibitory
inputs. The model cell fires when the membrane potential reaches
or exceeds a treshold T. Whenever the cell fires, the membrane
potential is reset to zero. For the present paper, we assume that
the total input patterns (the superposition of the excitatory input
spike trains and the superposition of the inhibitory input spike
trains) are Poisson processes with average firing rates equal to
P_e and P_i. In this case, the output firing pattern is a renewal
process and is characterized by the distribution of the intervals
between firings. Also, since P_e and P_i are equal to the sum of the
rates on the appropriate sets of input fibers, the values of P_e and
P_i are specified by the assumption that the firing rates of the
input fibers are the same as one would expect from auditory nerve
fibers with the same CFs. The predictions shown in the present
paper were generated assuming that the potential could not go below
the negative of the threshold value.

Model cells of this general type have been considered by several other investigators, e.g., simulated by Fetz and Gerstein
(1963), treated analytically by Stein (1965), and used by Molnar
and Pfeiffer (1968) in a purely excitatory model of this type to
describe the patterns of firing of cells in the cochlear nucleus.
Furthermore, Guinan et al. (1972a, b) have suggested that a model

cell of this type might be appropriate to describe the mechanism of
LSO neurons.

Our approach to the computation of the interarrival time
distribution of the model cell is taken from Molnar and Pfeiffer
(1968). Specifically, the probability distribution of the membrane
potential is calculated as a function of time and from this the
interarrival time distribution of threshold crossings is calculated.

Some properties of this model cell's response can be compared
with physiological responses from LSO cells with minimal assumptions
about the auditory nerve patterns. In particular, the shapes of the
interval histograms can be determined for various combinations of
the model parameters as a function of the input rates P_e and P_i.
The interval histograms from cells in the LSO (see Guinan et al.,
1972a, b) are generally unimodal and are conveniently characterized
(Guinan et al., 1972a, b) in terms of the relative sizes of their
mode (the interval between firings that occurs most frequently
-- the value of the interval for which the interval histogram has
its maximum value) and width defined to be the difference between
the intervals for which the histogram is equal to one tenth of its
maximum value . Generally, the shape of the histogram becomes more
symmetric (Gaussian) as the mode becomes large relative to the width
and becomes more exponentially shaped as the mode decreases relative
to the width. In Fig. 1 we have plotted pairs of values for the
mode and width taken from the data of Guinan et al. (1972).

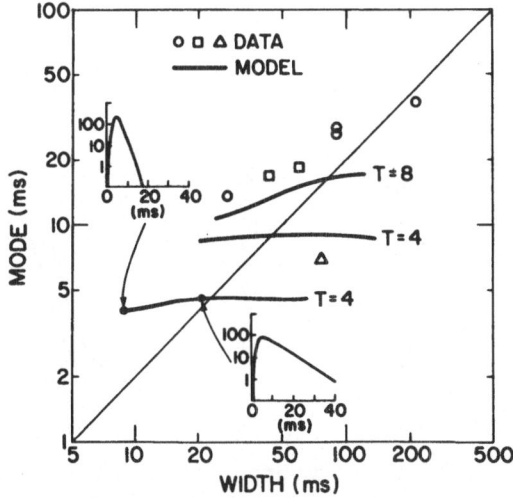

Fig. 1.   Interval histogram parameters: mode versus width (defined
          in text). Different symbols indicate different LSO cells
          (data from Guinan et al., 1972). Curves are contours of
          mode-width combinations calculated for a model (cont'd)

(Fig. 1 cont´d) neuron as a function of the inhibitory
rate P_i for three sets of values for the other parameters.
Note that the predicted contours can be shifted along the
diagonal by scaling the input rates and the decay rate of
the exponential in the model by a common factor. Insets:
Interval histograms calculated for the model neuron
corresponding to the points indicated by the arrows on the
lowest contour.

In the same figure we show contours of mode-width pairs that
are generated from our model cell as the rate of inhibitory inputs
is increased from zero (on the left end of each contour) to a value
equal to the excitatory rate (on the right end of each contour).
The three contours are for three different sets of model parameters.
The two insets illustrate the histograms generated by the model at
the two points indicated on the lowest contour. These contours and
the histograms illustrate the fact that as the inhibitory rate
increases in the model, the histogram shape tends to become more
exponential in appearance. The same general trend is seen in the
empirical data from the two cells for which multiple stimulus conditions were run. Considering that the predicted contours can be
shifted along the diagonal by scaling the rate parameters in the
model, we find the general agreement encouraging. We believe that
the contours for T 8 would provide a better fit to the data shown.

For some properties of the model cell´s response, it is necessary to specify the dependence of the input rates on the stimulus
parameters. The importance of this effect is illustrated for the
output rate of firing as a function of level in Fig. 2.

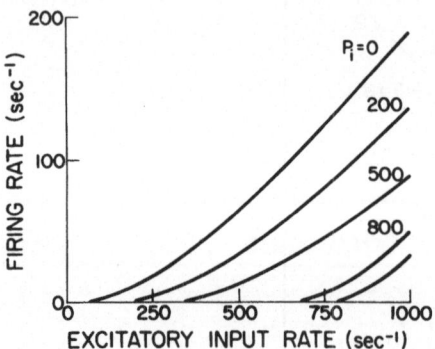 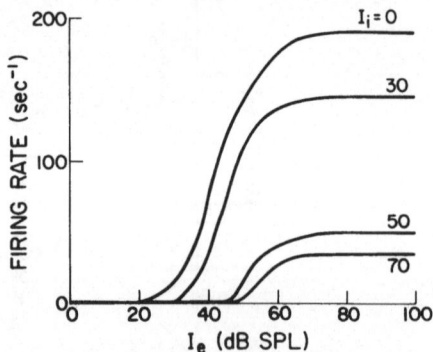

Fig. 2. Rate of firing of model neuron versus excitatory input
         level with inhibitory input level as a parameter. Left:
         Input rates P_e and P_i used as measures of input levels.
         Right: Stimulus sound pressure as the measure of input
         level; in this case, a single saturating monotonic curve
         was postulated to relate P_e and P_i to the sound (cont´d)

(Fig. 2 cont'd) pressure levels I_e and I_i.

On the left graph in Fig. 2, the output rate of the model neuron is plotted as a function of the excitatory input rate with the inhibitory input rate as a parameter. On the right graph in Fig. 2, the output rate is plotted as a function of the excitatory stimulus level in dB SPL with the inhibitory stimulus level in dB SPL as a parameter. The values plotted in the right graph were generated by assuming a common, saturating function to relate the firing rate to the stimulus level and then using the curves from the left graph. The curves on the right graph have many properties in common with the empirical curves shown by Boudreau and Tsuchitani (1970, Fig. 39). As another example, consider the rate-level functions for stimulation with ipsilateral (excitatory) tones alone. The rate-level functions saturate at progressively lower firing rates as frequency increases above the CF of the cell, whereas the saturation rates for tones below CF are approximately independent of frequency (Boudreau and Tsuchitani, 1970, Fig. 26). This might also be due to the transformation from the stimulus to the excitatory input rate in the model: some of the input fibers that respond and contribute to the input firing rate at CF would not be stimulated by a higher-frequency tone (because of the steepness of the high-frequency slopes of the tuning curves) so that the input excitatory rate would saturate at a lower value for frequencies higher than CF.

In order to evaluate the potential of this model to predict psychophysical performance in tasks such as interaural intensity discrimination, additional assumptions must be made. We have not pursued this evaluation as yet. One prediction that follows relatively directly from the hypothesized structure is that interaural intensity discrimination from a reference condition with a large interaural intensity difference would be poor relative to the equal intensity reference case as long as the overall intensity was randomized (so that the levels must be compared interaurally to achieve good performance).

Our conclusion is that there are no basic deficiencies yet apparent in these preliminary comparisons and therefore the model should be pursued further.

ACKNOWLEDGMENT

This work was supported by U.S. Public Health Service (NIH Grant No. NS10916).

REFERENCES

Boudreau, J. C. and Tsuchitani, C., 1970, Cat superior olive S-segment cell discharge to tonal stimuli, in: "Contributions to sensory physiology", D. Neff, ed., Academic Press, New York.

Colburn, H. S. and Durlach, N. I., 1978, Models of binaural interaction in: "Hearing", vol. IV of Handbook of Perception, E. C. Carterette and M. P. Friedman, eds., Academic Press, New York.

Colburn, H. S. and Latimer, J. S., 1978, Theory of binaural interaction based on auditory-nerve data. III. Joint dependence on interaural time and amplitude differences of discrimination and detection, J. Acoust. Soc. Am., 64: 95-106.

Fetz, E. E. and Gerstein, G. L., 1963, An RC model for spontaneous activity of single neurons, Quarterly Progress Report, M.I.T. Research Laboratory of Electronics, 71: 249-257.

Goldberg, J. M. and Brown, P. B., 1968, Functional organization of the dog superior olivary complex: An anatomical and electrophysiological study, J. Neurophysiol., 31: 639-656.

Goldberg, J. M. and Brown, P. B., 1969, Response of binaural neurons of dog superior olivary complex to dichotic tonal stimuli: Some physiological mechanisms of sound localization, J. Neurophysiol., 32: 613-636.

Goldberg, J. M., 1975, Physiological studies of auditory nuclei of the pons, in: "Handbook of Sensory Physiology", Vol. V/2, Springer, New York.

Guinan, J. J., Guinan, S. S. and Norris, B. E., 1972, Single auditory units in the superior olivary complex, I. Responses to sounds and classifications based on physiological properties, Intern. J. Neurosci., 4: 101-120.

Guinan, J. J., Norris, B. E. and Guinan, S. S., 1972, Single auditory units in the superior olivary complex. II. Locations of unit categories and tonotopic organization, Intern. J. Neurosci., 4: 147-166.

Hausler, R., Marr, E. M. and Colburn, H. S., 1979, Sound localization with impaired hearing, J. Acoust. Soc. Am., 65: S133.

Jeffress, L. A., 1948, A place theory of sound localization, J. Comp. Physiol. Psychol., 41: 35-39.

Masterton, B., Jane, J. A. and Diamond, I. T., 1967, Role of brainstem auditory structures in sound localization. I. Trapezoid body, superior olive, and lateral lemniscus, J. Neurophysiol., 30: 341-359.

Molnar, C. E. and Pfeiffer, R. R., 1968, Interpretation of spontaneous spike discharge patterns of neurons in the cochlear nucleus, Proc. Inst. Elec. Electron. Engr., 56: 993-1004.

Stein, R. B., 1965, Theoretical analysis of neuronal variability, Biophys. J., 5: 173-194.

PSYCHOPHYSICAL AND NEUROPHYSIOLOGICAL DATA ON THE SOUND SOURCE

PERCEPTION

J. A. Altman

Laboratory of Hearing Physiology
I. P. Pavlov Institute of Physiology
Academy of Sciences of the U.S.S.R.
Leningrad, 199164, U.S.S.R.

INTRODUCTION

Directional hearing has been studied intensively for a long time. As yet, only a few papers have dealt with the characteristics of the sound movement perception. It was shown in free field investigations that the perceived minimal shift of the moving sound source amounted to 2-4 degrees (Harris and Sergeant, 1971). When the velocity of the sound source movement was from 60 to 360 degrees/s, the detection threshold for the sound source movement rose from 5 to 20 degrees/s (Perrot and Musicant, 1977).

In the experiments with dichotic stimulation signals perceived as a moving fused auditory image (FI) were used. It was shown that dichotic click train stimulation with gradually changing interaural time delay evoked a sensation of FI movement (Altman, 1968). Similar signals were used for estimation of the FI movement detectability with variation of interaural time or intensity differences (Blauert, 1972; Grantham and Wightman, 1978).

The present paper presents data on perception of the sound source movement obtained in our laboratory in recent years in psycho-physical, behavioral and electrophysiological experiments.

FI MOVEMENT PERCEPTION BY NORMAL SUBJECTS

Click train with linearly diminished (to zero) or increased (from zero) interaural time or intensity differences (ΔT´s, ΔI´s) presented binaurally to the listeners through earphones with prac-

tically the same frequency characteristics. This stimulation evoked
a sensation of the FI movement from the ear to the head midline or
backwards. Trained subjects with normal hearing were used in the
experiments. The click train intensity amounted to 45-60 dB above
the subject s threshold in each ear. All the subjects participated
in about 50-150 trials during each experimental session, the total
number of sessions amounted to 5-18 in different experiments.

a) Limits of Perception of the FI Movement

A number of stimulus parameters such as the click repetition
rate, click train duration, click number, the range of ΔT- or Δ I-
changes were of importance for the sensation of the FI movement. To
find threshold values limiting perception of the FI movement signal
certain parameters were varied at random in a given experiment and
the subject had to answer whether the FI was localized at a certain
region within the head or whether the FI movement was perceived
(Viskov, 1975). Five to eight listeners were used in these experiments.

The lower limit of the click rate for the sensation of the FI
movement was within the range of 4-12 clicks/s as measured at 1.4 s
click train duration and the range of Δ T-changes of 700 μs. The
probability level of the FI movement sensation vs click rate was a
usual S-shaped psychometrical function and the 0.5 probability level
corresponded to the click rate of 7.6 s (average for all the subjects).

The lower limit of the click train duration for the sensation
of the FI movement was about 0.12 s on average with a 0.5-probability
level. At click train durations of 0.02, 0.04, 0.1 and 0.15 s the
probability of FI movement sensation amounted to 0.07, 0.12, 0.31
and 0.88 respectively.

The minimal click number needed for perceiving the FI movement
as low as two clicks in the binaurally presented train in the case
the click interval exceeded 0.1 s.

The data described allow to suggest a certain time mechanism
essential for the appearance of the FI movement sensation. With the
signal used a critical time interval of about 0.12 s was found to
be characteristic for the appearance of the FI movement sensation.
This critical time interval was shown in three independent experiments: i) with the lower limits of the click rate, 7.6/s, ii) with
the click train duration, 0.12 s and iii) with the minimal click
interval in a two-click train, above 0.1 s. It may be assumed that
this critical interval is within a period during which a short-term
sensory imprint of the localization of the unmoved stimulus is
formed. If this were true it may be supposed further that forming
of the FI movement sensation may be conditioned by wiping out this

imprint. Such a wiping out would deprive the auditory system of the possibility of the point-to-point localization of the sound source. Instead the sensation of a new quality would arise, i.e. the sensation of the FI movement. Another mechanism forming the FI movement sensation may be based on the process of temporal summation. To what extent such explanations will prove useful in free-field sound source movement perception is a matter of further investigations.

The lower limit of gradually changing interaural intensity differences still evoking the sensation of the FI movement was 6-9 dB through the click train of 1.5 duration. It exceeded greatly the Δ I threshold values for lateralization of the "unmoving" sounds (tenth of dB). It should be noted that the lowest values for lateralization of "moving" sounds resulted from gradually changed Δ T values were also higher (by about 100 μs) than those for lateralization of the "unmoving" sounds (about 10 μs). Such differences in threshold perception of moving and unmoving sounds were also observed in free-field experiments (Harris and Sergeant, 1971; Perrot and Musicant, 1977). It seems probable that these phenomena are connected with the necessity for the auditory system to follow to changing position of the moving sound for a certain period, while this is not necessary when localizing a stationary sound source.

b) Discrimination of Perceived FI Movement Velocity

On the basis of the experiments described above it was assumed that with the signal used the subjects will be able to perceive differences in the velocities of FI movement. The next task was to define difference limens (DL's) for velocity of the FI movement (Altman and Viskov, 1977). Four trained subjects were presented binaurally with two click trains separated by an interval of 1.5 s. In the main experimental series the click train intensity was 50 dB above the subject's threshold for each ear. The train duration amounted to 3 s, the click rate was 40/s. The interaural time difference increased linearly from zero to a maximal value of 0.1 to 3.4 ms (ΔT_{max}). The first and second click trains differed only in ΔT_{max} which was constant for the first train and varied at random for the second one. The value of ΔT_{max} determined the velocity of the Δ T change throughout the stimulus duration and thus the velocity of FI movement. The DL's for velocity of the FI movement were determined using two-alternative forced choice task. The subject was asked to compare FI movement of the two click trains and to report which of the trains evoked FI movement at a greater velocity. The difference in ΔT_{max} values expressed as velocity of FI movement in degrees/s , which corresponded to a 0.75 probability of correct responses was taken as the DL.

In order to express the velocity of the FI movement obtained from ΔT_{max} the following assumptions were made: i) FI movement may be considered as uniform curvilinear with the trajectory ap-

proximated by the arc of circle. Then the angular velocity of FI
movement may be defined as the ratio of the arc value S covered
with the FI movement to the time T of this movement: $\omega = S/T$. ii)
Under conditions of complete lateralization (i.e. complete FI shift
from the head midline to the leading ear) the arc value S equals
90°. These assumptions were discussed in more detail in a previous
publication (Altman and Viskov, 1977).

With the signal used the absolute velocity DL´s ($\Delta\omega$) defined
as a function of the velocity of FI movement (ω) displayed a nearly
linear relationship $\Delta\omega/\omega$. With a ω increase from 14° to 140°/s, the
$\Delta\omega$ changed from 10.8 to 19.3°/s at an average.

Velocity DL´s of the FI movement were also measured with two
additional click repetition rates (20 and 60 clicks/s) and two
additional signal intensities (30 and 70 dB above the subject´s
threshold). It was established that both the click repetition rate
and the intensity of the sound signals, within the range used, did
not produce a significant effect on DL values as compared with the
values given above.

It is of interest that DL´s for velocity of the free-field
sound source movement (Perrot and Musicant, 1977) proved lower than
DL´s defined in our experiments. The reason may be that in the free-
field sound localization the auditory system may use simultaneously
not only time but also intensity and spectral cues for estimating
interaural differences in stimulation.

c) Estimation of Perceived FI Movement Velocity in Subjective Scales

To study subjective scaling for estimation FI movement velocity, two well trained listeners were used in experiments by V. P.
Romanov (1980). Binaurally presented click train was used, the FI
movement was evoked by means of a gradual change of the interaural
time differences. The subject was asked to adjust the velocity of
FI movement to value ω_d, which would be perceived as twice as high
(or twice as low, ω_h) as compared with the velocity of the initial
standard stimulus (doubling or halving experiments). The click
train duration amounted to 11 s, with the click rate 40/s and intensity of 60 dB.

In the doubling experiments it was found that within the range
of velocity ω_o from 2.7 to 16.2 degrees/s $\omega_d > 2\omega_o$ whereas at higher
values of ω_o, up to 25.8 degrees/s the equality $\omega_d = 2\omega_o$ was
presened.

In the halving experiment with a velocity ω_o, ranging from 5.4
to 37.3 degrees/s, the experimental data could be approximated by

a linear equation $\omega_h = 0.5\omega_o$ and thus a linear subjective scale for estimating the velocity of FI movement could be stablished.

The difference between the subjective scales obtained in the two types of experiments is probably connected with the hysteresis effect (Stevens, 1957). According to the latter, the stimulus sequence is a determinant for results in scaling experiments. Different phenomena of this effect are described by Stevens (1957) for scales of equal halving.

The data obtained in the scaling experiments presented also some additional information concerning the perception of the velocity differences for FI movement. The range of lower velocity values was studied and it was found that for the value of 5.4 degrees/s the difference of about 2.7 degrees/s was perceived twice as low. Thus the DL for this velocity value should be less than 2.7 degrees/s.

FI MOVEMENT PERCEPTION FOLLOWING CENTRAL DISORDERS OF THE AUDITORY SYSTEM

a) Behavioral Experiments in Animals

In the above experiments some basic characteristics of FI movement perception in normal subjects were revealed. To study neurophysiological mechanisms underlying FI movement perception it seems essential to ascertain the role of different parts of the auditory system ensuring this perception.

As many published data show, destruction of the auditory cortex in mammals interferes strongly with the animal's ability of localizing stationary sound sources (see Erulkar, 1972). It was of importance to define also the localizing deficit in animals with ablated auditory cortex when tested with signals modelling a directional sound movement. This part of the investigation was performed on dogs by I. V. Kalmykova using conditioned reflex method.

Through earphones fixed at the dog's right and left ears the click train with gradually changing Δ T values was administered binaurally. Train duration of 3 s, click repetition rate of 40/s and intensity of 60 dB were used. The animals were trained to raise one leg with "FI movement" in one direction and the other leg with the "FI movement" in the opposite direction. Electric shock reinforcement was used throughout the training period.

The animals were easily trained to differentiate between two opposite directions of the "FI movement". However, following the unilateral ablation of the AI, AII and Ep areas of the auditory cortex, the probability of correct responses declined from 0.91 to 0.7. Following bilateral ablation of the defined cortical areas

the probability of correct responses diminished to a 0.5 level,
i.e. the animals were practically unable to differentiate between
the opposite directions of the "FI movement".

It was also established that the animals differentiated between moving and unmoving FI´s in case the velocity of the moving FI
amounted to 7 degrees/s or exceeded this value. Following bilateral
ablation of the auditory cortex this value increased greatly, up to
26 degrees/s.

Thus the animal experiments showed that like the free-field
localization and lateralization of stationary FI´s, the discrimination of the direction and velocity of FI movement needed the intact auditory cortex.

b) FI Movement Perception in Patients Treated with Unilateral Electroshock Therapy

The experiments on dogs allowed to suggest that also in man
the normal function of the auditory cortex is necessary for perception of the sound movement. With man, however, a pronounced hemisphere specialization should be taken into account. It was shown
in many investigations that the normal function of the left hemisphere is necessary mainly for speech recognition, whereas normal
activity of the right hemisphere is connected with music, different
kinds of non-speech sounds and prosodic speech characteristics
(Hécaen, 1969). So far data on the hemisphere specialization in
directional hearing and particularly in perception of sound source
movement are not known.

Specialization of human brain hemispheres in perception of
sound source movement was studied in 6 patients with normal hearing
but suffering from schizophrenia or endogenic depression and treated with unilateral electroshock therapy (Altman et al., 1979a).
Unilateral electroshock results in a temporary functional disorder
of the stimulated hemisphere. FI movement perception was estimated
before and following electroshock stimulation with the help of
dichotically presented click train of 4 s duration, 20/s repetition
rate and 40 dB intensity above threshold. FI movement sensation was
evoked from gradually ΔT change within $0 - 2.2$ ms range through
the click train presentation. The patient was asked to show the
point on the head surface where the sensation of FI movement was
initiated and the trajectory of the FI movement up to the endpoint.
The path covered by the moving FI was measured and presented in
degrees.

It was established that in left hemisphere functional disorder
following left-sided shock no substantial changes in the perception
of FI movement were observed. However, in right hemisphere functional
disorder following right-sided shock the perception of the FI move-

ment was significantly changed: the trajectory of FI movement within the right half of the head was greatly shortened so that FI movement was perceived only within an area of about 40 degrees near the right ear region. As to the FI moving within the left half of the head (prior to the shock), its trajectory shifted to the right and partly occupied the right half of the head, so that the area of about 30 degrees near the left ear became "vacant" following right-sided electric shock.

These results show a pronounced specialization of the brain hemisphere in FI movement perception which may be connected with normal activity of the right hemisphere.

c) FI Movement Perception in Patients with Damaged Temporal Brain Regions

The data presented implied the possibility of testing the FI movement perception for the purpose of diagnostics (Altman et al., 1979b). 17 patients with left-sided temporal lobe damage and 13 patients with right-sided damage (with the diagnosis of focal temporal epilepsy in 24 cases and brain tumor in 6 cases) were tested with the above dichotic click train stimulation evoking the sensation of FI movement on account of ΔT changes. FI movement trajectories were estimated as described above.

Different disorders in FI movement perception were observed: shifts of initial or end points of FI movement, shortening of the trajectory movement, complete inability to perceive the FI movement. A certain difference was established between patients with left- and right-sided damage. In patients with left-sided damage, disturbances of FI movement perception were observed in about 50% of cases, both in left- and right-sided FI movements. Meanwhile in patients with the right-sided damage, on the other hand, asymmetrical changes of FI movement perception took place: the distortions were observed in 54% of cases with left-sided FI movement and only in 15% with the right-sided one. These results proved useful for the diagnosis of "silent" right hemisphere damages.

It seems of importance that these data, though differing from those obtained with unilateral electroshock therapy, showed some common features with them. The similarity was mainly in asymmetrical changes of the right- and left-sided FI movement perception with right hemisphere disorders.

ELECTROPHYSIOLOGICAL DATA ON FI MOVEMENT PERCEPTION

This part of the investigation was performed by means of extra-cellular recording of the neuron impulse activity from the main centers of the auditory system (Altman, 1975) and from the cerebellum (Altman and Radionova, 1973; Altman et al., 1976). Similarly, as visual neuronal detectors (Hubel and Wiesel, 1959), neurons responding by a specialized reaction to sounds, modelling directional sound movement may be found in the auditory system. To distinguish such neurons the following criterion was used: different (asymmetrical) responses of neuron to sounds "moving" in opposite directions, i.e. responses which could not be predicted on the basis of the neuron reactions to "unmoving" sounds of different shifts from the midline.

The click train dichotic stimulation described above was used to simulate the sound source movement. In cats anaesthetized with chloralose-urethane mixture (30 and 500 mg/kg) no neuronal detectors were found at the superior olive level (21 neurons were studied). Meanwhile at the inferior colliculus level such neuronal detectors with specialized reaction to sounds "moving" in a certain direction were found in 13% of all neurons tested (79 neurons). In the medial geniculate and auditory cortex regions the percentage of neuronal detectors rose to 22 and 23%, respectively (of 50-52 neurons studied at each of the higher auditory levels). The percentage of neuronal detectors found by Sovijärvi (1973) in the cat auditory cortex with a free-field sound-source movement was even higher (32%).

A special feature of neurons of the higher auditory centers was a long-lasting after-discharge (its duration often amounted to 5-7 s or even more) which was connected with a certain direction of sound "movement".

It seems natural to suggest that neuronal detectors are of special importance for sound movement perception. Neuronal after-discharges which are specifically connected with a certain direction of sound source "movement" and are formed in higher auditory centers may be also responsible for a special role of these centers in perception of the sound source movement. Thus, a rough deficit of localization ability resulting from cortical damage both in man and animals may be connected with damage of the detector system as well as the system of delayed responses.

All the data presented above obviously show that formation of the moving FI, estimation of this movement and response to the "moving" sound stimulus (both in man and animals) requires memorizing some properties of this stimulus. The delayed after-discharges in neurons of the higher auditory centers may be probably looked upon as a correlative of the signal sensory imprint and of initial stages of forming the short-term memory necessary for the subsequent localizing behavior.

As is well known sound localization is closely connected with movements (the orienting reaction to sound, for instance). Therefore it was of interest to study brain structures which received information concerning sounds and were connected with movements. The cerebellum was chosen for this purpose: it is known to receive auditory afferent stimuli from the inferior colliculus and auditory cortex through the pontine nuclei up to the auditory cerebellar area (mainly the VI-VII-th vermal lobuli of Larsell). That information about the sound source localization can be transfered to the cerebellar auditory area was shown both with the evoked potential (Wolfe, 1972) and single unit activity recordings (Altman and Radionova, 1973; Aitkin and Boyd, 1975; Altman et al., 1976). As to the cerebellar reaction to the stimulus modelling sound source movement, it was found that 9 of 29 neurons studied (31%) showed the detector features in relation to a certain direction of the sound "movement". This portion is comparable with those for the higher auditory centers. These data seem essential for the mechanisms of sound localization and particularly for estimation of its movement.

For sound source localization, a model of a real external acoustic field is needed within the brain. Impulse flow processing within the classic centers of the auditory system may be probably connected with such model construction. In addition to this processing a reference level is necessary for estimating the sound source position. This reference level may be probably the body scheme which should be also presented within the brain. Then, for sound localization, the image (the model) of the external acoustic field should be compared with the body scheme within the brain centers. Extensive varying connections between different brain structures are obviously capable of such a comparison. However, a most economical and fast way of comparison would be a combination of the external acoustic field image and the body scheme in the same common brain structure. It is known that the body scheme is well presented within the cerebellar cortex (together with some other brain structures). On the other hand, as shown in the present work, the external acoustic field model, sound movement including, is also well presented within the same structures as far as sound localization parameters are concerned.

Data on brain hemisphere specialization in perception of sound source movement give additional support to the above assumption. The fact is that it is the right parietal lobe where the body scheme is presented (Hécaen, 1969). Therefore right hemisphere disorders following unilateral electroshock or following different damage would interfere with normal combining of the body scheme and the image of the external acoustic field. This would result in distortion of sound movement perception, as it was described above.

ACKNOWLEDGMENTS

The collaboration of my colleagues was invaluable throughout the work and their contribution is greatly acknowledged.

REFERENCES

Aitkin, L. M. and Boyd, J., 1975, Responses of single units in cerebellar vermis of the cat to monaural and binaural stimuli, J. Neurophysiol., 38: 418-429.
Altman, J. A., 1968, Are there neurons detecting direction of sound source motion?, Exp. Neurol., 22: 13-25.
Altman, J. A., 1975, Neurophysiological mechanisms of auditory localization, UCLA, Los Angeles.
Altman, J. A. and Radionova, E. A., 1973, Responses of neurons from the auditory area of the cerebellar vermis to monaural and binaural stimulation, Bull. Exp. Biol. Med., 4: 11-15. (in Russian).
Altman, J. A., Bechterev, N. N., Radionova, E. A., Shmigidina, G. N. and Syka, J., 1976, Electrical responses of the auditory area of the cerebellar cortex to acoustic stimulation, Exp. Brain Res., 26: 285-296.
Altman, J. A. and Viskov, O. V., 1977, Discrimination of perceived movement velocity for fused auditory image in dichotic stimulation, J. Acoust. Soc. Amer., 61: 816-819.
Altman, J. A., Balonov, L. J. and Deglin, V. L., 1979a, Effect of unilateral disorder of the brain hemisphere function in man on directional hearing, Neuropsychologia, 17: 295-301.
Altman, J. A., Rosenblum, A. S. and Lvova, V. G., 1979b, Perception of moving auditory images by patients with temporal lobe damages, Human Physiol., 5: 55-63. (in Russian).
Blauert, J., 1972, On the lag of lateralization caused by interaural time and intensity differences, Audiology, 11: 265-270.
Erulkar, S. D., 1972, Comparative aspects of spatial localization of sound, Physiol. Rev., 52: 237-360.
Grantham, D. W. and Wightman, F. L., 1978, Detectability of varying interaural temporal differences, J. Acoust. Soc. Amer., 63: 511-523.
Harris, J. D. and Sergeant, R. L., 1971, Monaural/binaural minimum audible angle for a moving sound source, J. Speech Hear. Res., 14: 618-629.
Hécaen, H., 1969, Aphasic, apraxic and agnostic syndromes in right and left hemisphere lesions, in: "Handbook of Clin. Neurol.", vol. 3-4, F. G. Finken and G. W. Brujn, eds., pp. 291-311, Elsevier, Amsterdam-N.-Y.
Hubel, D. H. and Wiesel, T. N., 1959, Receptive fields of single neurons in the cat's striate cortex, J. Physiol., 148: 574-591.
Perrot, D. R. and Musicant, A. D., 1977, Minimum auditory movement

angle: binaural localization of moving sound sources, J. Acoust. Soc. Amer., 62: 1463-1466.

Romanov, V. P., 1980, Scaling of the sound source movement perception in dichotic stimulation, Human Physiol., 6: 712-714.

Sovijärvi, A. R. A., 1973, Single neuron responses to complex and moving sounds in the primary auditory cortex of the cat. Academic dissertation, University of Helsinki, Helsinki.

Stevens, S. S., 1957, On the psychophysical law, Psychol. Rev., 64: 153-181.

Viskov, O. V., 1975, The perception of the fused auditory image movement, Human Physiol., 1: 371-376. (in Russian).

Wolfe, J. N., 1972, Responses of the cerebellar auditory area to pure tone stimuli, Exp. Neurol., 36: 295-309.

SESSION VII
NEURAL CODING OF SPEECH AND COMPLEX STIMULI
Chairmen: J. S. Buchwald and K. Sedláček

INFORMATION PROCESSING IN NEURONAL POPULATIONS OF THE HUMAN BRAIN DURING LEARNING OF VERBAL SIGNALS

N. P. Bechtereva and Yu. D. Kropotov

Institute of Experimental Medicine
Academy of Medical Sciences
Leningrad, U.S.S.R.

Application of stereotaxic method for diagnosis and therapy of different brain disorders opened new ways in the investigation of mechanisms of human mental activity. The complex approach to study these problems had been developed, including recording and processing of different physiological data, e.g. EEG, subcorticogram, slow potentials, pO_2, local brain tissue impedance, multiunit activity, etc. (Bechtereva, 1971). The previous investigations (Bechtereva et al., 1977) proved specific changes of some of these parameters during different psychological test administrations.

The aim of this paper is to present some neurophysiological correlates of verbal signal learning in patients with chronically implanted electrodes.

METHODS

Six patients with parkinsonism and epilepsy served as subjects. The golden electrodes were stereotaxically implanted into different nuclei of the thalamus, strio-pallidar system and different cortical regions. The electrodes had a diameter of 100 microns and their active surface varied between 0.01 and 0.15 square millimeters. The number of implanted electrodes varied from 32 to 54.

The psychological task, presented to the patient, was as follows. He had to memorize a given sequence of numbers, repeating them orally several times during the so-called memorization stage. This varied from 30 seconds to 2 minutes.

Completing the memorization stage the patient had to perform backward counting deduction of three from arbitrary three-digit number during 20 - 40 seconds. This stage was labelled as an interference stage. Then the patient was asked to reproduce orally the remembered numbers. Each task consisted of 10 - 20 trials, depending upon the patient's ability to cooperate. The simple repetition of well-known sequence of numbers (e.g. 1,2,3,...) served as control.

The multiunit activity was recorded during the test administration. Its peak-to-peak amplitude in approximatelly 50% of cases exceeded 75 uV. (The noise level of the amplifiers when their inputs were shorted, did not exceed 15 uV). Thus 96 cases were analysed. The corresponding data were tape recorded and processed off-line by computers MINSK-32 and PLURIMAT S. The poststimulus time histograms were plotted (on an X-Y plotter) and then statistically analyzed.

RESULTS AND DISCUSSION

The above mentioned approach gave us the possibility to classify the dynamics of neuronal discharge frequency in different regions into the following groups.

1. The neuronal populations of the first group changed their activity at the beginning of either the memorization or interference stages only. These changes lasted from 1 to 4 seconds and were correlated with the new series of numbers presentation (see Fig. 1 - at the bottom).

Their character was either an activation or inhibition type and they were observed in various thalamic nuclei, striopallidar system and premotor cortex.

2. The neuronal populations of this group were active during the whole trial. Their discharge frequency did not change during the memorization or interference stages. In some cases the difference between experimental and control tests was observed. The neurons belonging to this group were mainly localized in centrum medianum, ventrolateral thalamic nuclei and the premotor cortex.

3. The most interesting were the populations of the last group. They revealed the delayed changes of neuronal spike activity, occurring several seconds after the onset of the trial. These neurons belonged to the centrum medianum, to the ventrolateral thalamic nuclei and the premotor cortex. The character of the changes represented either activation or inhibition and we shall call them sensitization-like (S) and habituation-like (H) activity. No corresponding changes were detected in the control trials. The neuronal populations of the H-type became active during the first seconds of every trial and then their activity rapidly decreased

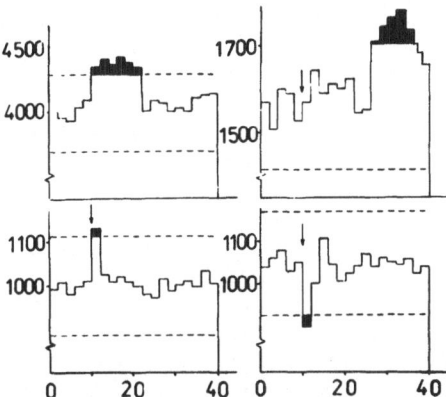

Fig. 1.   The examples of neuronal populations dynamics are given.
          Horizontally: time in s, vertically: the total discharge
          frequency. The arrow indicates the trial onset. The hori-
          zontal dashed lines indicate 99% confidence limits.

(Fig. 1, top-left). On the contrary in the neuronal populations
with S-type activity their discharge frequency increased only with
some delay after the trial onset (Fig. 1, top-right).

     The latency of frequency alterations depended upon the amount
of numbers to be learned. For example, in cases of memorizing 3 two-
digit numbers the latency was about 20-30 seconds, whereas in cases
of learning the 4 two-digit numbers it was approximately 30-40
seconds.

     The frequency dynamics of these neuronal populations were
compared with the learning curves, derived from psychological tests
when the time of memorization stage varied. Analysing the results
of such tests it was possible to obtain a curve "percentage of
correct responses vs duration of memorization stage".

     Our investigations revealed that the learning curves were
closely related to poststimulus-time histograms for neuronal populations of described type. Moreover, it was shown that the delayed
changes of neuronal spike activity were accompanied by the rather
rapid decrease of brain tissue impedance, measured in corresponding
sites of the brain.

One more point must be noted here. The neuronal populations of S-type and H-type had different dynamics during the recall stage after interference. The activity of H-type neuronal populations did not restore the frequency level, developed during memorization stage. On the contrary, the S-type populations had a tendency to restore the level, developed during learning of verbal signals.

This fact and some others, not mentioned here, give us an opportunity to conclude that S-type neuronal populations seem to be responsible for the long-term storage.

Based on these empirical results a mathematical model of a neuronal network was suggested together with S. V. Pakhomov. It consists of inhibitory and excitatory formal neurons, connected together by lateral connections. The equations for the membrane potential dynamics were derived for each neuron. These equations described the spatial and temporal integration of postsynaptic potentials, the existence of threshold, refractory period and some other features of real neurons. The influence of sensitization and habituation on the model behaviour was studied. We were able to suggest that sensitization resulted in the increase of signal-to-noise ratio and therefore it might be responsible for the long-term changes of synaptic efficiency.

REFERENCES

Bechtereva, N. P., 1971, "Neurophysiological aspects of human mental activity", Medicina, Leningrad (in Russian, reprinted in English).
Bechtereva, N. P., Bundzen, P. V. and Gogolitsyn, Yu. L., 1977, Brain codes of mental activity, Nauka, Leningrad, (in Russian).

A COMPARISON OF THE RESPONSES EVOKED BY ARTIFICIAL STIMULI AND

VOCALIZATIONS IN THE INFERIOR COLLICULUS OF SQUIRREL MONKEYS

Judith A. Manley and Peter Müller-Preuss

Max-Planck-Institute for Psychiatry

Munich, F. R. G.

Although our present understanding of the auditory system comes primarily from studies using easily defined acoustic parameters, an organism's acoustic environment consists primarily of constantly changing complex stimuli. Our study was undertaken to compare the responses of inferior colliculus (i. c.) neurons to simple artificial stimuli and to more complex species-specific vocalizations.

Squirrel monkeys were implanted with a reclosable chronic recording cylinder. During the acute recording sessions the monkeys were awake, but restrained in a primate chair in a sound-attenuated room. Tungsten microelectrodes were pushed directly through the dura. Histological data, together with the presence during recording of background neural activity ("swish ; Aitkin, 1979) indicate that most of the 215 units included in this study were located in the central nucleus, but a few may have been located beyond the dorso-lateral edge of the nucleus. White noise, clicks, tone bursts (300 ms) and 8 tape-recorded monkey calls were used as stimuli. Two of the calls had a tonal quality, 3 had FM components and 3 contained fairly broad-band noise. All stimuli were presented at 70^{+}_{-} 5db SPL peak pressure.

Cells in the i. c. of squirrel monkeys are overall more responsive and therefore less selective to all types of acoustic stimuli than cells in the auditory cortex of these monkeys (Winter and Funkenstein, 1973; Newman and Wollberg, 1973; Manley and Müller-Preuss, 1978; Müller-Preuss and Manley, in preparation). The responsiveness of i. c. cells and the distribution of various response types are shown in Fig. 1. Almost all the cells responded at this intensity to noise (97 %) and tones (97%) and the majority also to clicks (74%). All of the 125 cells tested with all 8 calls responded

to at least one, 90 % responded to 5 or more and 60 % to all 8 calls.
The auditory cortex cells that we tested under the same conditions
with the same 8 calls were slightly less selective, with 41% responding to all 8 calls. Individual vocalizations evoked a response in
68-92 % of the i. c. units, a slightly higher percentage than in the
cortex. In contrast to the cortex, all the i. c. cells which responded to calls also responded to noise and tones, i. e. there were no
highly selective cells which responded only to vocalizations.

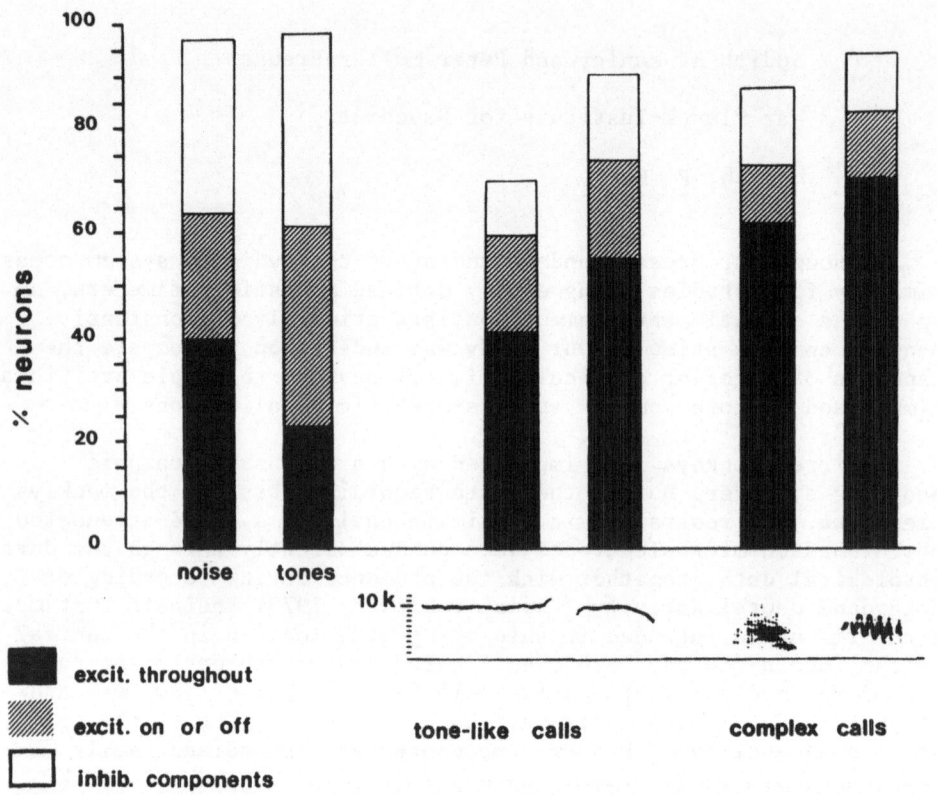

Fig. 1.   The height of each bar shows the percentage of the total
          number of neurons tested which responded to the indicated
          stimulus types. A frequency vs. time spectrogram is given
          for each of the representative calls. The black portion
          of each bar shows the proportion of cells giving tonic
          excitatory responses; shading represents phasic cells
          and the white portion all cells with inhibitory responses
          or with complex responses containing inhibitory and exci-
          tatory components.

Individual cells responded with a variety of patterns to each of the stimuli. Of the 208 cells responding to noise, 27 % responded with phasic excitation (onset and/or off), 40 % with tonic excitation, 9% with only inhibition and 25% with both excitatory and inhibitory components. Most of these latter responses consisted of phasic or tonic excitation followed by inhibition at the cessation of the stimulus.

Consistent with the lack of specificity to vocalizations, most units responded over a broad frequency range to tones. Of 158 cells tested (most over more)than 6 octaves , the average frequency response range at 70 $\pm$5dB was 4.1 octaves. The same basic response patterns were elicited by tones as by noise, however tone responses more commonly contained phasic and/or inhibitory components. Phasic excitatory responses were elicited over the greatest part of the response range in 39% of the cells responding to tones (n=163), tonic excitation in 23%, inhibition alone in 10% and complex excitatory-inhibitory patterns in 28%. In a number of cells phasic excitatory responses were elicited by tones and sustained responses by noise. In other cells inhibitory components were elicited over all or part of the response range to tones, but did not appear in the response to noise. Broad-band stimuli apparently elicits more tonic firing and obscures inhibitory inputs which may be evoked over only a part of the response range. In 40% of the units the response pattern varied with frequency. Tonic excitation tended to occur more often at low frequencies and phasic excitation at high frequencies. Coupled with this was a tendency for inhibitory responses and responses with inhibitory components to occur more often at high frequencies.

In comparing the responses of i. c. neurons to vocalizations with the responses to noise and tones, several trends emerged. Cells which responded to all 8 calls had on average only a slightly broader frequency response range than the more selective cells. Likewise frequency response range was not always a good indicator of the calls to which a unit would respond. Some units responded to either more (18%) calls than one would predict from the responses to tones and noise or to fewer (21%) calls than one would predict.

In contrast to artificial stimuli, vocalizations elicited a higher percentage of excitation persisting throughout the duration of the stimulus (see Fig. 1). Many of these "tonic" responses clearly followed frequency or amplitude changes within the complex calls, however the tone-like calls also elicited a higher proportion of "tonic" responses than either noise or tones. Expressed as a percentage of the responses elicited (rather than a percentage of neurons tested as in Fig. 1), excitation throughout the various calls occurred in 61-78% of the responses (noise, 40%; tones, 23%). Conversely neural firing only at the onset and/or cessation occurred less often in response to calls than to noise or tones. Vocalizations, especially complex ones, appear to provide the i. c. neurons with continuous frequency or amplitude changes eliciting sustained firing. The presence of broad spectral energy probably contributes to this effect. Our study of the cortex indicates that the proportion of "tonic" responses to calls is more nearly similar to the proportion elicited by noise and tones. Inhibition was also less prominent in the collicular responses to vocalizations than to noise or tones. The proportion of pure inhibitory responses elicited by the various calls and by noise and tones was similar, but the proportion of complex responses with excitatory and inhibitory components was much less for calls (3-13%) than for noise (25%) or tones (28%). In this respect there was no difference between the tone-like calls and the complex calls. Again the presence in the complex calls of frequency or amplitude changes coupled with the presence of broader spectral energy may provide sustained excitatory input which obscures much of the inhibitory input to these cells. Nevertheless this would not appear to adequately explain the small proportion of inhibitory components elicited by tone-like calls. In the cortex the proportion of responses with inhibitory components appears to be more nearly similar for vocalizations and artificial stimuli.

ACKNOWLEDGMENT

Supported by a Deutsche Forschungsgemeinschaft grant to Detlev Ploog.

REFERENCES

Aitkin, L., 1979, The auditory midbrain, Trends in Neurosciences, Dec. 1979: 308-310.

Manley, J. and Müller-Preuss, P., 1978, Response variability of auditory cortex cells in the squirrel monkey to constant acoustic stimuli, Exp. Brain Res., 32: 171-180.

Newman, J. and Wollberg, Z., 1973, Multiple coding of species-specific vocalizations in the auditory cortex of squirrel monkeys, Brain Res., 54: 287-304.

Winter, P. and Funkenstein, H., 1973, The effect of species-specific vocalizations on the discharge of auditory cortex cells in the awake squirrel monkey (Saimiri sciureus). Exp. Brain Res., 18: 489-504.

ACOUSTIC PROPERTIES OF CENTRAL AUDITORY PATHWAY NEURONS

DURING PHONATION IN THE SQUIRREL MONKEY

P. Müller-Preuss

Max-Planck-Institute for Psychiatry

Munich, F. R. G.

The present work is based on the assumption that the complex nature of acoustic communication within certain vertebrates, such as primates, requires control circuits between structures involved in phonation and audition. This means that interactions between those structures exist, and that acoustic communication is not only carried out through genetically preprogrammed processes. Control of an individual s own vocal output can take place via the ear (auditory feedback), via somatosensory structures for example in the larynx (proprioceptive or tactile feedback), or via neuronal circuits in the brain itself (central control). An example for the importance of auditory feedback is the dramatic consequence of deafness in humans when learning to speak (Seeman, 1969). As far as central control circuits are concerned, several studies have shown that vocal activity can influence peripheral structures of the auditory pathway, such as the middle ear (Suga and Jen, 1975), and also central stations, such as the lateral lemniscus (Suga and Shimozawa, 1974) and the inferior colliculus (Schuller, 1979).

The present study was undertaken for the purpose of investigating relationships between brain structures involved in phonation and audition on the single unit level in primates, that is, squirrel monkeys (Saimiri sciureus). From this complex of problems, we concentrated in particular on the following two questions: 1. Does the auditory system receive neuronal information from structures involved in phonation before vocalizations are emitted by the larynx? 2. Does the auditory system process selfproduced vocalizations differently from vocalizations coming from the environment?

In an attempt to answer these questions, the activity of single neurons in parts of the central auditory pathway was recorded extracellularly during phonation. The particular levels within the auditory system were the inferior colliculus, the medial geniculate body and the primary and secondary auditory cortex. Because it is difficult to evoke vocalizations from these monkeys in a controlled behavioral situation (as it would be in conditioning experiments), our main efforts were directed at eliciting vocal activity through electrical brain stimulation. But we also recorded cell activity during spontaneously uttered calls. In the case of brain stimulation, we chose those stimulation points in the central grey matter which also elicited vocalizations after the end of stimulation. This was necessary to avoid masking effects of the stimulus artifact at the recording site. Stimulus duration was 1 - 1.5 s, pulse length 1 ms, intensity up to 400 microAmps, and frequency 30 Hz. The stimulation of individual electrodes consistently evoked the same call, differing only in amplitude. So the auditory pathway was stimulated during phonation by either electrically elicited or spontaneously uttered vocalizations. In order to have a comparable acoustic stimulus without the effects of phonation, these self-produced vocalizations were recorded on an endless-loop tape and played back to the monkey. The auditory input during playback of the self-produced vocalizations was similar to the original input during phonation, even given changes caused by bone conduction. To characterize the intensity/ response-relationships of the cells, the playback calls were presented at different sound intensities in the range of 40 to 80 dB SPL. Finally, to test the general acoustic responsiveness of the neurons, various species-specific vocalizations and artificial sounds, such as white noise, were used. The time available for observing the unit activity before phonation was, in the case of spontaneously uttered calls, unlimited, in the case of electrically elicited calls, limited by the stimulus artifact to 100 ms to 800 ms. In all three structures under investigation, we did not detect a change in spontaneous activity before the onset of the vocalizations.

A comparison of the neuronal responses to self-produced vocalizations during phonation versus the responses to the playback on self-produced calls showed the following effects: In all three structures, acoustic stimulation via the loudspeaker evoked in most cells a more or less excitatory response. In the cortex, less than half of these cells responded with very similar patterns to the self-produced calls and the playback calls. In more than half of these cells the responses during phonation were clearly weaker or totally absent in comparison to the response to the playback calls. This inhibitory effect was found mainly during electrically elicited calls but also during spontaneously uttered vocalizations. In additional experiments, where the acoustic responsiveness of the cells in the interval between the end of stimulation and the begin of phonation was tested, it could be shown that the inhibitory effect was not caused by the electrical stimulation itself.

The situation in the medial geniculate body is very similar to that in the cortex. We found many cells which did not respond during phonation, but clearly responded to the playback call. And there were also cells in the thalamus which did not differentiate between self-

Phonation   **Play-back**

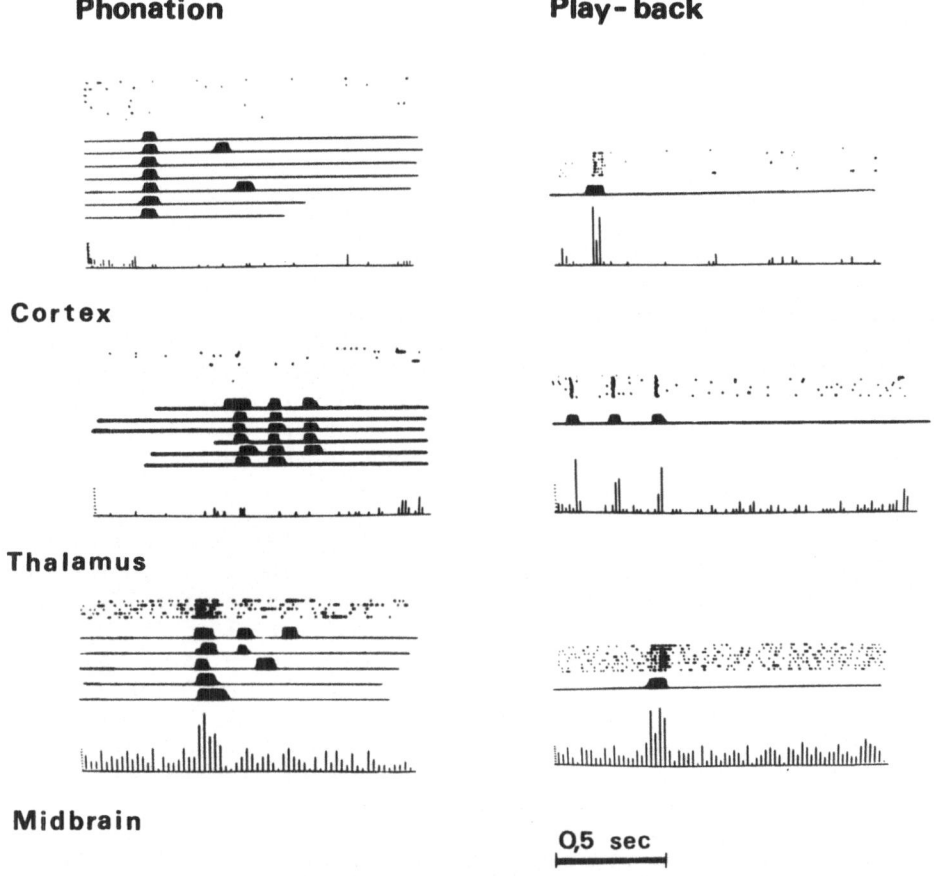

Cortex

Thalamus

Midbrain

0,5 sec

Fig. 1.   Shown here are three examples of auditory pathway unit
           responses to phonation (left) and to the playback of those
           self-produced calls (right). In the upper part of each
           section the dot display represents the unit activity, in
           the center the half amplitude courses of the calls are
           shown, in the case of phonation for each particular trial.
           In the lower part PST-histograms are delineated. Note the
           inhibitory effect in the cortical and thalamic example,
           where no response to phonation but a clear excitation to
           the playback calls can be seen. In contrast, the cell shown
           below displays similar response patterns to phonation and
           to the playback call, which is typical for midbrain cells.

produced and loudspeaker-transmitted calls. Examples of those cells
showing the inhibitory effect are given in the center and upper part
of Fig. 1. The response properties in the inferior colliculus are
quite different. In this structure, we did not find the inhibitory
effect during phonation. All cells we recorded in the inferior
colliculus show similar responses to phonation and to playback calls.
An example of such midbrain cells is shown in the lower part of Fig.
1.

Concerning the auditory cortex, we did not find a difference
between primary and secondary fields. In the absence of microlesions
along the electrode penetrations, we are at the moment unable to
decide if the differentiating neurons lie within definite cortical
layers. As there were no cells which showed a change in spontaneous
discharge rate before the onset of a vocalization, it is concluded
that the three auditory structures do not receive at this time
information about the activity of brain areas involved in phonation.
That responds to our first question in that manner that, at its cortical, thalamical and midbrain level, the auditory system does not
get a message from the motor structures during the time of "neuronal
preparation or generation" of vocalizations in the sense of "corollary discharges" (Evarts, 1972).

The fact that a considerable amount of cortical and thalamical
cells do not respond to self-produced vocalizations is an indication
that the auditory cortex and the auditory thalamus receive a specific picture about the individual s own vocal activity. It is suggested that the inhibitory effect during phonation is caused by
activated vocalization structures and might be the result of corollary discharges given from motor structures to parts of the auditory
system which lie higher than the midbrain. Concerning the functional
significance of such an inhibitory and differentiating mechanism,
we like to propose the following considerations: Parts of the auditory cortex and thalamus receive full neuronal information about the
organism s vocal output comparable to the information it gets about
other calls. This information might play a role in modulating and
learning calls, or in maintaining learned vocalizations constant.
Other subunits of the auditory cortex and thalamus are not fully
in use during phonation and therefore free for processing essential
acoustic signals coming from the environment during this time. In
general, this inhibitory mechanism might be part of a multi-modal
control circuit system which enables an organism to adapt its vocal
activity to the requirements of the acoustic environment.

ACKNOWLEDGMENT

Supported by Deutsche Forschungsgemeinschaft.

REFERENCES

Evarts, E. V., 1972, Feedback and corollary discharges: A merging
    of the concepts, Neurosc. Res. Symp. Summ., 6: 86-112.
Seeman, M., 1969, "Sprachstörungen bei Kindern", VEB Verlag, Berlin.
Schuller, G. J., 1979, Vocalization influences auditory processing
    in collicular neurons of the CF-FM Bat, Rhinolophus ferrume-
    quinum, J. Comp. Physiol., 132: 39-46.
Suga, N. and Shimozawa, T., 1974, Site of neural attenuation of
    responses to self-vocalized sound in echolocating bats, Scien-
    ce, 183: 1211-1213.
Suga, N. and Jen P. H. - S., Peripheral control of acoustic signals
    in the auditory system of echolocating bats, J. exp. Biol.,
    62: 277-311.

SELECTIVITY OF AUDITORY NEURONS FOR VOWELS AND CONSONANTS IN THE FOREBRAIN OF THE MYNAH BIRD

G. Langner, D. Bonke and H. Scheich

Institut für Zoologie
Technische Hochschule Darmstadt
Schnittspahnstrasse 3, Darmstadt B.R.D.-6100

Animals are able to categorize certain classes of speech sounds in a way similar to man (Kuhl and Miller, 1975; Burdick and Miller, 1975). Further evidence for accurate processing of speech signals is provided by the ability of certain birds, like parrots and mynahs, to imitate human voices. The aim of this investigation was to analyze neuronal mechanisms in the auditory neostriatum of "talking" mynahs pertaining to processing of vowels and consonants (Langner et al., 1979).

Single units were recorded in the field L of 7 adult mynahs which were awake and had chambers implanted in the skull for stereotaxic recordings. Methods for the implant and for closed-field acoustic stimulation were as described for another bird (Scheich et al., 1977). The telencephalic field L is a layered and tonotopically organized auditory projection area in the neostriatum (Bonke, D. et al., 1979). Discrimination of vowels was analyzed by stimulation with 9 natural vowels from a german speaker. Mechanisms of vowel selectivity were analyzed with 5 synthetic vowels composed of 2 formants F_1 and F_2 with 3 harmonics each which were multiples of a fundamental frequency varying from 120 Hz to 220 Hz in steps of 20 Hz. The formants could be presented separately. In addition to the vowels in a number of units responses to systematic variations of synthetic stop consonants were studied. In a series of 9 stimuli only the transient in front of the second formant was varied so that the first 4 stimuli were identifiable as /ba/, /bɛ/ or /bi/ and the last 4 stimuli as /ga/, /gɛ/ or /gi/. The synthetics were produced by the laboratory of Prof. Libermann, New Haven.

Among 250 units 132 (53 %) were responsive to at least 1 vowel. 20 % may be considered selective because they responded to only 1

or 2 vowels, 12 % gave a good response to only 1 vowel. The selective response of many units for a distinct vowel may be explained
by one excitatory input at one formant frequency and one or several
inhibitory inputs which are removed from the frequency of the other
formant. In some cases the isointensity response (IR) to pure tones
(65 dB SPL) revealed two separate excitatory frequency bands
corresponding to F_1 and F_2 (Fig. 1). The formants of the vowel /e/
with f_o = 120 Hz matched best the two excitatory bands of unit
N 140. The complete vowel ($F_1 + F_2$) elicited a stronger response than
either formant alone. There was no response to the vowel /o/ although F_1 of /o/ equals F_1 of /e/ and thus is excitatory.

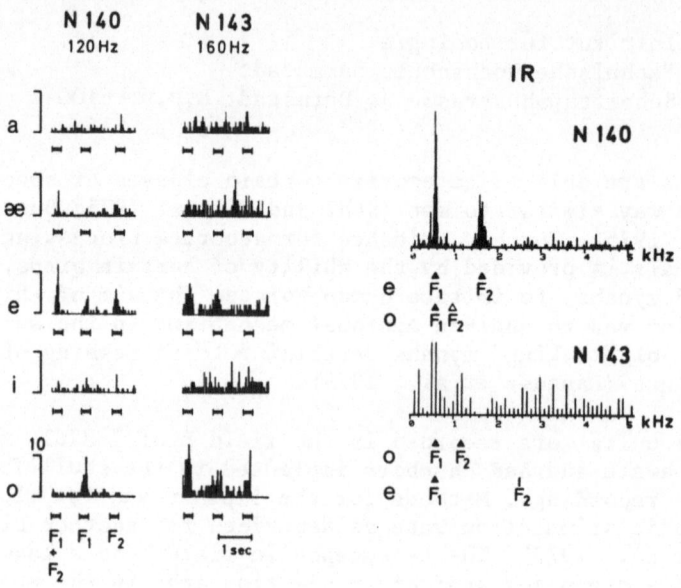

Fig. 1.   Two excitatory bands. In each of the two columns on the
left side of the figure the responses of 2 units to 5
synthetic vowels ($F_1 + F_2$) and to the formants F_1 and F_2
alone are illustrated by PST-histogram (15 ms binwidth,
10 repetitions). Time marks for the stimuli are given at
the bottom. The frequencies of the fundamentals below the
unit number at the top hold for all vowels in a column. On
the right side the isointensity responses (IR) to pure
tones at 65 dB SPL are plotted together with the positions
of the formants of some of the vowels.
+ stands for excitatory, - for inhibitory and | for no
effect of the formant. Both units have two excitatory
inputs corresponding to the two formants of the vowels
/e/ in N 140 and /o/ in N 143.

However F_2 of /o/ lies between the two excitatory bands and therefore is probably inhibitory. In unit N 143 the two formants of /o/
match the excitatory frequency bands and the response to the complete vowel is stronger than to either formant alone.

For several units it is possible to define selective areas
in the formant plane. For 4 units these areas are symbolized as
shaded areas in Fig. 2. Plus and minus symbols along the F_1- and
F_2-axis indicate excitatory and inhibitory responses of the formants
alone. In the graphs series of vowels are illustrated with varying
fundamental frequency. Each excitatory response to a vowel is
indicated by a plus, an inhibitory effect by a minus and no identifiable effect by a broken minus. All investigated units seemed to
have boundaries of their selective areas which were functions of

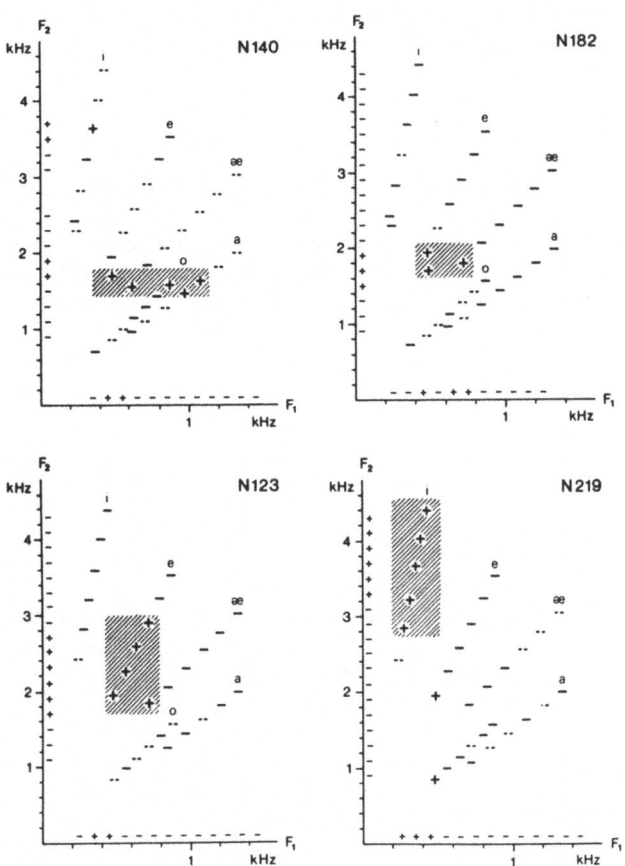

Fig. 2.   Selective areas of forebrain units. The responses of 4
          selective units are symbolized in the plane of formants.
          Plus and minus along the axes indicate in what (cont´d)

(Fig. 2 cont'd) frequency range F_1- and F_2-sounds presented alone yielded excitatory or inhibitory responses. Series of responses to complete vowels are illustrated with varying fundamental frequency f_o so that F_1 and F_2 which are defined as multiples of f_o shift through a range of values. Excitatory response of a vowel was indicated by +, inhibitory by -, no identifiable effect by --. The shaded areas give the selective range of the units in the plane of formants.

one formant frequency alone independent of the other formant frequency, e.g. the boundaries seemed to be parallel to the axes. This fits well with some psychophysical results on boundaries of synthetic two-formant vowels by Mushnikov and Chistovich (1971) and of four-formant vowels by Slepokurova (1972). With one exception these boundaries also have been found to be parallel to the formant axes. Extremely parallel boundaries at least to the F_2-axis were the result of recent detailed psychophysical studies in our laboratory with two formant vowels similar to those used in the present neurophysiological study in the forebrain of the mynah (B. Hose et al., in prep.).

42 units were examined for their discrimination for 3 consonant-vowel continua /ba/ - /ga/, /bɛ/ - /gɛ/, /bi/ - /gi/ . Among some units which were able to discriminate between /ba/ and /ga/ and /ga/ or /bɛ/ and /gɛ/ were 2 units which responded in a categorical fashion. They preferred all 4 variations of /ba/ resp. /ga/ against all 4 variations of /ga/ resp. /ba/.

ACKNOWLEDGMENT

Supported by the Deutsche Forschungsgemeinschaft SFB 45.

REFERENCES

Bonke, D., Scheich, H. and Langner, G., 1979, Responsiveness of units in the auditory neostriatum of the Guinea Fowl (Numida meleagris) to species-specific calls and synthetic stimuli. I Tonotopy and functional zones, J. Comp. Physiol., 132: 243-255.

Burdick, C. K. and Miller, J. D., 1975, Speech perception by the chinchilla: Discrimination of sustained /a/ and /i/. J. Acoust. Soc. Am., 58: 415-427.

Kuhl, P. K. and Miller, J. D., 1975, Speech perception by the chinchilla: voiced-voiceless distinction in alveolar plosive consonants, Science, 190: 69-72.

Mushnikov, V. N. and Chistovich, L. A., 1971, O sluchovom opisaniji glasnovo, I. Priznaki, razlichajushtshie /i/ i /e/, in: "Analiz retshevich signalov chelovekom, Nauka, Leningrad.

Scheich, H., Langner, G. and Bonke, D., 1979, Responsiveness of units in the auditory neostriatum of the Guinea Fowl (Numida meleagris) to species-specific calls and synthetic stimuli. II. Discrimination of Iambus-like calls. J. Comp. Physiol., 132: 257-276.

Scheich, H., Langner, G. and Koch, R., 1977, Coding of narrow-band and wideband vocalizations in the auditory midbrain nucleus (MLD) of the Guinea Fowl (Numida meleagris), J. Comp. Physiol., 117: 245-265.

Slepokurova, N. A., 1972, O procedure raspoznavanija stacionarnich glasnyh, in: "Sensorinyje sistemi. Voprosi teorii i metodov issledovanija vosprijatija rechevich signalov", Nauka, Leningrad.

SOME ASPECTS OF FUNCTIONAL ORGANIZATION OF THE AUDITORY NEOSTRIATUM (FIELD L) IN THE GUINEA FOWL

D. Bonke, B. A. Bonke, G. Langner and H. Scheich

Institut für Zoologie, Technische Hochschule Darmstadt

Schnittspahnstrasse 3, Darmstadt, B.R.D.-6101

There has been great interest in the last years in understanding basic principles of neuronal processing of simple and complex sounds, i.e. pure tones and species-specific vocalizations, related to the functional organization of auditory nuclei. Most investigations have focused on the mammalian auditory system, where a tonotopic organization of most nuclei in the auditory pathway could be shown.

The aim of this investigation was to study the structural and functional organization of field L in the auditory neostriatum, which is the primary auditory projection area in birds comparable to the auditory cortex in mammals. The spatial distribution of the terminals of the ascending fibers within field L has been desribed by Karten (1968) with the fiber degeneration technique. In our study we used tritiated aminoacids for terminal labelling. Injections into the diencephalic nucleus ovoidals show moderate labelling in three laminae, which are called L_1, L_2 and L_3 in dorso-ventral sequence. The heaviest labelling is found in the intermediate layer L_2, which shows a clublike expansion medio-ventrally. L_2 is the primary projection layer in field L (Bonke, 1979a). The layering bears some similarities to the structural organization of the auditory cortex in mammals. AI has two intermediary neighbouring layers (III and IV) which receive the input of most of the ascending fibers which project to AI (Aitkin, 1976).

We have studied a large number of neurons in these 3 layers by systematic electrode penetrations in a horizontal grid of 1.5 by 2.5 mm with a stepwidth of 0.2 mm while stimulating with pure tones and species-specific calls.

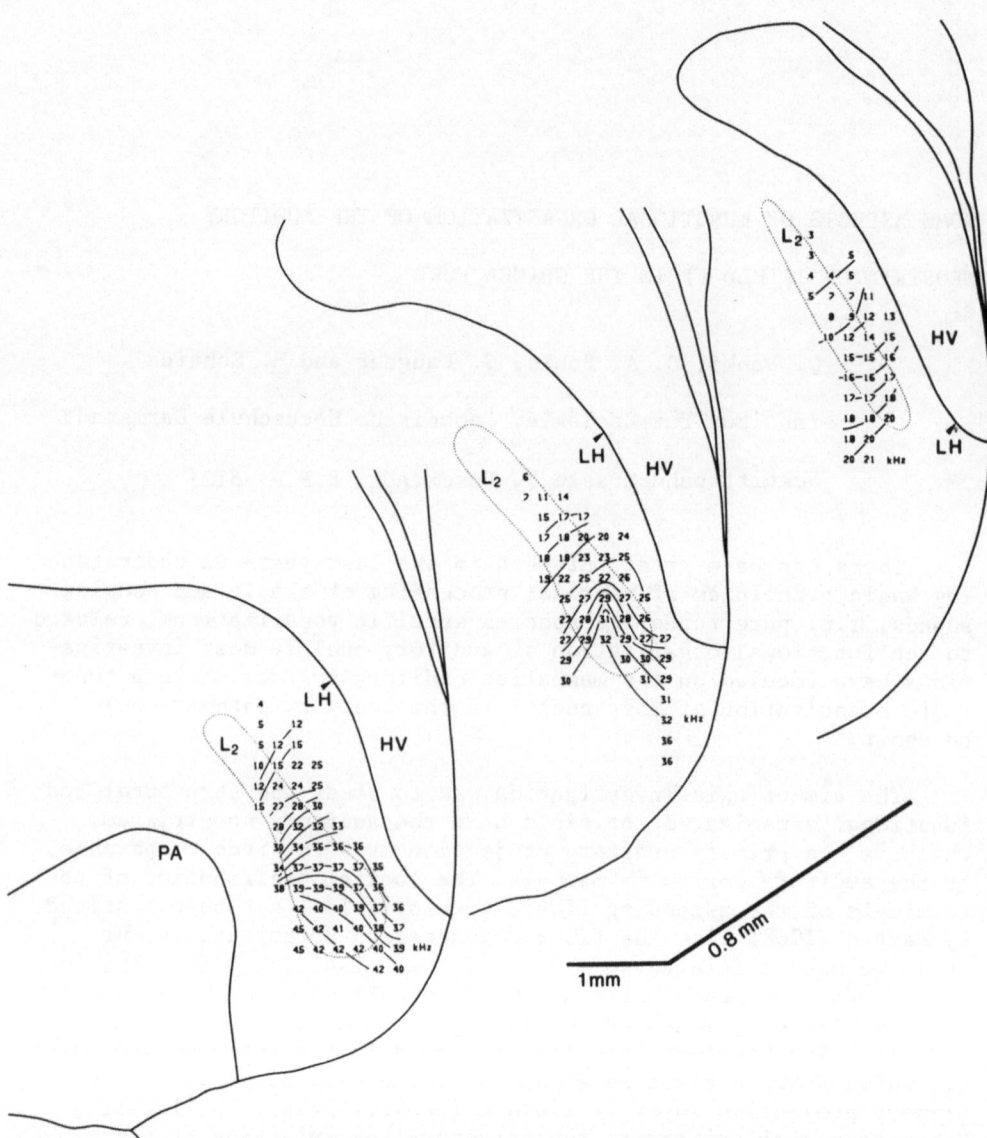

Fig. 1. Best frequency map of field L reconstructed in three
        different frontal planes, all 800 μm apart. Loci with the
        same best frequency are connected by isofrequency lines.
        HV hyperstriatum ventrale, LH lamina hyperstriatica, L₂
        layer L₂ of field L, PA paleostriatum.

Neuronal responses show a close correspondence to the anatomically defined layers L_1 - L_3. Neurons in L_2 show high response activity to pure tones and a high level of spontaneous activity. The most common type of units in L_2 shows tuning curves with best frequencies between 300 Hz caudally and 6 kHz rostrally with bands of inhibition on either side of the exitatory band and varying responsiveness at other frequencies. More uncommon and less frequent L_2 units show very wide, only unilateral or no bands of inhibition.

Neurons in L_1 and L_3 show less spontaneous activity and tone response. Comparable results can be shown by autoradiography of 2-deoxyglucose labelling in layers L_1 - L_3 (Scheich, 1979a). Neurons in field L_1 - L_3 are tonotopically organized. Fig. 1 shows three reconstructions of best frequency responses in three different frontal planes, 800 μm apart. Loci which show response to the same frequency are connected by isofrequency lines. In the dorsal part of field L isofrequency lines are crossing layer L_2 at right angle. In the ventral part of field L, which has the club shaped expansion, isofrequency lines are bent downwards. This gives the idea that functional organization may undergo some change. See results of 2-deoxyglucose labelling of this region (Scheich and Maier, this vol.). Isofrequency lines can be drawn in two different planes. This reveals that one single frequency is represented two dimensionally, i.e. in an isofrequency plane.

In addition neurons have been tested with species-specific calls in order to identify organizational aspects related to the discrimination of those calls which usually cover a wide frequency range. Four calls from the Guinea fowl repertoire (Maier, 1977) have been selected for stimulation: 1. the kecker, 2. the ee-trill, 3. the iambus call and 4. the tremolo. The frequency range of these calls is overlapping. From neuronal recordings at systematically varying places in field L during presentations of these calls PST histograms were calculated. The reconstruction of neuronal activity according to their x, y and z position showed that wide band calls activate large areas in all three layer of field L. The highest activity corresponds to layer L_2. This layer shows a clear shift of the activity distribution when stimulated with different calls. The shift can be explained by the tonotonic responsiveness of the area and by the different energy distribution of the calls.

Fig. 2 shows the power spectra of three calls and the responsiveness of 502 neurons from L_2 during presentation of the calls. To build up the histogram the integrated and normalized response of each neuron to a call is accumulated on that place of the frequency axis where each neuron had its best frequency. There is a close correspondence between energy peaks in the power spectrum of calls and the neuronal responsiveness. These results indicate that the energy peaks, in other words the formants of calls, determine the activity pattern of L_2 units (Scheich et al., 1979b).

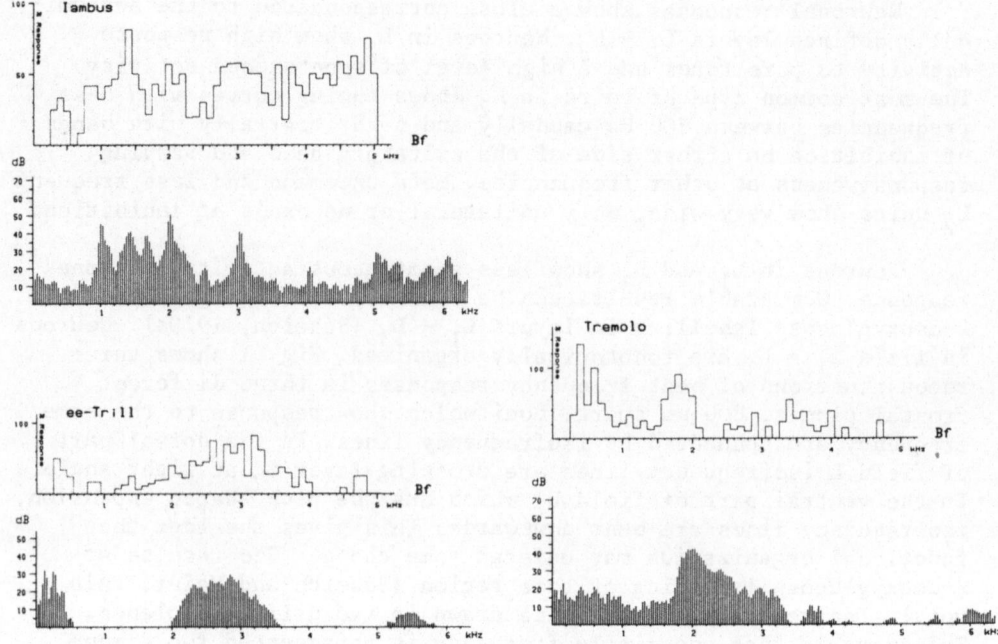

Fig. 2. Comparison of call responsiveness of 502 L_2 units to three
         different Guinea fowl calls (iambus, ee-trill, tremolo)
         with the best frequency of these units.

ACKNOWLEDGMENT

Supported by the Deutsche Forschungsgemeinschaft SFB 45.

REFERENCES

Aitkin, L. M., 1976, Tonotopic organization of higher levels of
    the auditory pathway, in: International review of physiology,
    neurophysiology II., R. Porter, ed., University Park Press,
    Baltimore.
Bonke, B. A., Bonke, D. and Scheich, H., 1979a, Connectivity of the
    auditory forebrain nuclei in the Guinea fowl (Numida meleagris).
    Cell Tissue Res., 200:101-121.
Bonke, D., Scheich, H. and Langner, G., 1979b, Responsiveness of
    units in the auditory neostriatum of the Guinea fowl (Numida
    meleagris) to species-specific calls and synthetic stimuli.
    I.Tonotopy and functional zones, J. Comp. Physiol., 132: 243-
    255.

Karten, H. J., 1968, The ascending auditory pathway in the pigeon (Columba livia). II. Telencephalic projections of the nucleus ovoidalis thalami, Brain Res., 11: 134–153.

Meier, V., 1977, Vocal communication in the Guinea fowl (Numida meleagris), Proc. XVth Int. Ethological Conference, Bielefeld, Section II, 48.

Scheich, H., Bonke, B. A., Bonke, D. and Langner, G., 1979a, Functional organization of some auditory nuclei in the Guinea fowl demonstrated by the 2-deoxyglucose technique, Cell Tissue Res. 204: 17–27.

Scheich, H., Bonke, D. and Langner, G., 1979b, Tonotopy and analysis of wide band calls in field L of the Guinea fowl, Exp. Brain Res. Suppl. 2: 94–109.

14-C-DEOXYGLUCOSE LABELING OF THE AUDITORY NEOSTRIATUM

IN YOUNG AND ADULT GUINEA FOWL

H. Scheich and V. Maier

Institut für Zoologie
Technische Hochschule Darmstadt
Schnittspahnstrasse 3, BRD-6101 Darmstadt

Auditory areas in the forebrain of the guinea fowl have proven to be excellent substrates for a functional analysis with the deoxyglucose technique in non-anesthetized birds. The method yields sufficient resolution to appreciate differential labeling after stimulation with tones and more complex patterns including species-specific calls (Scheich et al., 1979a). The aim of the present report is to describe some differences of stimulus induced labeling between young birds at various ages after hatching and adult birds. The emphasis hereby is placed on the tonotopic organization of these forebrain areas as revealed with this anatomical method.

A total of 25 birds between 2 and 150 hours and around 3 weeks after hatching were injected with $[14^C] - 2 -$ deoxyglucose ($40\mu Ci/100g$) into the pectoral muscle immediately before the 45 min exposure to acoustic stimulation. In 7 adult birds the vena cutanea ulnaris was cannulated with a polyethylene tube and deoxyglucose was administered ($18\mu Ci/100$ g) in three aliquots 1, 15 and 30 min after the onset of acoustic stimulation. This difference in the amount of injected deoxyglucose yielded comparable labeling in chicks and adults.

Field L in the caudal neostriatum is a primary auditory projection area which receives input from the thalamic n. ovoidalis (Karten, 1968; Bonke et al., 1979). As demonstrated with injection of labeled aminoacids into n. ovoidalis the projection to field L is not uniform and allows the distinction of three laminae, L_1, L_2 and L_3 in dorsoventral sequence, among which L_2 receives the bulk of the input terminals (Bonke et al., 1979). The lamination is related to physiological properties of units. Bonke, D. et al. (1979) have shown that L_2 harbours chiefly "simple" units with high

spontaneous activity, one best frequency and one or two inhibitory
bands. These units have little discriminative power for species-
specific calls while L_1 and L_3 units may be highly selective respon-
ders to calls (Scheich et al., 1979b). The units usually have low
spontaneous activity and more complex tuning properties, if they
respond to tones at all.

Field L is a tonotopically organized flat structure. In the
plane of L_2 where neurons have clear best frequencies isofrequency
contours can be plotted in approximately rostro-caudal direction
(Bonke, D. et al., 1979). In addition in the rostral and middle
thirds of the field isofrequency contours cut across all three laminae. Thus units with the same frequency input are in a two-dimensional plane, perpendicular to L_2 and in rostro-caudal direction.

Local differences in the physiological properties of units
correspond to the functional organization revealed with the deoxyglucose technique. In control birds which were restrained in a dark
sound-proof chamber without acoustical stimulation the lamina L_2 is
labeled throughout its extent. In frontal sections the pattern of
labeling corresponds in shape to a dumb-bell (Fig. 1). Spontaneous
labeling may correspond to the high spontaneous activity of units
or of the input terminals. When adult birds are stimulated with
a tone of 1 ms duration at 1 ms intervals a stripe pattern of labeling in the autoradiograms of frontal sections of the brain can be
followed from caudal to rostral (Fig. 1). The tone-activated stripe
is most distinct in the caudal part of the field (Fig. 1, C). In
the middle third of the field the tone-activated stripe cuts across
all three laminae and may form a cross pattern together with the
labeling in L_2 or may shrink to a dark spot in L_2 (Fig. 1, B). Each
frequency produces a spot or stripe in a specific position along L_2.
Low frequencies are represented most dorsally. Stimulation with a
harmonic tone of 1 kHz fundamental frequency produces at least 3
stripes corresponding to 1, 2 and 3 kHz in dorso-ventral sequence.
The distance between the stripes is roughly equal and measures
800 μm. From these and other pure tone experiments tonotopy is evident and likewise that the band between 0.5 and 3 kHz covers most
of the area of field L.

A tone-activated stripe in field L is visible in chicks already
2 hours after hatching (Fig. 2). Though the pattern of labeling
appears somewhat blurred and the area of L_2 is smaller in the chick
brain the overall features as described above are present. There are
two major differences however.

Firstly the cross pattern formed between L_2 and the tone activated stripe is also present in the rostral half of the field where
in the adult the stripe has shrunk to a spot in L_2. Among several
possible interpretations one is that in the chick the dominant
frequency input which is responsible for the best frequency of neu-

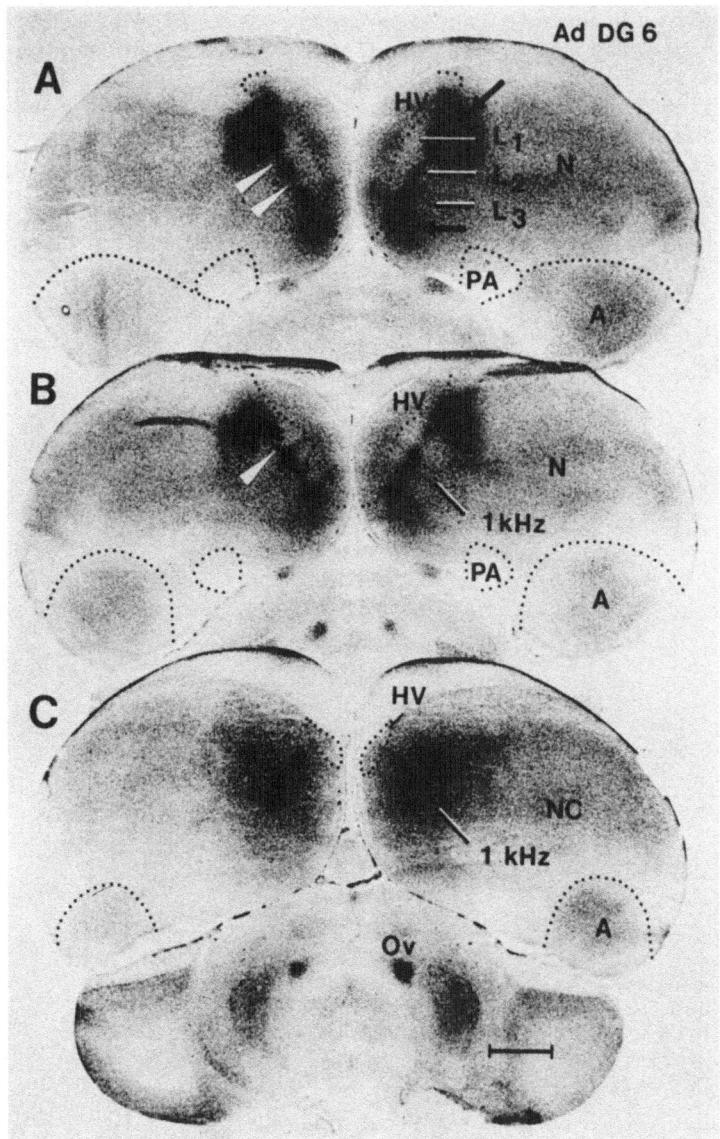

Fig. 1. Autoradiographs of deoxyglucose-labeled forebrain of the
Guinea Fowl in three different frontal planes. The stimulus
that produced the labelling was an 1 kHz pure tone. A section at the level of the rostral third of field L. Note
layers L_1, L_2 and L_3 of field L. L_1 extends up to the lamina
hyperstriatica, the dotted line between field L and the
hyperstriatum ventrale (HV). The black arrows mark the
swellings at the dorsal and ventral edges of the dumb-bell
shaped layer L_2. White arrows mark band of reduced (cont´d)

(Fig. 1 cont'd) labeling in L_2.
B Section at the level of the middle third of field L. The
marker indicates the 1 kHz-activated labeling in L_2. C Section from the caudal third of field L. The marker indicates
the 1 kHz-activated stripe-pattern. Scale = 2 mm. A archistriatum, N neostriatum, NC neostriatum caudale, PA palaeo-
striatum, Ov n. ovoidalis.

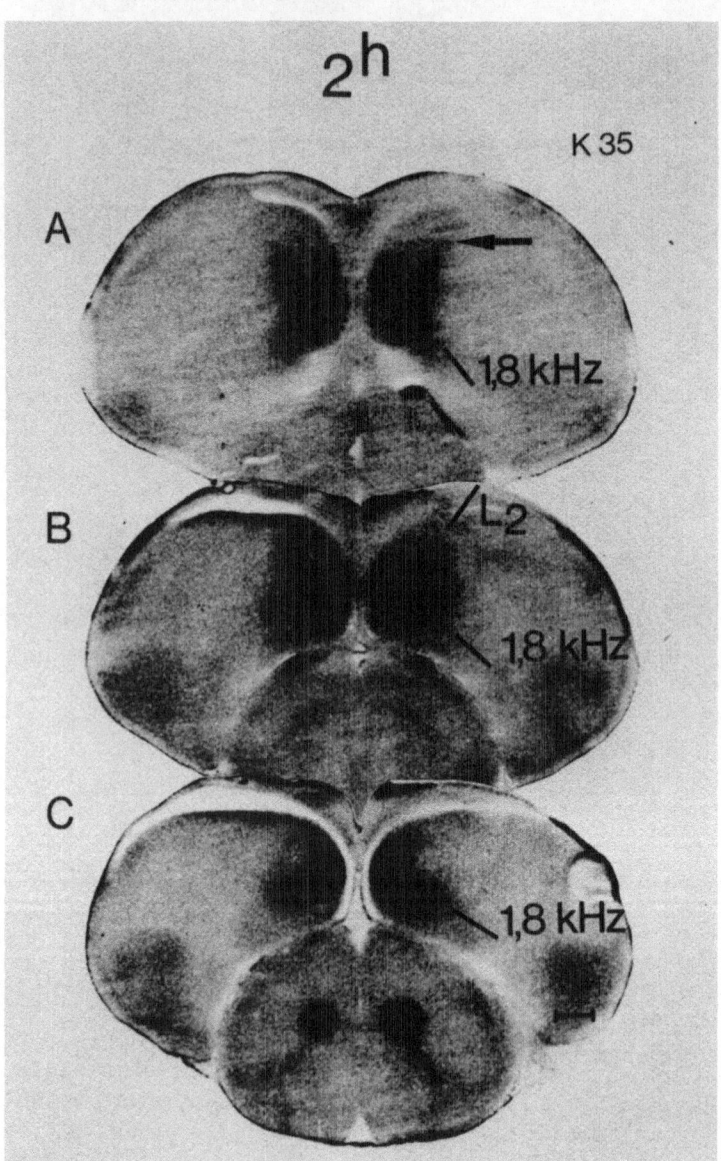

Fig. 2. Autoradiographs of sections through field L in (cont'd)

(Fig. 2 cont'd) a chick of two hours of age corresponding to the levels in Fig. 1. Note the cross pattern formed by the labeling of L_2 and the tone-activated stripe in A and B. The black arrow indicates the position where in the adult the dorsal swelling of L_2 is present.

rons within an isofrequency plane exerts rostrally a stronger influence in L_1 and L_2 than is the case in the adult. In the adult L_1 and L_2 neurons there have a high discriminative power for spectrally complex patterns (Bonke, D. et al., 1979). This property may not be fully developed in the young chick.

The second difference concerns inhibition in L_2. In the adult in the rostral and middle thirds of the field the tone-activated spot of labeling in L_2 is flanked by a more or less distinct reduction of spontaneous labeling of L_2 (Fig. 1, white arrows). This corresponds to lateral bands of inhibition seen in most L_2 neurons (Bonke, D. et al., 1979). In chicks of one day of age these interruptions in the labeling of L_2 are not present. It may be argued that the less sharply defined labeling in the proportionally small chick brain may blurr this feature. Since the lateral reduction of labeling is clearly present in some chicks of several days of age which have hardly larger brains there remains the possibility that inhibitory mechanisms are not fully developed in young chicks. An increasing power with age of inhibitory mechanisms may also be responsible for the reduction of tone activation in L_1 and L_2 neurons in the anterior parts of field L. Thus neurons in these laminae which selectively respond to complex patterns like species-specific calls may reach this level of selectivity by higher efficiency of inhibitory mechanisms.

The third difference concerns the labeling at the dorsolateral edge of the dumb-bell of L_2 (Fig. 1). In the adult this area is always labeled largely independent of the acoustic stimulus used. Judging from its location the area may represent the low frequency end of the tonotopic map. At present these two aspects are not easily reconciled. It is interesting to note that the labeling of this area is age dependent. Labeling is not observed in one day old chicks (Fig. 2).

ACKNOWLEDGMENT

Supported by the Deutsche Forschungsgemeinschaft SFB 45.

REFERENCES

Bonke, B. A., Bonke, D. and Scheich, H., 1979, Connectivity of the auditory forebrain nuclei in the Guinea Fowl (Numida meleagris),

Cell Tissue Res., 200: 101-121.

Bonke, D., Scheich, H. and Langner, G., 1979, Responsiveness of units in the auditory neostriatum of the Guinea Fowl (Numida meleagris) to species-specific calls and synthetic stimuli. I. Tonotopy and functional zones, J. Comp. Physiol., 132: 243-255.

Karten, H. J., 1968, The ascending auditory pathway in the pigeon (Columba livia). II. Telencephalic projections of the nucleus ovoidalis thalami, Brain Res., 11: 134-153.

Scheich, H., Bonke, B. A., Bonke, D. and Langner, G., 1979a, Functional organization of some auditory nuclei in the Guinea Fowl demonstrated by the 2-deoxyglucose technique. Cell, Cell. Tissue Res, 204: 17-27.

Scheich, H., Langner, G. and Bonke, D., 1979b, Responsiveness of units in the auditory neostriatum of the Guinea Fowl (Numida meleagris) to species-specific calls and synthetic stimuli. II. Discrimination of Iambus-like calls, J. Comp. Physiol., 132: 257-276.

INTEGRATION OF VOCO-AUDITORY CENTERS IN SONG BIRDS

Nozomu Saito and Masao Maekawa

Department of Physiology
Dokkyo University Medical School
Mibu, Tochigi 321-02, Japan

Among the problems of neural processing in auditory systems of animals, one of the most interesting is their perceptual strategy in sound communication. Avian vocalizations are developed along their own strategies of auditory perception (song ontogeny). The song ontogeny will be referred to the characteristic integration of the neural substrate of vocalization and hearing.

The concept of song ontogeny was classified into two types: the ontogeny of Type I is to develop normal song by referring to an auditory template which is given by a conspecies as an external model. Type I requires not only auditory feedback but also an auditory template. The conspecific call of Type II does not require any learning, but birds vocalize their calls on a genetically inherited basis (Nottebohm et al., 1972). Neither auditory feedback nor template is necessary.

Evolutionary trends increase the enlargement of two striata (hyperstriatum and archistriatum) of the birds of Type I, both of which contain vocal centers, hyperstriatum ventrale caudalis (HVc) and robustus archistriatalis (RA) (Karten et al., 1967; Nottebohm et al., 1976; Stokes et al., 1974). HVc of canary (Type I) has been defined by behavioral study (Nottebohm et al., 1976). In the present study, an attempt to produce vocalizations from canaries was made by electrical stimulation of HVc in the unrestrained posture without anesthesia. Stimulating pulse (0.1 ms width) trains with less than 100 μA of intensity, were applied at 10 Hz. In comparing the electrically-elicited vocalization with voluntary vocalization, left-sided stimulation of HVc caused the canary to vocalize with patterning of the phrase structure. The sonagram resembled the voluntary version. When the right HVc was stimulated the phrase structures were quite different and became labile. The spectra of sound elements were very diverse. The other vocalization nucleus, RA, of the canary and mynah telencephalon is clearly recognized, being surrounded by a distinct spherical boundary of thick layers of fibers. Electrical stimulation of RA elicited a pattern of song which consistently resembled a voluntary version.

Anterograde projections of the tracts of vocalization are observed from RA to the hypoglossal nerve (n.XII, motor nerve of the syrinx-vocal cord) by autoradiographic technique using ^{14}C-deoxyglucose intravenous application of 3μCi/10g body weight . Label of the ^{14}C-deoxyglucose consumed was examined with serial sections of mynah brains during the vocalization elicited by application of electrical train pulses to RA. Effective stimulation confined to RA permitted the production of a pattern of label above background. The extents of heavy labelling at the stimulation site, RA and adjacent labelled sites were examined. The label occurs in the long ipsi- and contralateral projections of tractus occipito-mesencephalicus (OM) and tractus thalamofrontalis medialis (TFM) to the brain stem, reaching n.XII (Fig.1). A high density of label is observed through all of ipsilateral OM. A small contingent density of bilateral label appears to occupy a portion where OM and TFM merge together just caudal to commissura anterior (CoA). A similar density of label occurs bilaterally in nucleus intercolliculus (ICo) and in medullary auditory nuclei(Fig. 1,d). A much higher density of label in OM reappears towards ipsilateral n.XII.

The anterograde pathways of the hemispheric RA travel through ipsilateral OM, although bilateral projections are observed in the midbrain including ICo and TFM. It is puzzling that label of both sides occurs contingently yet it occurs ipsilaterally at the level of n.XII. It is most probable that auditory responses of mynahs to their own electrically-elicited vocalizations would cause bilateral labelling of medullary auditory nuclei (refer to Scheich et al., 1979).

The deoxyglucose-labelled regions in the thalamus apparently include the unilateral auditory nucleus, OV (Fig. 1,b). A question arises whether RA connects directly or indirectly with OV. The question was examined using labelling with horseradish peroxidase (HRP) which was evidently accumulated in the cell somata by retrograde axonal transport. A 10 nl solution of HRP was injected into RA of the mynah and the canary. Leaving the injection site, retrograde label reaches two regions: one is the ipsilateral HVc, and the other one appears in the ipsilateral thalamus.

The latter is immediately adjacent to the auditory nucleus of the thalamus, nucleus ovoidalis (OV) (Fig. 2). The majority of labelled cells form a nucleus medial to OV. The tail of the nucleus continues ventrally and is interspersed in the fibers of OM, imme-

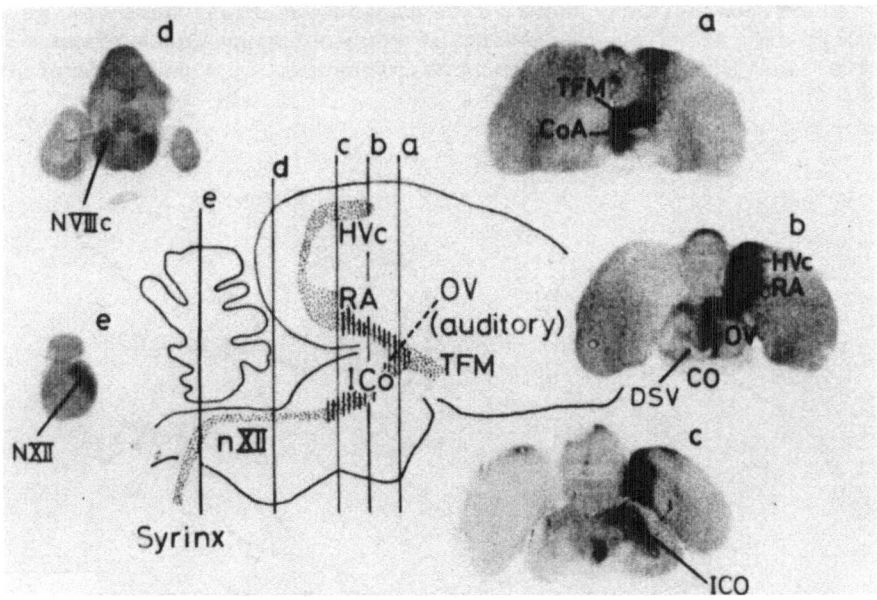

Fig. 1.   Autoradiographs of deoxyglucose-labelled brain of the
          mynah in five frontal planes. Right RA is stimulated to
          vocalize. Right brain is to the right. Vertical shaded
          area is tractus occipitomesencephalicus (OM).

Co, Chiasma opticum   NVIIIc, Nervus octavus,
CoA, Commissura anterior   pars cochlearis
DSV, Decussatio supraoptica   NXII, Nervus hypoglossus
     ventralis   OV, Nucleus ovoidalis
HVc, Hyperstriatum ventrale   RA, Nucleus robustus
ICo, Nucleus intercollicularis   archistriatalis
                                    TFM, Tractus frontalis-
                                         thalamicus medialis

diately underlying the caudal part of OV and the ovoidal tract
(TOV) which extend ventrally (Karten et al., 1968; Stokes et al.,
1974).

    There was no direct connection between RA and OV. However, an
evident anterograde projection to RA originated from the nucleus
immediately ventral to OV. The latter nucleus may be presumed to
connect RA and OV. An indirect connection of the voco-auditory
nuclei has been also demonstrated in the canary telencephalic
projection from field L to RA (Kelly et al., 1979). Accordingly,
voco-auditory integration in song birds seems to occur mostly via
an indirect connection, although only two species of song bird were
examined.

The voco-auditory neural integration described here may be
involved in the unique mechanism of song ontogeny which gives a
partial insight into the perceptual mechanism in sound communication.

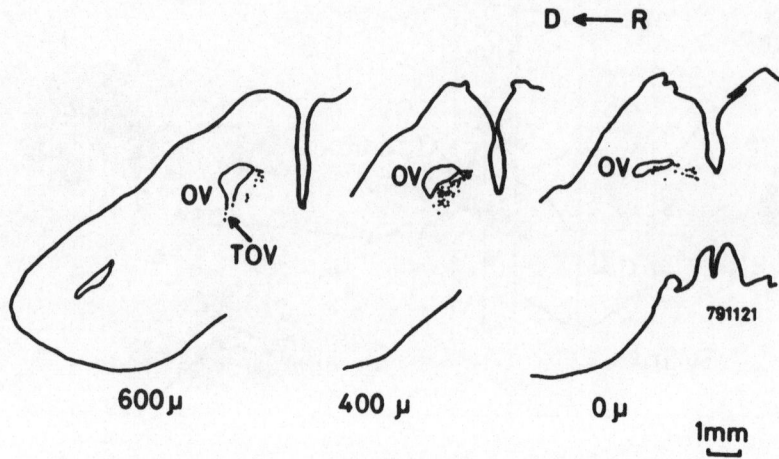

Fig. 2.   HRP-labelling of ipsilateral thalamic projection from left
          nucleus robustus archistriatalis (RA) of the mynah. Note
          the HRP labelled cells (dots) adjacent to nucleus ovoidalis
          (OV) and tractus ovoidalis (TOV). Arrow indicates rostro-
          caudal direction of the frontal sections.

REFERENCES

Karten, H. J., 1968, The ascending auditory pathway in the pigeon
     (Columbia livia) II. Telencephalic projections of the nucleus
     ovoidalis thalami, Brain Res., 11: 134-153.
Karten, H. J. and Hodos, W., 1967, "A Stereotaxic Atlas of the
     Brain of the Pigeon", The Johns Hopkins University press,
     Baltimore.
Kelly, D. B. and Nottebohm, F., 1979, Projection of a telencephalic
     auditory nucleus-field L- in the canary, J. Comp. Neurol.,
     183: 455-470.
Nottebohm, F., Konishi, M., Hillyard, S. and Marler, P., 1972,
     Ontogeny of acoustic behavior, in: "Auditory processing of
     biologically significant sounds", F. G. Worden et al., eds.,
     Neurosciences Res. Prog. Bull., 10: 31-49.
Nottebohm, F., Stokes, T. M. and Leonard, C. M., 1976, Central
     control of song in the canary, Serinus canarius, J. Comp.
     Neurol., 165: 457-486.

Scheich, H., Bonke, B. A., Bonke, D. and Langner, G., 1979, Functional organization of some auditory nuclei in guinea fowl demonstrated by the 2-deoxyglucose technique, Cell Tissue Res., 204: 17-27.

Stokes, T. M., Leonard, C. M. and Nottebohm, F., 1974, Telencephalon diencephalon, and mesencephalon of the canary, Serinus canaria in stereotaxic coordinates, J. Comp. Neurol., 156: 337-374.

RESPONSE PROPERTIES AND SPIKE WAVEFORMS OF SINGLE UNITS IN THE
TORUS SEMICIRCULARIS OF THE GRASSFROG (RANA TEMPORARIA) AS
RELATED TO RECORDING SITE

J. J. Eggermont, D. J. Hermes, A. M. H. J. Aertsen and
P. I. M. Johannesma

Department of Medical Physics and Biophysics
University of Nijmegen
Nijmegen, The Netherlands

INTRODUCTION

The torus semicircularis (T.S.) is the highest auditory nucleus
in Anurans from which routinely single unit recordings can be
obtained (Capranica, 1976). Its position in the midbrain has been
studied by Potter (1965a) who also distinguished various sub-nuclei.
The best frequencies (BF) of the units are reported to be bimodally
distributed in approximately a 1 : 1 ratio around a separation frequency of about 900 Hz (Potter, 1965b). A tonotopic organisation
does not seem to be present. Response latencies were reported in
the range of 8 - 150 ms at ventral parts of the torus (Potter,
1965a; Priot-Droy, 1977) which probably will affect spike waveforms. Therefore this study aims to relate spike waveforms, general
response properties and recording site in the torus in order to
describe the functional properties of the torus in lightly
anaesthetised grassfrogs.

METHODS

Lightly anaesthetised (0.025 - 0.03% of an MS-222 solution,
pH = 7) frogs showing a cornea reflex and feetpinch withdrawal
reflex were used. Topical anaesthesia was applied to the surgically
exposed tissue and the animal was placed in a sound proof room and
kept at a temperature of about 15° C. Stainless steel microelectrodes with an exposed tip of 15 μm and having a 2 - 14 MΩ resistance
were lowered through the optic tectum into the torus. The moments
of occurrence of single unit spikes were stored digitally and the
recorded signal was also stored on analog tape. Off-line cross-

correlation between spike moments and functionals of the digitally
stored stimulus ensemble was performed (Aertsen and Johannesma,
1979) to obtain a spectrotemporal receptive field of the units. BF
and latency were determined from the receptive field. At the end
of successful electrode tracks laesions were made by injecting
anodal currents of 0.3 μA for 10 seconds. After completion of the
experiment the animals were deeply anaesthetised and perfused
with Ringer followed by Bouin fixative. The brain was embedded in
paraffin and cut into 20 μm thick sections which were stained with
haematoxyline-eosine.

From the analog tape recorded electrode signals average spike
waveforms were obtained for most of the units and subjected to a
cluster analysis. This showed four distinct groups of waveforms
(Fig. 1) : Type I and Type II represented low voltage spikes (60 -
150 μV) while type III and type IV represented higher voltage
(≐ 300 μV) spikes. No correlation was found between spike waveform
and electrode impedance nor with particular animals.

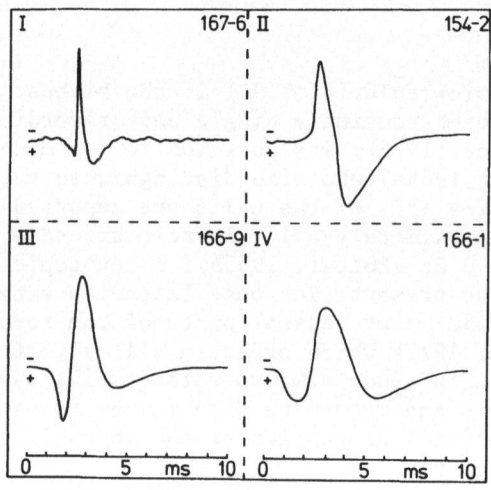

Fig. 1.   Spike waveforms in the torus semicircularis.

Sound was presented to the animal using closed sound systems
with a flat frequency characteristic (within 5 dB) up to 3000 Hz.
Left and right earphones had identical responses within 3 dB across
this frequency range.

RESULTS

Neural Response Properties

Recordings were made from 127 single units in the T.S. of 45
grass-frogs. From these units 45 showed a sustained response to
stationary noise but first order cross correlation between spike
moments and the noise signal did only produce results in two units
with BF below 300 Hz. Second order crosscorrelation between spike
moments and functionals of the noise or tonal stimuli resulted for
81 units in an unambiguous estimate of the BF. The distribution of
BFs suggests three groups of nerve fibers: two groups in the range
below 800 Hz and one group in the range above 800 Hz. The separation
at about 800 Hz results in a 1 : 1 ratio for low-plus middle-fre-
quency units and high-frequency units. In ten units double or multiple BFs could be discerned. No correlation between BG and spike
waveform was found.

Response latencies could be estimated accurately in 95 units
resulting in a multimodal distribution in the 12 – 138 ms range
(Fig. 2). Units responding in a sustained way to noise were nearly
invariably found in the short latency range (12 – 28 ms). No correlation was found between BF and latency. Type I waveform units in
majority were in the short latency range, whereas Type II units had
exclusively long latencies. Type III units were distributed uni-
formly across the entire latency range.

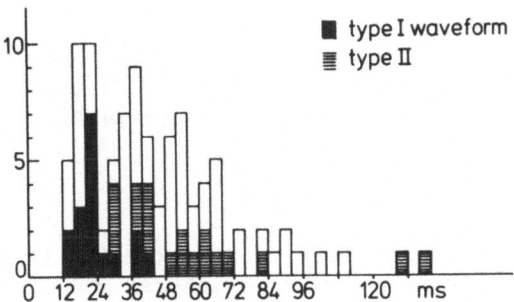

Fig. 2.   Response latency distribution for 95 units.

Recording Site

For classifying the recording sites in the torus the nomenclature and subdivisions as proposed by Potter (1965a) were adopted.
In addition it is suggested that the anteroventral part of the principal nucleus be a separate nucleus: the anteroventral nucleus. From

Table I.   Recording site, spike waveform and response properties.

	laminar n.	principal n.	antero-ventral n.	magno-cellular n.
Best frequency				
< 800 Hz	9	15	11	4
> 800 Hz	8	2	13	4
Multiple BF's	0	1	0	3
latency short	8	5	12	2
long	9	14	7	9
response to noise				
sustained	10	6	14	3
non-sustained	10	20	17	9
spike waveform				
I	8	3	6	1
II	1	2	11	6
III	10	16	13	7
IV	1	3	2	2
No. of units	20	28	32	17

the 127 units studied ten units could not be unambiguously attribu-
ted to one of the nuclei and ten other units were found outside the
T.S. The remaining 97 units showed the properties presented in
Table I.

DISCUSSION

The finding of a trimodal BF distribution in this study corro-
borates the observations in the peripheral auditory system in higher
Anurans that the fibers from the Amphibian papilla appear to split
up into two groups (Capranica, 1976). Fibers receiving this input
from the Amphibian papilla seem to dominate the caudal part of the
torus but evidence for a tonotopic organization could not be found.

In the anteroventral parts of the torus short latency units
were more abundant than in the caudo-ventral parts. In addition
short latency units and type I waveforms showed a large correspondence. This could both relate to the fact that incoming fibers
project upon the torus dominantly in the antero-ventral parts while
degenerating terminals were sparse in e.g. the magnocellular nucleus

(Rubinson and Skiles, 1975). The finding of type II waveforms nearly exclusively (17 out of 20) in the ventral parts of the torus suggests a correspondence to the cells, with numerous dendrites projecting in all directions, found by Potter (1965b) and Priot-Droy (1979) in this region. In the laminar and principal nucleus in contrast most cells show a single thick dendrite directed toward the caudocentral part of the torus. This correspondence of type II units with ventral parts of the torus forms the main reason for assigning a separate status to the anteroventral part of the principal nucleus.

ACKNOWLEDGMENT

This research project was supported by the Netherlands Organization for the Advancement of Pure Research (ZWO).

REFERENCES

Aertsen, A. and Johannesma, P., 1979, Spectro temporal analysis of auditory neurons in the grassfrog, in: "Hearing mechanisms and Spech",O. Creutzfeld, H. Scheich and Chr. Schreiner, eds., Suppl. II, Exp. Brain Res.

Capranica, R. R., 1976, Morphology and physiology of the auditory system, in: "Frog neurobiology",R. Llinas and W. Precht, eds., Springer, Berlin.

Potter, H. D., 1965a, Mesencephalic auditory region of the bullfrog, J. Neurophysiol., 28: 1132-1154.

Potter, H. D., 1965b, Patterns of acoustically evoked discharges of neurons in the mesencephalon of the bullfrog, J. Neurophysiol., 28: 1155-1184.

Priot-Droy, M. T., 1977, La typologie neuronique du tegmentum des Anoures, J. Hirnforschung, 18: 321-333.

Rubinson, K. and Skiles, M. P., 1975, Efferent projections of the superior olivary nucleus in the frog, Rana catesbeiana, Brain Behav. Evol., 12: 151-160.

CODING OF AMPLITUDE-MODULATED TONES IN THE MIDBRAIN AUDITORY

REGION OF THE FROG

N. Bibikov and Oxana Gorodetscaya

Acoustical Institute

Moscow, U.S.S.R.

Until recently it was generally accepted that the ability to reproduce temporal aspects of sounds deteriorated from the peripheral to central regions of the auditory pathway. This conclusion was made from experiments using gated tones and noises and from others dealing with sound waveform reproduction. But the investigation of response modulations produced by small amplitude modulations of sounds have shown that the situation is far more complicated. In auditory nerve fibers the frequency domain response functions for modulation are lowpass functions having cutoff frequencies in the range of 1000 Hz. Amplitude-modulated sounds with small amplitude modulation give rise to a modulation of a discharge rate in only a narrow range of sound intensities (Møller, 1981).

In the cochlear nuclei the frequency domain transfer functions usually have a distinct peak at a certain modulation frequency in the range 50-300 Hz. Throughout this dynamic range pronounced enhancement by sound modulations are observed (Møller, 1972). Thus there are transformations of the coding of small amplitude modulation from the auditory nerve to the cochlear nuclei.

The mechanisms for the further processing of small amplitude changes in the auditory system are more obscure. The studies of inferior colliculus and medial geniculate units deal generally with large indices of amplitude modulation. Only in bat has greater sensitivity to small amplitude modulations recently been demonstrated (Schuller, 1979).

We studied the neural coding of small amplitude changes in the midbrain auditory region of the frog. We well understood the possible difficulties of a direct comparison of our data from the frog with the results obtained in mammals. Our previous investigations have shown however that the principles of temporal coding in the auditory system of these vertebrates are rather similar (Bibikov, 1975).

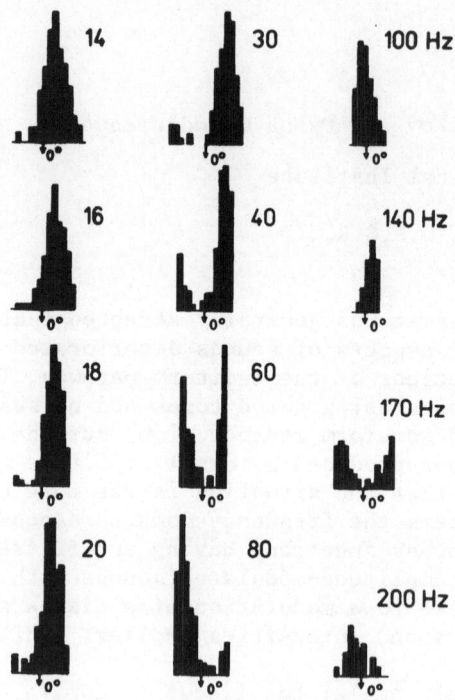

Fig. 1.   Cycle histograms of the activity of a unit in response to amplitude-modulated tones. The modulation frequency is indicated by numbers appended to each cycle histogram. The modulation of tone amplitude was 10% and its frequency (BF of the unit) was 1.6 kHz. The stimulus intensity was 20 dB above the unit's threshold.

We worked with immobilized lake frogs using a standard micro-electrode technique. A closed acoustic system with contralateral

stimulation was used. Best frequencies and response patterns to tone
bursts were recorded prior to stimulation with modulated sounds. All
tones were presented at the best frequency of the unit.

The great majority of the units located in the midbrain auditory region of the frog were incapable of generating a sustained
train of discharges in response to continuous pure tone, and units
responding tonically during our standard 300 ms tone bursts usually
ceased their firing after some seconds of stimulation. However even
a small amplitude modulation could change the response radically.
A 10% amplitude-modulated tone evoked a firing that lasted several
minutes with only moderate adaptation.

The discharge frequency of such units in response to sinu-
soidally-modulated tones is modulated over a wide range of sound
intensities. The shapes of the folded histograms synchronized with
the modulation waveform were usually similar to a half-rectified
sine wave (Fig. 1).

In a certain range of modulation frequencies a pronounced
enhancement of the modulation was observed. The folded histograms
were treated as vector distributions from which the synchronization
index phase angles were computed. In many cases during the presentation of a 10%-modulated tone the synchronization index substantially
exceeded 0.5 - the value corresponding to a 100% sinusoidal modulation.

For such units modulation transfer functions were obtained
using sounds with different frequencies of sinusoidal modulation.
(Fig. 2). On the basis of this characteristic the units could be
classified into three groups. In the first group synchronization was
maintained over a large range of modulation frequencies up to 200
Hz. The second group demonstrated lowpass filter characteristics
with cut-off frequencies in the range 15-50 Hz. Finally there are
some units which have a pronounced peak in their frequency characteristics.

The phase angles measured between the maximum of the modulation waveform and the folded histograms were usually linear functions of the modulation frequencies. Some regression lines are shown
in Fig. 3.

Usually the mean rate of discharge decreased markedly during
a long presentation of an amplitude-modulated tone. Our observations
show that these adaptive changes considerably increase the synchronization of the discharges with the modulation waveform. The
synchronization obtained during the first seconds of the sinusoi-
dally-modulated tone was substantially lower than it was after some
ten seconds. This suggests that adaptation increases neuron sensi-

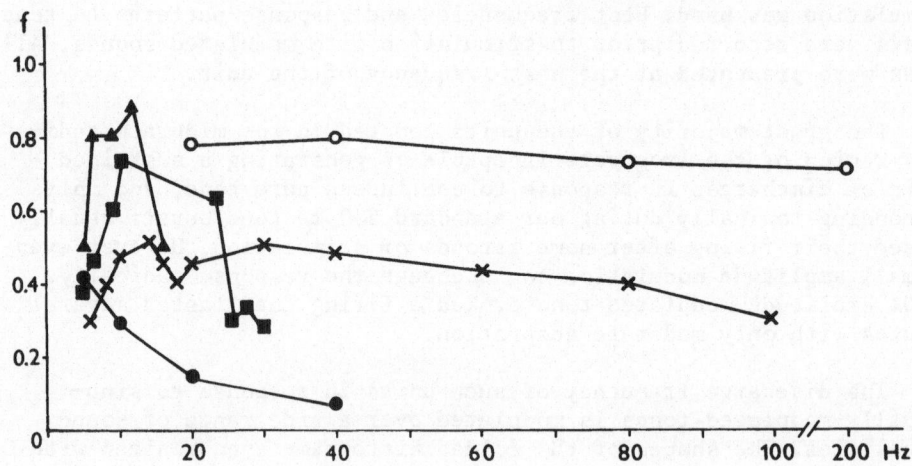

Fig. 2.   The dependence of synchronization indices on the frequency
          of modulation. Results from 5 representative units are
          shown.

tivity to a small amplitude modulation. It is of interest that
related phenomena have been observed in psychoacoustical experiments.
Recently Viemeister (1979) has explored this question using the
forced-choice procedure. He found that the threshold for amplitude
change detection at low modulation frequencies increased when a
continuous tone was substituted for an interrupted tone.

      In conclusion let us summarize the main distinctions in the
reactions of the medullary and mesencephalic auditory units to
amplitude modulated tones. 1. In the midbrain less than half the
units respond tonically throughout the long amplitude-modulated
tones. In cochlear nuclei such long responses occur in the great
majority of units. 2. The tonic midbrain units however respond to
modulated tones far more vigorously than to pure tones. In the
cochlear nuclei of the rat "it is a general finding that the mean
discharge rate is the same for modulated tones and for unmodulated
tones" (Møller, 1974). 3. The enhancement of the neuronal discharges
by small sinusoidal amplitude modulations is generally far more
pronounced in frog midbrain tonic neurons than in units located in
the mammalian cochlear nuclei. 4. In midbrain units the shapes of
the folded histograms are usually distorted. The responses to sinusoidally modulated sounds appear as half-wave rectified sine waveforms. In cochlear nuclei the shapes of these histograms are usually
sinusoidal. 5. In midbrain tonic neurons the adaptation processes
increase the synchronization of the discharges to the modulation

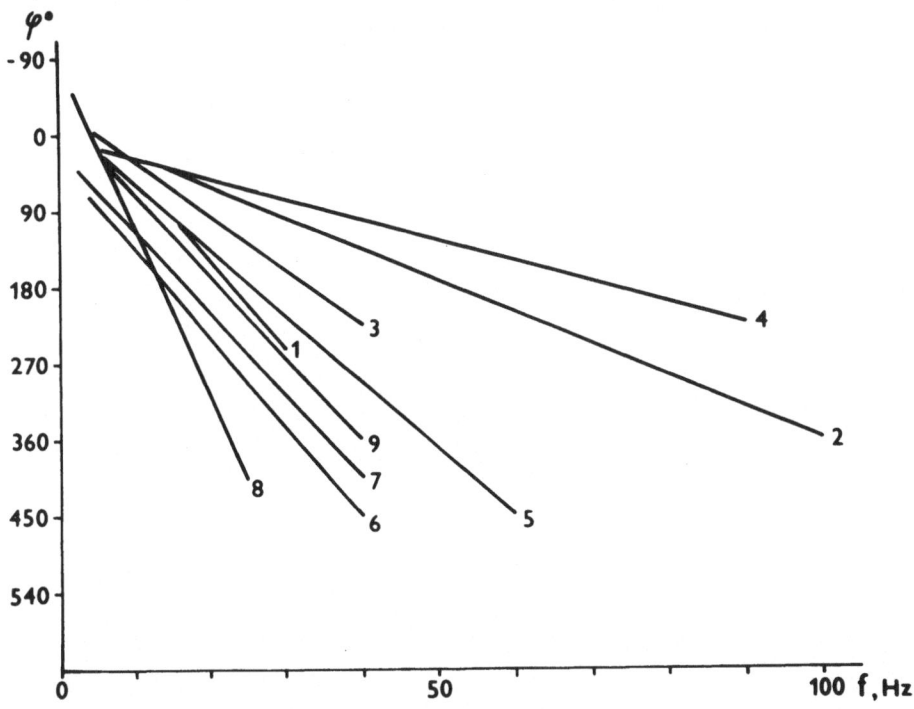

Fig. 3. The phase angle between the modulation of the sound and
         discharges as a function of the modulation frequency.
         Regression lines from 9 units are shown.

waveform. In cochlear nuclei this effect is far less pronounced and
there was "only a slight difference in the reproduction of the
modulation waveform for continuous tones and tones with a wide range
of interruption rates" (Møller, 1973).

REFERENCES

Bibikov, N., 1974, Encoding of the stimulus envelope in peripheral
        and central regions of the auditory system of the frog, Acust-
        ica, 31: 310-314.
Møller, A., 1973, Statistical evaluation of the dynamic properties
        of cochlear nucleus units using stimuli modulated with pseudo-
        random noise, Brain Res., 57: 443-456.
Møller, A., 1974, Responses of units in the cochlear nucleus to
        amplitude-modulated tones, Exp. Neurol., 45: 104-117.
Møller, A., 1981, This volume.
Schuller, G., 1979, Coding of small sinusoidal amplitude modulations
        in the inferior colliculus of "CF-FM" bat, Rhinolophus fer-

rumequinum, Exp. Brain Res., 34: 117-132.
Viemeister, N., 1979, Temporal modulation transfer functions based
    upon modulation thresholds, J. Acoust. Soc. Am., 66: 1364-1380.

SESSION VIII
DEPRIVATION AND DEVELOPMENTAL STUDIES
Chairmen: C. N. Woolsey and J. Mysliveček

EFFECTS OF EARLY AUDITORY STIMULATION ON CORTICAL CENTERS

J. Hassmannová, J. Mysliveček* and V. Nováková**

Dept. of Physiology, Medical Faculty of Hygiene,
Charles University, Prague
*Inst. of Hygiene and Epidemiology, Prague
**Inst. of Physiology, Czechosl. Academy of Sciences
Prague, Czechoslovakia

INTRODUCTION

The aim of the present study was to investigate if and how early stimulation induces functional and biochemical changes in the cerebral cortex of the rat. We have shown previously that early complex or visual stimulation increases the protein and ribonucleic acid (RNA) content in the occipital cortex of 4-week-old rats and enhances amplitudes of visual cortical evoked potentials (EP) of adult animals (Mysliveček and Štípek, 1979). Recently (Hassmannová et al., 1980), comparing the effects of early complex (visual, auditory and somesthetic-kinesthetic) stimulation with visual stimulation alone, we were able to reveal an increase of the total RNA content in the cortical neurons of projection areas of those modalities which were involved in the early stimulation, changes in the visual projection area being most significant. A decrease in the neuronal RNA content was found in the prefrontal cortex. The present communication deals with the effects of early auditory stimulation during the second postnatal two weeks of rat ontogeny.

METHODS

Rat pups were stimulated by intermittent acoustic stimuli (1 s tone pips, 3kHz/10s, 70dB) alone or combined with optic stimuli of the same frequency and exposure to somesthetic-kinesthetic stimulation (forced movements in an arena 30x30x30cm with a moving floor, speed 200mm/s). In total, there were 30 stimulation sessions à 30 min. The effect of the stimulation was studied twice: immedi-

ately after the stimulation period (28 days of age) and 4 weeks
later(in 56-day-old animals). EPs to auditory, visual and somesthetic stimuli were recorded in each of the corresponding projection

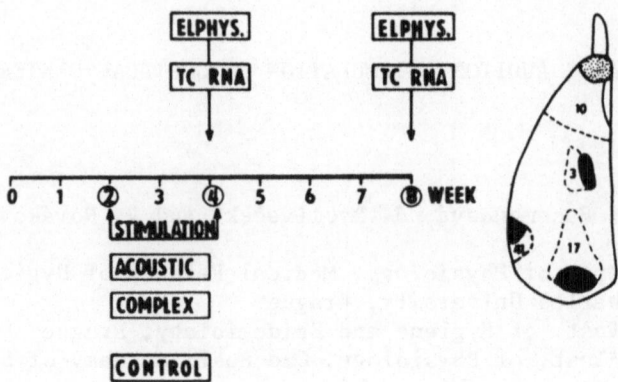

Fig. 1. Schedule of the experiment; on the right - sites of record-
         ing and specimen taking.

areas in awake immobilized rats at the stimulation rates of 0.4, 1,
4 and 8/s. At these sites, we also estimated the total RNA content
in the cortical neurons by means of an integrating microdensitometer;
this was also done in the prefrontal cortex (Fig. 1).

EVOKED POTENTIALS

EPs were averaged across groups (10 animals per group, 20 stimuli within individual stimulation rate in each animal, i.e. 200 stimuli in each averaged record). The peak latencies of the primary
positive-negative EP complex to click stimulation in the auditory
cortex of 4-week-old animals were 17.25 and 32.45 (on average) in
controls, 13.45 and 26.1 respectively in the group with early auditory stimulation, and 13.45 and 31.2 ms respectively in animals after
complex stimulation. This represents a latency decrease of more than
20% in animals after auditory stimulation relative to controls and
a smaller decrease in rats after complex stimulation. This could be
considered as a functional improvement of the projection areas, but
the integrated amplitudes of P_1, N_1 and P_2 waves in animals with
early auditory stimulation were decreased by about 40% at lower
stimulation rates and this was still more evident at the rate of
8/s. No decrease was observed after early complex stimulation. EP
amplitudes to compound visual-auditory stimuli relative to EP

amplitudes to auditory stimuli alone were changed in the following
manner: a decrease in 4-week-old control rats and in those stimulated by auditory stimuli during the 2nd two weeks, but an increase
in rats stimulated formerly with complex stimuli; at the age of 8
weeks an increase in both stimulated groups, whereas an unsignificant
decrease in controls.

TOTAL CONTENT OF RNA IN CORTICAL NEURONS

At the age of 4 weeks, the total RNA content was highly significantly increased in the acoustically stimulated animals in all
cortical projection areas studied. These changes are similar to those
we obtained (Hassmannová et al., 1980) after complex stimulation;
no changes were, however, recorded in the prefrontal cortex (Fig. 2).
At the age of 8 weeks the trend of changes in the projection areas
was principally maintained, only the statistical significance of
differences between the groups was decreased in the auditory and
somatosensory projection areas; the RNA content in the prefrontal
neurons was increased this time. The dynamics of changes in the
prefrontal cortex were different from those after complex stimulation.

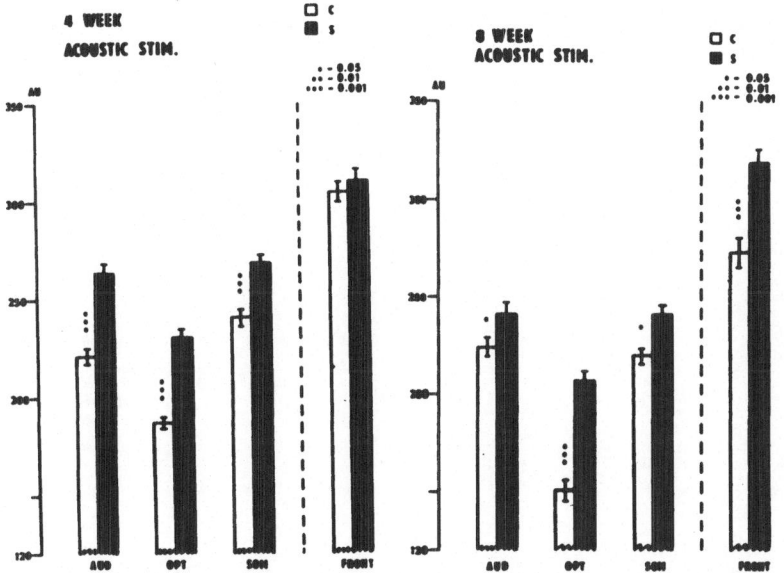

Fig. 2. Total RNA content (arbitrary units-AU) in the auditory
         (AUD), visual (OPT), somesthetic (SOM) and prefrontal
         (FRONT) cortical areas in control (C) and in acoustically
         stimulated animals at the age of 4 and 8 weeks. Significant
         differences between groups indicated by dots.

CONCLUSION

We want to emphasize only two facts: first that a non-continuous stimulation of the auditory system implemented within physiological limits during the second postnatal two weeks produces long-term changes of functional and biochemical parameters at least of cortical neurons. Second, contrary to visual stimulation inducing changes of EPs and RNA content only in the visual projection area, and moreover in the prefrontal cortex, auditory stimulation induces changes in the auditory, visual and somatosensory projection areas. The internal mechanisms of this multiple influence have to be revealed.

REFERENCES

Hassmannová, J., Mysliveček, J. and Nováková, V., 1980, Changes of the total ribonucleic acid content in the cortical neurons after early stimulation. IBRO Monographs, Raven Press. In press.
Mysliveček, J. and Štípek, S., 1979, Effects of early visual a and complex stimulation on learning, brain biochemistry and electrophysiology, Exp. Brain Res., 36: 343-357.

EFFECTS OF ACOUSTIC DEPRIVATION ON MORPHOLOGICAL PARAMETERS

OF DEVELOPMENT OF AUDITORY NEURONS IN RAT

J. Coleman

Department of Physiology
University of South Carolina
Columbia, South Carolina, U.S.A.

Acoustical experience, along with predetermined maturation
processes, appears to participate in the development of neurons of
the central auditory pathway. In cat, physiological investigations
have demonstrated a decrease in neural response latency with
maturation (Aitkin and Moore, 1975; Romand and Marty, 1975). Sound
deprivation during development was reported to produce changes in
binaural interactions and latencies of neurons in the rat inferior
colliculus (Silverman and Clopton, 1977; Clopton and Silverman,
1978). Morphological investigations in mice (Webster and Webster,
1977, 1979) and rats (Coleman and O´Connor, 1979) have further shown
that sound attenuation during early development leads to reduction
in cell size of central auditory neurons. The present investigation
was aimed at morphological parameters of normal development and the
effects of one form of acoustical deprivation during infancy.

Thirty two Sprague-Dawley rats from three litters each enclosed
with their mother served as subjects. Sixteen normally reared animals were sacrificed at one of the following ages: 10 days; 16 days;
24 days; or 36 days. Twelve rats were subjected to monaural ear
canal ligatures at 10, 16, or 36 days and sacrificed at 70 days
along with normal controls. Following cardiac perfusion with forma-
lin-saline and standard brain exposure to a sugar-formalin solution,
40 μm sections were cut and stained in cresyl violet. Total volumes
of the ventral and dorsal cochlear nuclei were calculated from
planimetric measurements at x20. In adult normal and sound-deprived
animals all neurons classified as large spherical cells (area III
of Harrison and Irving, 1965) were drawn under oil immersion at
x1000 using a camera lucida from a sample of two corresponding
representative sections in each anteroventral cochlear nucleus.
Neurons with visible nuclei and nucleoli were measured repeatedly

by planimetry for two-dimensional cell area calculations. Data from
all conditions was subjected to computer analysis of variance.

Results from volume measurements of the dorsal and ventral
cochlear nuclei at different developmental stages are shown in
Fig. 1. The dorsal cochlear nucleus of normal rat pups sacrificed
at 10 days of age prior to meatus opening shows a mean volume

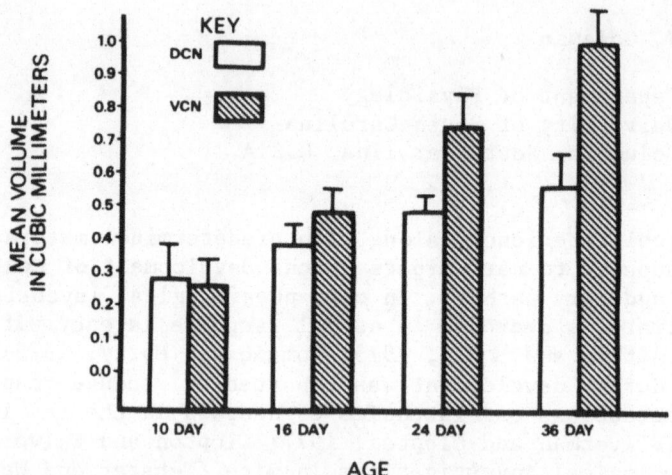

Fig. 1.   Volume of dorsal and ventral cochlear nuclei at different
          postnatal ages. S. E. mean indicated.

of 0.30 mm^3. This increases to 0.41 mm^3 at 16 days, 0.50 mm^3 at
24 days, and 0.58 mm^3 at 36 days. Differences in dorsal cochlear
nucleus volume between all age groups was significant beyond the
.05 level. A more dramatic developmental increase in volume occurs
in the ventral cochlear nucleus which contains the large spherical
cells, as well as other neuron types (e.g., small spherical cells,
globular cells, etc.). From 10 day old animals which show a mean of
0.29 mm^3 volume of the ventral cochlear nucleus increased dramati-
cally to 0.51 mm^3 at 16 days. Further large increases in volume
were observed at 24 days (0.76 mm^3) and 36 days (0.93 mm^3). Differences in all ventral cochlear nucleus groups were statistically
significant.

Neurons classified as normal adult large spherical cells had a mean two-dimensional area significantly larger than those of the deprived side in surgically ligated animals. The latter sample included 1,078 neurons. Similar differences were observed between the deprived and non-deprived sides of treated animals. Large spherical cells of rats deprived from 16 days were larger than those of the 10 day deprivation group, but significantly smaller than those of normals. However, cells of rats deprived from 36 days of age were not significantly smaller than cells from normals.

The ventral cochlear nucleus and the dorsal cochlear nucleus of rat appear to have different rates of growth. Total volume of the dorsal cochlear nucleus increases moderately from 10 to 36 days in the rat. Growth of the ventral cochlear nucleus is somewhat more striking, especially between 10 and 16 days. This period in rat is characterized by opening of the auditory meatus, onset of the Preyer reflex and recordable auditory nerve compound action potentials (Wada, 1923; Crowley and Hepp-Reymond, 1966; Bosher and Warren, 1971; Carlier, Lenoir and Pujol, 1979). Growth of the ventral cochlear nucleus may reflect the rapid development of auditory nerve activity input observed during this time (Carlier, Lenoir, and Pujol, 1979). Differential growth of perikaryl, fiber and glial elements within this period is under study.

As a consequence of ligature-induced monaural sound deprivation from 10 days mature rats exhibit a reduction in large spherical cell size. This supports previous observations of acoustical deprivation in rat (Coleman and O'Connor, 1978) and in mouse (Webster and Webster, 1977, 1979). In our previous work, ear ossicles were surgically removed and smaller samples of measurements were taken from the largest most distinct cells of the large spherical cell region of the anteroventral cochlear nucleus. In the present study, all neurons categorized as large spherical cells were measured in sample sections. In contrast, sound deprivation from 36 days of age produced no deficit in perikaryl size which suggests that some critical period exists for the impact of environmental acoustic stimuli on morphological development of large spherical cells. The morphological changes resulting from sound deprivation during development are clearly not as extreme, however, as the central degenerative effects produced by otocyst ablation observed in chicks (Parks, 1979). In rat, monaural acoustic deprivation was shown to have a striking effect on binaural electrophysiological responses of the inferior colliculus (Silverman and Clopton, 1977; Clopton and Silverman, 1978). Although appropriate behavioral data in rat is absent, monaural ear plugs in developing guinea pigs were shown to later produce more pronounced deficits in sound localization under binaural conditions than after rearing with binaural ear plugging (Clements and Kelly, 1978).

ACKNOWLEDGMENT

Supported by the Deafness Research Foundation and NIH
Ag-1571.

REFERENCES

Aitkin, L. M. and Moore, D. R., 1975, Inferior colliculus. II.
    Development of tuning characteristics and tonotopic organiza-
    tion in central nucleus of the neonatal cat, J. Neurophysiol.,
    38: 1208-1216.
Bosher, S. K. and Warren, R. L., 1971, A study of the electroche-
    mistry and osmotic relationships of the cochlear fluids in the
    neonatal rat at the time of the development of the endocochlear
    potential, J. Physiol. (London), 212: 739-761.
Carlier, E., Lenoir, M. and Pujol, R., 1979, Development of cochlear
    frequency selectivity tested by compound action potential
    tuning curves, Hearing Res., 1: 197-201.
Clements, M. and Kelly, J. B., 1978, Auditory spatial responses of
    guinea pigs (Cavia porcellus) during and after ear blocking,
    J. Comp. Physiol. Psychol., 92: 34-44.
Clopton, B. M. and Silverman, M. S., 1978, Changes in latency and
    duration of neural responding following developmental auditory
    deprivation, Exp. Brain Res., 32: 39-47.
Coleman, J. R. and O'Connor, P., 1979, Effects of monaural and
    binaural sound deprivation on cell development in the antero-
    ventral cochlear nucleus of rats, Exp. Neurol., 64: 553-566.
Crowley, D. E. and Hepp-Reymond, M.-C., 1966, Development of
    cochlear function in the ear of the infant rat, J. Comp.
    Physiol. Psychol., 62: 427-432.
Harrison, J. M. and Irving, R., 1965, The anterior ventral cochlear
    nucleus, J. Comp. Neurol., 124: 15-42.
Parks, T. N., 1979, Afferent influences on the development of the
    brain stem auditory nuclei of the chicken: otocyst ablation,
    J. Comp. Neurol., 183: 665-678.
Romand, R. and Marty, R., 1975, Postnatal maturation of the cochlear
    nuclei in the cat: a neurophysiological study, Brain Res.,
    83: 225-233.
Silverman, M. S. and Clopton, B. M., 1977, Plasticity of binaural
    interaction. I. Effect of early auditory deprivation, J. Neuro-
    physiol., 40: 1266-1274.
Wada, T., 1923, Anatomical and physiological studies on the growth
    of the inner ear of the albino rat, Am. Anat. Mem. No. 10.
Webster, D. B. and Webster, M., 1977, Neonatal sound deprivation
    affects brain stem auditory nuclei, Arch. Otolaryngol., 103:
    392-396.
Webster, D. B. and Webster, M., 1979, Effects of neonatal conductive
    hearing loss on brain stem auditory nuclei, Ann. Otol. Rhinol.
    Laryngol., 88: 684-688.

BEHAVIORAL AND ANATOMICAL STUDIES OF CENTRAL AUDITORY DEVELOPMENT

Jack B. Kelly

Department of Psychology
Carleton University
Ottawa, Canada, K1S 5B6

Behavioral studies of the development of hearing indicate that central auditory processes associated with sound localization are in a state of functional readiness soon after the onset of peripheral sensitivity to sounds. In human infants, the ability to localize sounds in space can be demonstrated shortly after birth (Mendelson and Haith, 1976; Muir and Field, 1979; Turkewitz, Birch, Moreau, Levy and Cornwell, 1966; Wertheimer, 1961). Also, in the highly precocial guinea pig, directed approach responses to sound are seen as soon as the first four days after birth (Clements and Kelly, 1978). Furthermore, in guinea pigs, it has been shown that even at this early stage of development binaural cues are important for maintaining reliable approach responses toward a sound source. Monaural ear blocks reduce directional responding to a chance level even over periods of prolonged testing. From this observation, one can infer that central structures involved in the integration of binaural cues and the execution of appropriate spatial responses are already active. Central mechanisms for sound localization are present in the infant very early in development.

More recent studies have shown that these conclusions may apply also to altrical mammalian species once the much later development of the peripheral receptor has been taken into account (Potash and Kelly, 1978, 1980). For example, in the rat directional responses to auditory stimuli appear as early as the first week after the onset of cochlear responses to moderately intense sounds. During this period approach responses can be elicited by continuous playback of tape-recorded vocalizations made by rat pups together in their home cage. Again, the importance of binaural integration can be demonstrated by ear blocking studies. Monaural ear blocks result in a reduction of approach responses to a chance level, whereas

binaural ear blocks have virtually no effect on these responses.
Similar results have been obtained with gerbils.

Auditory approach responses of the type described for the
guinea pig, rat and gerbil are complex, and involve an element of
motivation as well as motor skill for execution. To pinpoint the
onset of sound localization more accurately, it was considered
desireable to look at a simpler response, or, at least, to address
the issue of sensory development using a different response. Peter
Judge and I have employed a simple test based on head orientation
to investigate the onset of sound localization in the rat. Infant
rats are placed inside a small cage made of wire mesh located
inside the quiet environment of a sound attenuated room. Two loudspeakers are positioned on the left and right of the cage. After
the rat pup has settled down in this situation and is in a position
approximately 90° to the axis between the two speakers, a single
520 millisecond noise burst (2-32 kHz) is presented from left or
right loudspeaker. The behavioral response of the rat is observed
by closed-circuit television with a camera mounted directly above
the test cage. From videotaped records, orientation responses are
scored in terms of the percentage of responses toward the sound
source, and the magnitude of error in orientation. Typically, ten
trials are given over a two day period. Fig. 1 shows the results
of a study with groups of rat pups at 8, 11, 14, 17 and 20 days of

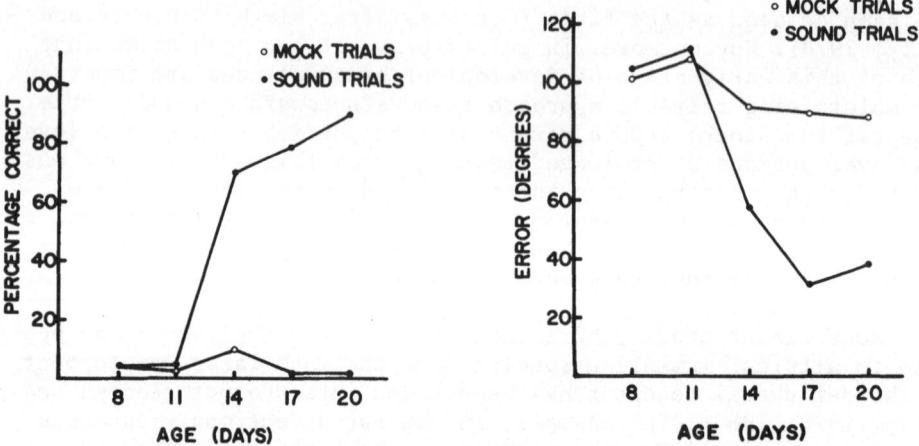

Fig. 1.   Head orientation scores for rat pups at different ages.
          Average percentage of trials with orientation toward the
          sound source is shown on the left. The magnitude of error
          in orientation is shown on the right. Six animals were
          used in each age group.

age. Clearly the first indication of directional head orientation
to sound is on day 14. The onset of sound localization is between
days 11 and 14 which corresponds closely to the onset of cochlear
responses to sounds of moderate intensity.

These behavioral studies have been paralleled by anatomical
studies of the developing auditory system using (^{14}C) -2-deoxyglu-
cose autoradiography(Kelly and Skeen, 1979). Rat pups of different
ages including 8, 11, 14, 17 and 20 days were given intraperitoneal
injections of (^{14}C) -2- deoxyglucose (40 μCi/100 g body weight),
and were subjected to 45 minutes of noise (2-32 kHz) at 80 dB SPL
from an overhead loudspeaker. To prevent adaptation the noise was
gated in pulses of 70 milliseconds and was presented at a rate
which varied from 5/s to 10/s. Following auditory stimulation the
animals were sacrificed, and their brains were removed and frozen.
Thirty micron sections were taken in the frontal plane throughout
the auditory system from cochlear nucleus to auditory cortex, and
were exposed to Kodak SB-5 and Lo´dose mammography films. The
resulting autoradiographs reflect the pattern of metabolic activity
in the central nervous system (Kennedy, DesRosiers, Jehle, Reivich,
Sharpe and Sokoloff, 1975; Plum, Gjedde and Samson, 1976).

Stimulated animals were compared with non-stimulated, ear
plugged animals to determine the amount of stimulus induced metabolic activity in central auditory structures. Generally, in older
animals increased metabolic activity was seen in brain stem structures as a result of noise stimulation. Differences between stimulated and non-stimulated cases were particularly apparent in the
cochlear nucleus and inferior colliculus. Stimulus induced activity
was much less obvious in the medial geniculate and auditory cortex.

The development of central auditory activity becomes apparent
in comparisons of animals of different ages. Fig. 2 illustrates
the response seen in the caudal inferior colliculus of animals at
days 8, 11, 14 and 20. The activity in the inferior colliculus of 8
and 11 day rat pups (a and b) is low and is not noticeably greater
than that seen in non-stimulated animals. Also, no differences were
found in other brain stem structures. Stimulus induced activity is
first seen in the inferior colliculus of 14 day animals (c). Only
part of the inferior colliculus is affected by stimulation at this
age, which may be related to restricted sensitivity to sound in
peripheral structures. By days 17 and 20 (d) a broad band noise
stimulus results in a more evenly distributed response in the
inferior colliculus. The pattern is similar to that seen in adult
(60 day) rats under the same stimulus conditions.

In conclusion, both behavioral and anatomical studies agree
that there is an onset of central auditory function in the rat
around days 11-14 postnatally. The beginning of central auditory
processing in turn closely corresponds to the development of

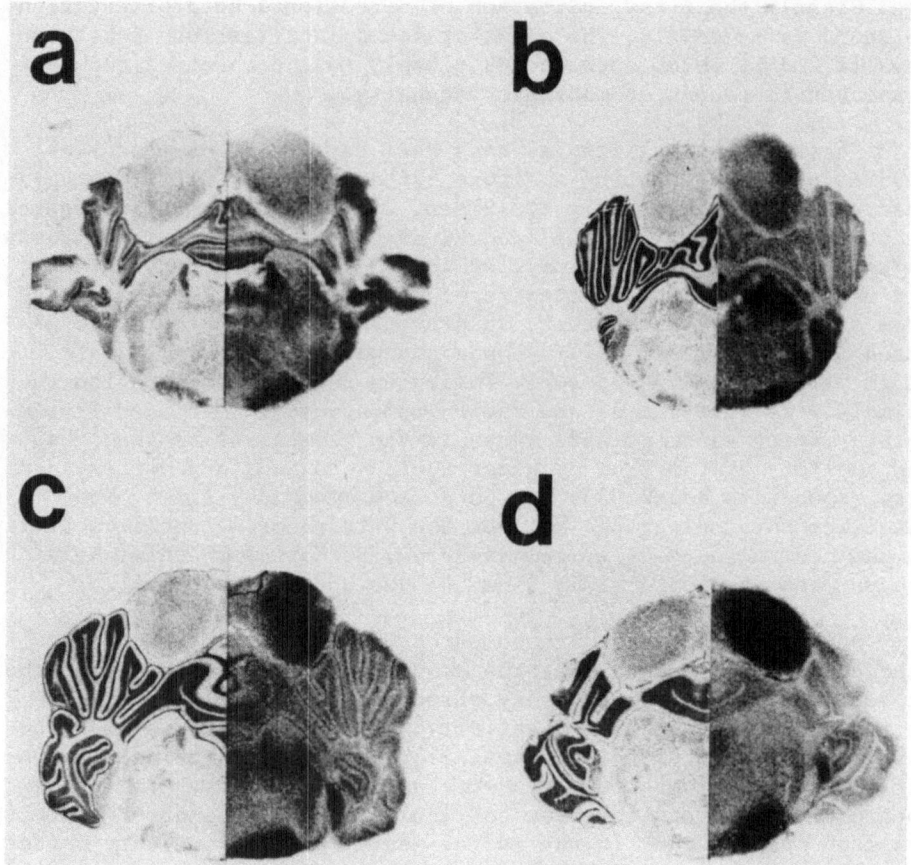

Fig. 2. Deoxyglucose response in the inferior colliculus is shown
for 8, 11, 14 and 20 day rat pups (a, b, c and d). For
each frontal section, the deoxyglucose autoradiograph
appears on the right, and a cresylviolet stain of the
histological section from which the autoradiograph was
produced is shown on the left.

peripheral sensitivity to sound. At this time structures such as
the inferior colliculus which are known to be important in sound
localization are already responsive to auditory stimulation.

ACKNOWLEDGMENTS

    This work was supported by the National Science and Engineering
Research Council of Canada.

REFERENCES

Clements, M. and Kelly, J. B., 1978, Auditory spatial response of young guniea pigs (Cavia porcellus) during and after ear blocking. J. Comp. Physiol. Psychol., 92: 34-44.

Kelly, J. B. and Skeen, L. C., 1979, Development of central auditory function in the rat; A (^{14}C) deoxyglucose study, Neurosci. Abstr., 5: 23.

Kennedy, C., DesRosiers, M. H., Jehle, J. W., Reivich, M., Sharpe, F. R. and Sokoloff, L., 1975, Mapping of functional neural pathways by autoradiographic survey of local metabolic rate with (^{14}C) - deoxyglucose, Science, 187: 850-853.

Mendelson, M. J. and Haith, M. M., 1976, The relation between audition and vision in the human newborn. Monographs of the Society for Research in Child Development, 41, (Serial No. 167).

Muir, D. and Field, J., 1979, Newborn infants orient to sounds, Child Development, 50: 431-436.

Potash, M. and Kelly, J. B., 1978, Early development of sound localization by rats and gerbils, J. Acoust. Soc. Am., 64: S-86.

Potash, M. and Kelly, J. B., 1980, The development of directional responses to sounds in the rat (Rattus norvegicus), J. Comp. Physiol. Psychol., (in press).

Turkewitz, G., Birch, H. G., Moreau, T., Levy, L. and Cornwell, A. C., 1966, Effect of intensity of auditory stimulation on directional eye movements in the human neonate, Anim. Behav., 14: 93-101.

Wertheimer, M., 1961, Psychomotor coordination of auditory and visual space at birth, Science, 134: 1962.

INPUT - DEPENDENT 2-DEOXY-D-GLUCOSE UPTAKE IN THE CENTRAL

AUDITORY SYSTEM OF RANA TEMPORARIA

H. Flohr, R. Ammelburg, H. Kortmann and W. Elsen

Department of Neurobiology
University of Bremen
2800 Bremen, F.R.G.

The (^{14}C)-deoxyglucose - technique developed by Sokoloff et al. (1977), can be used to quantify regional cerebral energy metabolism in specific discrete structures of the brain. It had been used successfully to analyse the relationship between sensory input and patterned neuronal activity in different systems. For instance, Sokoloff (1975) and Kennedy et al. (1975) studied the metabolic activity in the visual system of the rhesus monkey with intact binocular vision as well as after uni- and bilateral eye occlusion. They found that the pattern of 2-DG uptake closely coincided with the changes in functional activity expected from known properties of the visual system. They were able to demonstrate ocular dominance columns in the striate cortex originally described by Hubel and Wiesel (1962 and 1972) by means of electrophysiological and autoradiographic studies on the visual cortex of the cat and the macaque monkey. Hubel et al. (1978) used the 2-DG technique to demonstrate orientation columns in the monkey's visual cortex. Through olfactory stimulation with specific odours it has been shown that specific spatial patterns of neuronal activity participate in coding the sensory information. In studies on the vestibular system of the rat Sharp (1976) was able to demonstrate that vertical rotation produces a specific differential metabolic activation in the vestibular nuclei and several areas of the cerebellum. Some of the highest rates of local cerebral metabolic activity have been found in the structures of the auditory system of rats and monkeys (Sokoloff, 1977). Bilateral auditory deprivation by occlusion of both external auditory canals in rats reduces glucose utilization bilaterally in the cochlear nuclei, inferior colliculi, lateral

lemnisci and superior olivae. Unilateral auditory deprivation
depresses the activity asymmetrically; for example, the activity
of the ipsilateral cochlear nucleus is reduced to 75% of that of
the contralateral. Recently Scheich et al. (1979) used the technique
to demonstrate stimulus dependent activity patterns for pure tones,
harmonic tones and species specific calls in the auditory system
of the guinea fowl. There was a good correspondence between the
observed spatial DG uptake pattern and previous results obtained
with microelectrode recordings.

In the present study the 2-DG uptake in the structures of the
auditory system of <u>Rana temporaria</u> was studied. Measurements were
made under normal conditions and after uni- and bilateral destruction of the inner ear. The anuran auditory system is relatively well
known. Its organization is basically similar to that of reptiles,
birds and mammals. However it represents an example of a peripheral
specialization; that is, the sensory organ is sensitive to only a
small set of species specific signals to be identified by central
structures. Therefore these systems offer an attractive model for
the analysis of possible spatial factors involved in the processing
of auditory information.

METHODS

Both male and female adult frogs were used. The destruction of
the inner ear was performed according to Ewald (1892). "Acute"
effects were studied 3-7 days, "chronic" effects 60-80 days after
the operation. $(1-{}^{14}C)$-2-deoxy-D-glucose (53mCi/mmole, New England
Nuclear) was used. 16,7 µCi/100g body weight suspended in sterile
saline solution were injected into the dorsal lymph sac. During the
45 minute post-injection incubation the animals were exposed to
normal laboratory noise. Following this incubation period, they
were sacrificed immediately. The brain was quickly removed and
frozen in isopentane to -160°C. 12 µm thick frozen sections of the
brain were cut with a WKF Kryotom WK 1150 and picked up with slides.
The sections were freeze-dried at -84°C in a 10^{-6}mbar vacuum for
12 hours using a Leybold Hereus freeze-dryer GT 1, after which they
were brought into contact with emulsion-coated slides (Ilford G 5)
at -20° C and exposed for 20-24 days at -20°C in Clay-Adams boxes
containing drierite. After separating them from the sections the
films were developed (Kodak D-19 developer) and fixed in the usual
way. The sections were stained with cresylviolet. The autoradiographs were evaluated quantitatively using a television image
analyzer (Cambridge Instruments, Quantimet 720) equipped with a
densitometer module. The areas to be analyzed were delineated by
an imageeditor. The detected area and its total integrated optical
density were measured. The mean density was derived.

RESULTS

The structures of the auditory system in <u>Rana temporaria</u> are
characterized by a relatively high 2-DG uptake. Fig. 1a shows a
transverse section through the medulla oblongata of a normal animal.
As can be seen the activity corresponding to the dorsal nucleus and
to the superior olive is symmetrically distributed. Fig. 1b shows
a typical section through the midbrain of a normal animal. The
central region of the torus semicircularis shows a high, symmetrically distributed 2-DG uptake mainly in the areas of the nucleus
principalis and the nucleus magnocellularis. After bilateral auditory deprivation by destruction of both inner ears, the activity
is markedly depressed throughout the system (Fig. 1c and 1d). The
Figs. 1e and 1f show sections through the medullary nuclei and the
midbrain after acute unilateral destruction of the right inner ear.
The metabolic activity is asymmetrically distributed and reduced
in the ipsilateral dorsal nucleus, the contralateral superior olive
and the contralateral torus semicircularis. The DG distribution in
chronically deafferented animals shows the same asymmetries. Fig. 2
summarizes the quantitative data obtained for the different groups.

COMMENTS

The organization of the anuran auditory system is shown
schematically in Fig. 3. Fibers from the amphibian and basilar
pappilae terminate in the ipsilateral dorsal medullary nucleus.
From this nucleus fibers project principially to the bilateral
superior olivary nuclei and to the contralateral dorsal nucleus.
The nuclei of the torus semicircularis receive their main input
from the ipsilateral superior olivary nucleus. The present results
fully agree with this concept of the structure of amphibian auditory pathways. They reconfirm the usefulness of the DG technique
in mapping functional neuronal pathways in the central nervous
system. The level and distribution of the metabolic activity seen
in the auditory system as revealed by the autoradiographs are
clearly input-dependent. Comparable relationships have been observed
in other sensory systems of various species. However it cannot be
concluded that the activity patterns merely reflect the relative
electrical activities during the timespan in which DG is available.
On the basis of calculations on the metabolic costs of the electrical activity in neurons this seems to be rather improbable (Creutzfeldt, 1975). The exact mechanism of this close coupling between
input and energy metabolism in sensory systems is not yet understood.

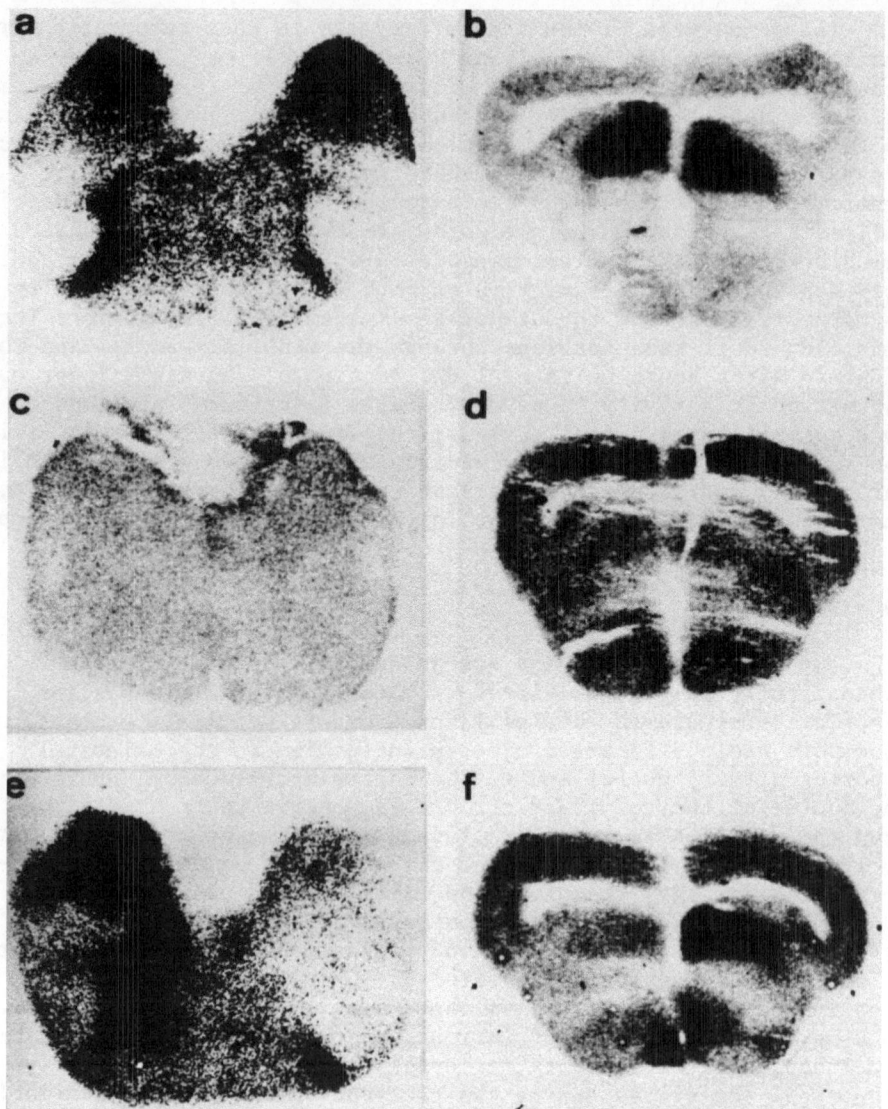

Fig. 1. The figure shows DG-autoradiographs of transsections
through the central auditory regions of the medulla
oblongata and the mesencephalon. a) nucleus dorsalis and
b) torus semicircularis of an animal with intact inner
ears. c) and d), equivalent areas of an animal after
destruction of both inner ears. e) and f), equivalent
areas of an animal after unilateral destruction of the
right inner ear.

Fig. 2.

Relative DG distribu-
tion in the auditory
system of the frog.
Column N shows the
relative left/right
distribution in the
control group (intact
animals). Columns A
and C show the uptake
of the deafferented
in percent of the in-
tact side for acutely
and chronically deaf-
ferented in percent
of the intact side for
acutely and chronical-
ly deafferented ani-
nals respectively.

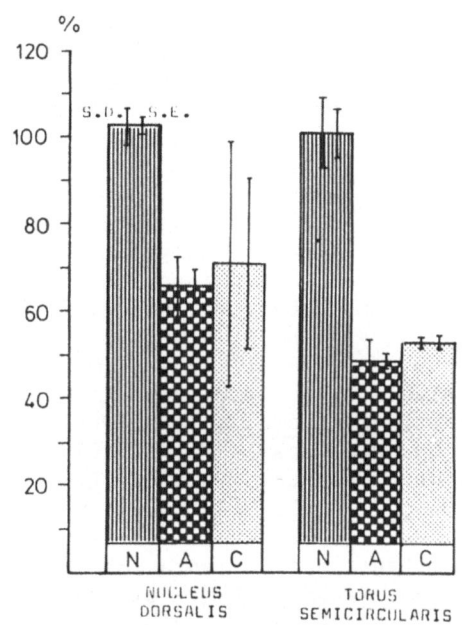

Fig. 3.
Main auditory pathways
and nuclei of the frog
(modified after Feng 1975)

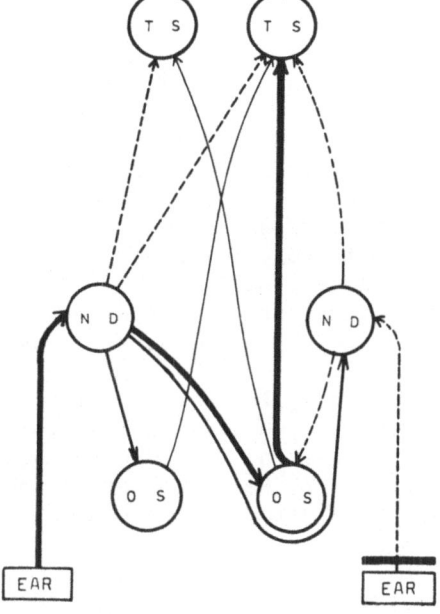

REFERENCES

Capranica, R. R., 1976, Morphology and physiology of the auditory
system, in: "Frog neurobiology", Llinás, R. and Precht, W.,
eds., Springer, Berlin.

Creutzfeldt, O. D., 1975, Neurophysiological correlates of different
functional states of the brain, in: "Brain work", D. H. Ingvar
and N. A. Lassen, eds., Kopenhagen.

Ewald, J. R., 1892, Physiologische Untersuchungen über das Endorgan
des Nervus octavus, Bergmann, Wiesbaden.

Feng, A. S., 1975, Sound localization in anurans: An electrophysiological and behavioral study, Thesis, Ph. D. Degree, Cornell
University, Ithaca, New York.

Hubel, D. H. and Wiesel, T. N., 1962, Receptive fields, binocular
interaction and functional architecture in the cat´s visual
cortex, J. Physiol., 160: 106-154.

Hubel, D. H. and Wiesel, T. N., 1972, Laminar and columnar distribution of geniculo-cortical fibers in the macaque monkey,
J. Comp. Neurol., 146: 421-450.

Hubel, D. H., Wiesel, T. N. and Stryker, M. P., 1978, Anatomical
demonstration of orientation columns in macaque monkey,
J. Comp. Neurol., 177: 361-379.

Kennedy, C., des Rosiers, M. H., Jehle, J. W., Reivich, M., Sharpe,
F. and Sokoloff, L., 1975, Mapping of functional neural path-
ways by autoradiographic survey of local metabolic rate with
(^{14}C)deoxyglucose, Science, 187: 850-853.

Krnjević, K., 1975, Neuronal metabolism and electrical activity,
in: "Brain work", D. H. Ingvar and N. A. Lassen, eds., Kopenhagen.

Potter, H. D., 1965, Mesencephalic auditory region of the bull frog,
J. Neurophysiol., 28: 1132-1154.

Sharp, F. R., 1976, Activity related increases of glucose utilization associated with reduced incorporation of glucose into its
derivates, Brain Res., 107: 663-666.

Sharp, F. R., Kauer, J. S. and Shepherd, G. M., 1975, Local sites
of activity related glucose metabolism in rat olfactory bulb
during olfactory stimulation, Brain Res., 98: 596-600.

Scheich, H., Bonke, B. A., Bonke, D. and Langner, G., 1979, Functional organization of some auditory nuclei in the guinea fowl
demonstrated by the 2-deoxyglucose technique, Cell Tissue Res.,
204: 17-27.

Sokoloff, L., 1975, Influence of functional activity on local
cerebral glucose utilization, in: "Brain work",D. H. Ingvar
and N. A. Lassen, eds., Kopenhagen.

Sokoloff, L., 1977, Relation between physiological function and
energy metabolism in the central nervous system, J. Neurochem.,
29: 13-26.

Sokoloff, L., Reivich, M., Kennedy, C., des Rosiers, M. H., Patlak,
C. S., Pettigrew, K. D., Sakurada, O., Skinohara, M., 1977,
The (^{14}C)deoxyglucose method for the measurement of local

cerebral glucose utilization: Theory, procedure and normal values in the conscious and anaesthetized albino rat, J. Neurochem., 28: 897–916.

Wiesel, T. N., Hubel, D. H. and Lam, D. M. K., 1974, Autoradiographic demonstration of ocular-dominance columns in the monkey striate cortex by means of neuronal transport, Brain Res., 79: 273–279.

PLASTIC CHANGES IN THE INFERIOR COLLICULUS FOLLOWING

COCHLEAR DESTRUCTION

Ikuo Taniguchi

Department of Physiology
Dokkyo University School of Medicine
Mibu, Shimotsuga, Tochigi 321-02, Japan

Local neuronal activity in the central nervous system is closely coupled to the rate of local glucose consumption. If functional alterations occur locally in the central nervous system in response to peripheral damage, an autoradiographic technique using ^{14}C deoxyglucose (DG) (Sokoloff et al., 1977) can monitor the nature of the induced change in activity, and provide an estimate of its magnitude. In the present study plastic changes in the inferior colliculus (IC) following cochlear destruction were studied by the DG technique.

Immature ICR mice were divided into four groups with different cochlear lesion histories: (1) with both cochleas intact (2) with one cochlea destroyed for 1 day to 8 months; (3) with both cochleas destroyed for 1 to 40 days; (4) with one cochlea destroyed for 8 months, and the remaining cochlea destroyed for 1 day before the experimental manipulation. Cochlear destruction was made at 3 weeks of age under pentobarbital anesthesia. DG was administered through the tail vein as a pulse (2 μCi/10 g body weight) to the unanesthetized animal. After injection, each animal was stimulated by noise bursts whose intensity was fixed at 60 dB SPL. Forty-five minutes after the DG injection, the brain was quickly removed and frozen brain sections 30 μm thick were cut. The sections were then exposed to X-ray film for 2 days, and the film was developed. The optical density of the autoradiographs was measured by a densitometer. The optical density for the IC was expressed relative to that for the opposite IC or the commissure of IC.

After unilateral cochlear destruction, the IC contralateral to the lesion exhibited markedly less labelling than did the ipsilateral side. This result agrees with previous neuroanatomical

(Osen, 1969) and DG autoradiographic (Sokoloff, 1976) studies
indicating that the IC is predominantly innervated by the crossed
afferent fibers. After a few days, however, the optical density
in the contralateral IC gradually increased and attained the normal
level about 1 month later as shown in Fig. 1. These results are
consistent with those in a previous study (Taniguchi and Saito,
1978).

We made the assumption that the ipsilateral input might begin
to contribute significantly to activation of the IC after destruction of the contralateral afferent input which is normally predominant. If the activity of the IC contralateral to the lesion is thus
enhanced by the ipsilateral input, destruction of the remaining
cochlea should result in a remarkable decrease in DG uptake. Un-
expectedly, the result did not follow the assumption. The IC ipsilateral as well as contralateral to the new lesion did not exhibit

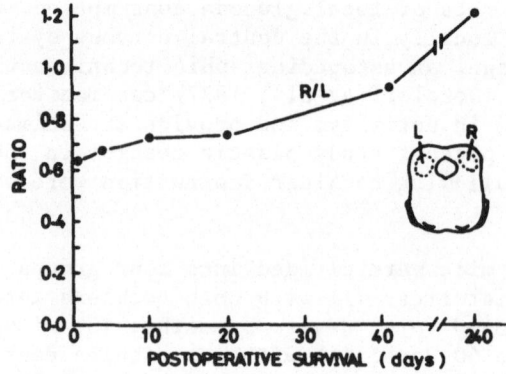

Fig. 1.   The time course of increase in DG uptake in the right IC
           relative to that in the left (R/L) after destruction of
           the left cochlea.

marked decrease in DG uptake. The whole area of the IC was diffusely
labelled. The central region of the IC contralateral to the new
lesion, however, is less densely labelled than the surrounding
zones which may receive the somatosensory input (RoBards, 1979).

The observations described above led us to investigate the
possibility that the IC may recover normal levels of metabolic

activity even if auditory inputs are completely removed. This was
confirmed by the autoradiography made after simultaneous destruction
of both cochleas. The autoradiograph on the 1st postoperative day
showed remarkable decreases in DG uptake on both sides of the IC.
On subsequent days, however, DG uptake increased symmetrically. On
the 20th postoperative day DG uptake in the IC reached essentially
the normal level bilaterally as shown in Fig. 2. The time course
of gradual increase in DG uptake was quite similar to that observed
after unilateral cochlear destruction. Apparently the neurons of
the IC become active even without their auditory afferent input.

These results suggest some possible mechanisms. First, spontaneous activity in the IC neurons increases by some type of dis-
inhibition or supersensitivity. Recently, Silverman and Clopton
(1977) demonstrated that inhibition typically evoked in the IC by
stimulation of the ipsilateral ear was absent in the monaurally
deprived rats. Abnormally enhanced excitability in the IC neurons

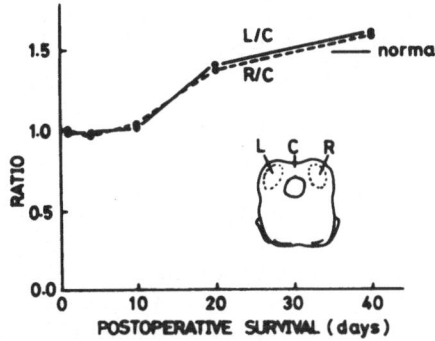

Fig. 2.   The time course of increase in DG uptake in the right and
          the left IC relative to that in the commissure of IC (R/C,
          L/C) after bilateral cochlear destruction.   A horizontal
          bar indicates the ratio measured on the autoradiograph of
          the normal animal.

has also been demonstrated in acoustic priming (Urban and Willott,
1979). To explain extreme susceptibility to audiogenic seizures
in the primed animals, many investigators have proposed supersensitivity after deprivation or disuse of auditory afferent input
to the higher neuronal structures. Second, reinnervation by col-

lateral sprouting in the IC cannot be ruled out. The external
nucleus of the IC receives both auditory and somatosensory input
(Aitkin et al., 1978). If the synaptic sites in the IC are trans-
synaptically degenerated by cochlear destruction, the somatosensory
afferent sprouts may to occupy the evacuated sites, thereby causing
restoration of DG uptake in the IC after cochlear destruction.

The mechanisms described above are not exclusive mutually. To
define the mechanism, further studies are needed.

REFERENCES

Aitkin, L. M., Dickhaus, H., Schult, W. and Zimmermann, M., 1978,
    External nucleus of inferior colliculus: auditory and spinal
    somatosensory afferents and their interactions, J. Neurophy-
    siol., 41: 837-847.
Osen, K. K., 1969, The intrinsic organization of the cochlear nuclei
    in the cat, Acta Otolaryngol., 67: 352-359.
RoBards, M. J., 1979, Somatic neurons in the brain stem and neocor-
    tex projecting to the external nucleus of the inferior colli-
    culus: an anatomical study in the opossum, J. Comp. Neurol.,
    184: 547-566.
Silverman, M. S. and Clopton, B. N., 1977, Plasticity of binaural
    interaction, I. Effect of early auditory deprivation, J. Neuro-
    physiol., 40: 1266-1274.
Sokoloff, L., 1976, Auditory stimulation, in: "Neuroanatomical Func-
    tional Mapping by the Radioactive 2-Deoxy-D-Glucose Method",
    Plum, F., Gjedde, A. and Samson, F. E., eds., Neurosci. Res.
    Prog. Bull., 14: 489.
Sokoloff, L., Reivich, M., Kennedy, C., DesRosiers, M. H., Patlak,
    C. S., Pettigrew, K. D., Sakurada, O. and Shinohara, M.,
    1977, The ^{14}C deoxyglucose method for the measurement of
    local cerebral glucose utilization: theory, procedure, and
    normal values in the conscious and anesthetized albino rat,
    J. Neurochem., 28: 897-916.
Taniguchi, I., and Saito, N., 1978, Plastic reorganization in the
    inferior colliculus of the immature mouse studied by ^{14}C de-
    oxyglucose method, Proc. Japan Acad., 54: 496-499.
Urban, G. P. and Willott, J. F., 1979, Response properties of
    neurons in inferior colliculi of mice made susceptible to
    audiogenic seizures by acoustic priming, Exp. Neurol., 63:
    229-243.

DEVELOPMENTAL CHANGES OF AUDITORY EVOKED RESPONSES IN NORMAL AND

KANAMYCIN TREATED RATS

S. Matsuura and T. Tokimoto

Department of Physiology
Osaka City University, Medical School
Osaka, Japan

Auditory function is well developed at the time of birth in most animals, but in such animals as mice and rats, hearing is not functioning at birth and the ear is completed postnatally. In this study carried out on rats, auditory evoked cortical and brain stem responses were recorded to elucidate differences in the response to sound stimuli between early postnatal periods and the adult ages, and also to demonstrate the presence of a specially susceptible period to aminoglycosidic antibiotics at the time of initial appearance of auditory function.

Rats from different litters were used at various ages from birth to adulthood. Cortical and brain stem responses (CR and BSRs) were recorded through a small pin- or screw-type electrode which was pricked into the skull and permanently fixed with dental cement or aron alpha. Kanamycin (daily dose: 400 or 200 mg/kg) was applied to some groups of rats according to the experimental schedule. The injected animals, especially younger ones which received larger daily dose, showed a slower increase in the body weight. Some died during the course of the experiment (less than 10 % in number). Within a few days after the cessation of kanamycin treatment, however, the body weight of most of the younger animals recovered almost to the same level as the control.

Auditory responses were recorded under light anesthesia (25 mg/kg pentobarbitone or ether only) at an interval of a few days. Younger animals mostly recovered from the anesthesia within a few hours and began to suckle. The data were discarded when they did not begin to suckle several hours after the recording. Tone pips of various intensities and frequencies were delivered through a loud-speaker placed about 30 cm from the front of the animal. Frequency of the stimulation was about 1/s and 5/s for CR and BSR, respectively. One hundred and 999 responses were averaged in CR and BSR, respectively. The sound intensity was monitored throughout the experiments and expressed with reference to the sound pressure level (2×10^{-5} Pa). In some experiments, an electrical stimulus was applied to the eighth nerve fibers with a pair of tungsten wire electrodes (tip diameter; 0.1 mm, interpolar distance; about 0.3 mm) inserted into the cochlea from a small perforation near its apex to evoke simulated cortical responses.

In normal rats, auditory evoked response most commonly appeared on the 12th day after birth, and every normal rat examined in this study showed initial sign of both CR and BSR between 11th and 14th days after birth, while the lowest intensity of the sound necessary to evoke the responses (more than 80 dB SPL at the level of the ears) was much higher than that required for mature animals. The early response showed characteristic signs of immaturity in being low in amplitude and long in latency in both CR and BSR. Especially CR showed an immature pattern different from those recorded in the adult animals. The immature pattern changed rapidly to a more mature response type during the first week. The development thereafter became much slower. For example, CR in response to a constant frequency and intensity of sound reached an amplitude more than 80 % of the adult responses within 3 weeks after birth. Though the threshold intensity did not show much difference depending on the frequency used in every rat examined on the initial days of the appearance of the responses, the threshold intensity decreased with the lapse of days, especially to the sound between 4 kHz and 12 kHz than to the lower frequencies of the sound. (Frequencies higher than 12 kHz were not applied because of the limitation of the equipment used.) The threshold intensity for sound of 4 - 12 kHz lowered below 50 dB SPL within 3 weeks after birth.

The latencies of the BSR as measured for the first two or three waves usually reached the values comparable to the adult ones within 2 weeks after their initial appearance, but changes in the latencies of cortical evoked potentials took a longer time to reach the adult values.

In the kanamycin-treated rats, the development of the auditory responses were different depending on the schedule of administration of kanamycin in the early neonatal period. In a group of rats in which kanamycin (400 mg/kg) was daily applied from the first day of the birth to the 10th day, the development of the CR and BSR was not different from that observed in the control rats without kanamycin-treatment, though slightly smaller responses were sometimes observed as compared with the control group. When the drug administration was stopped 3 - 4 days before the initial appearance of the auditory responses, even this small difference in the response size was no more apparent.

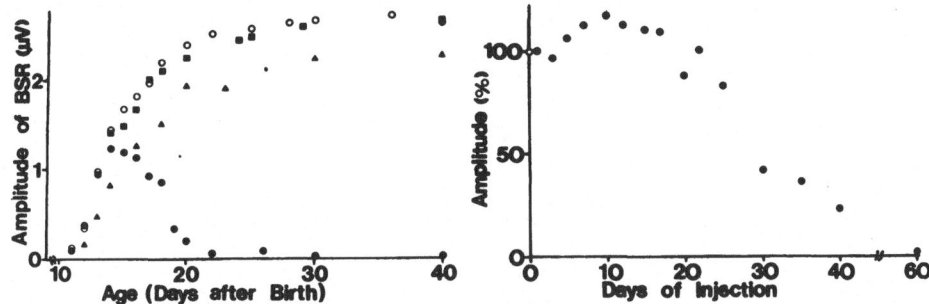

Fig. 1.   Sample records of BSR and CR in normal and kanamycin-
          treated rats. A: BSR in normal rat. B: BSR in a rat treated
          with kanamycin (400 mg/kg) from the 11th to 20th day after
          birth. C: CR in normal rat. Each five records of A, B and
          C were obtained on the same rat. D and E: Records from a
          normal rat of 8 days of age. CR was induced not by sound
          (D), but by an electrical stimulus (E). F and G: Similar
          to D and E, but obtained from an older rat to which kana-
          mycin was applied from 11th to 20th day after birth. Figures
          on each record indicate the days after birth. Arrows show
          the time of electrical stimulus.

     On the contrary, in the group of rats to which kanamycin (400
mg/kg) was administered from the 11th to the 20th day after birth
CR and BSR developed apparently normally up to the 15th day or so,
i.e., the fourth day following the start of the drug application,
but the responses declined thereafter rapidly and were almost
abolished on the 20th day or so. Almost no recovery was observed
after the end of the drug administration. When the drug application began after a few days following the start of the initial rapid
phase of development of the responses or much later, the same daily
dose of kanamycin for ten days had almost no effect on the responses. In the adult ages, auditory evoked responses showed almost
no sign of decrease until the 20th day of the kanamycin application.

     In the experiment of electrical stimulation of the eighth
nerve, CR could be recorded as early as on the 8th day after birth
when no responses could be elicited with the sound stimulus of any
intensity. The electrically induced cortical potential at this
early stage of development was low in amplitude and long in latency,
and appeared with a pattern different from the sound-evoked CR of
adult animals. At the 20 th day of age, electrically stimulated
cortical response was comparable with the response produced by the
sound stimulus in its amplitude and response pattern. The CR evoked
by a supramaximum electrical stimulus increased in amplitude with
the development of the auditory system and usually became larger

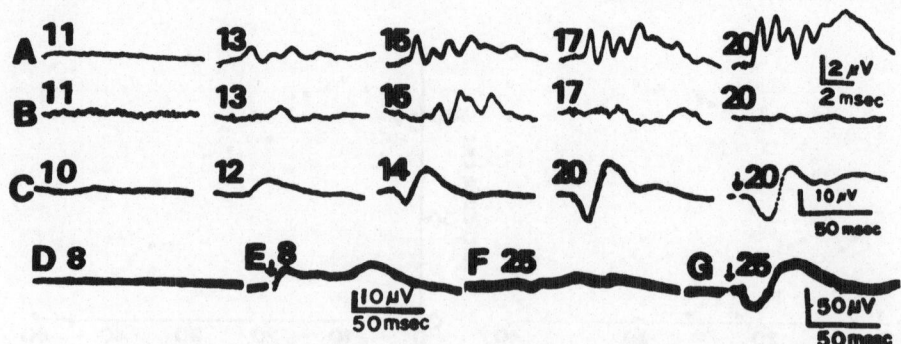

Fig. 2. Amplitude of BSR in normal and kanamycin-treated rats. A:
Results from early postnatal periods. Open circle, normal
rats; triangles, cases in which kanamycin was applied from
birth to the 10th day; filled circles, kanamycin applied
from the 11th to the 20th day; squares, kanamycin applied
from the 14th to the 23th day. Each plot shows average of
more than three measurements. B: Amplitude of the BSR as
expressed in % of the control BSR size. Abscissa is days
of kanamycin administration (daily dose, 400 mg/kg). Each
circle shows the average of more than three measurements.
The average amplitude before kanamycin application was
taken as 100 %.

than the sound-evoked ones in response to a high intensity of sound
(100 dB SPL). The simulated CR in response to an electrical stimulus
delivered to the auditory nerve was not different from the control
response without kanamycin-treatment even in the rats which became
almost unresponsive to sound by kanamycin applied for 10 days from
the 11th to the 20th day after birth, demonstrating the resistance
of the cochlear afferent fibers to kanamycin.

In the electronmicroscopy, evident signs of damage were detected on the organ of Corti in the rat which received kanamycin (daily
dose; 400 mg/kg) for 4 days starting from the 11th day after birth.
There was pronounced damage on almost every outer hair cell in the
rats treated with kanamycin for 7 - 10 days starting from the some
day. A near-total or complete loss of the outer hair cells was
observed in the basal turn, showing a clear correlation with the
functional loss as revealed in the CR and BSR, while the inner hair
cells seemed to be well preserved.

The depressive action on the sound evoked responses of kanamycin, was drastic when applied at the period of initial appearance
of hearing and quite different from the effects observed at other
stages of development. Although direct evidence to localize the site

of action of the antibiotics on the hair cells are still lacking, the rapid appearance of ototoxic effects during the critical period seems to indicate that the antibiotics may easily reach the site of action on the hair cells during that special period. Presumably the drug may interfere with the receptor mechanism by combining with the superficial structures (Wersäll and Flock, 1964) located on the hair-bearing surface of the sensory hair cells by entering the endolymphatic space (Matsuura et al.,1971; Konishi, 1979) through structures such as the stria vascularis which has just started functioning (Kikuchi and Hilding, 1965) or the external sulcus which reaches full maturity only about 15 days after birth (Bosher and Warren, 1971). According to this explanation, it is not necessary to assume an especially elevated susceptibility on the part of hair cells at a certain developmental stage. Further study is necessary for clarifying the mechanism underlying the critical period.

REFERENCES

Bosher, S. K. and Warren, R. L., 1971, A study of the electrochem-
     istry and osmotic relationships of the cochlear fluids in the
     neonatal rat at the time of the development of the endocochlear
     potential, J. Physiol. (London), 212: 739–761.
Kikuchi, K. and Hilding, D., 1965, The development of the organ of
     Corti in the mouse, Acta Otolaryngol., 60: 207–216.
Konishi, T., 1979, Effects of local application of ototoxic anti-
     biotics on cochlear potentials in guinea pigs, Acta Otolaryngol.
     88: 41–46.
Matsuura, S., Ikeda, K. and Furukawa, T., 1971, Effects of strepto-
     mycin, kanamycin, quinine and other drugs on the microphonic
     potentials of goldfish sacculus, Jap. J. Physiol., 21: 579–590.
Wersäll, J. and Flock, A., 1964, Suppression and restration of the
     microphonic output from the lateral line organ after local
     application of streptomycin, Life Sci., 3: 1151–1155.

SESSION IX
AUDITORY PROSTHESES,
PHYSIOLOGICAL BACKGROUND
Chairmen: W. D. Keidel and S. Tichý

PHYSIOLOGICAL BACKGROUND OF HEARING PROSTHESES

Wolf D. Keidel

Institute of Physiology and Biocybernetics
University of Erlangen-Nürnberg
Universitätsstrasse 17, D-8520 Erlangen, F.R.G.

There have been many efforts to help completely deaf people with prostheses which would allow them to communicate verbally with other people. All the requirements for such a type of auditory prosthesis have so far not been fulfilled, however. One of the main reasons for this failure is the great complexity of the auditory stimuli used by nature for conveying spoken information from one subject to another. The language phonemes, as they have been labeled, are not only remarkably complex with respect to their frequency distribution, but they also change in time and intensity with respect to the different components of the sound spectra used as stimuli for speech communication. One of the first to clearly show the complexity of auditory stimuli of this type was Licklider (1951). He gave an example of a spoken word in a three-dimensional plot where the abscissa indicates time, the rear ordinate the frequency, increasing from front to rear, and the third ordinate, perpendicular in direction, the intensity of each component. Such a figure looks like the skyline of Manhattan and makes very clear the extreme complexity of a spoken phoneme or word. Obviously, therefore, the difference limen for intensity (Δi) and frequency (Δf) has to be small enough to enable human speech communication (Zöllner and Keidel, 1963).

Meyer-Eppler (1950) published a plot of the density of the information flow within the conventional graph of performance of the ear where the abscissa is the logarithm of frequency and the ordinate the intensity of sound. Here, it can be clearly seen that only in a relatively restricted range of sound frequencies, which are also used in conventional telephone systems, namely between 300 and 3000 Hz, the density of the information flow via the ear toward the human auditory cortical level is significantly increased.

389

This is shown in Fig. 1 by the great number of points of equal information content. Similarly, the concentration for the highest

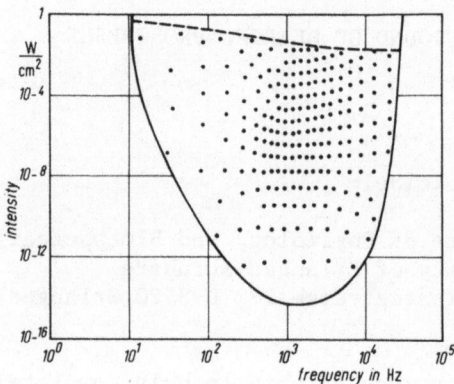

Fig. 1.   Information density of speech in the hearing area. Each
           point represents a certain amount of information (according
           to Meyer-Eppler, 1950).

information flow with respect to intensity is also concentrated in a relatively small range between about 40 dB and, for normal speech, not more than 90 dB SPL (re 20 μPa = 2×10^{-4} dyn/cm^2). This shows that the most interesting part of the auditory system is not the capability of the ear to convey information, but rather the small area in this plot defined as above.

Now, what does the auditory system do to feed this sort of complex auditory sound into the cortical level? I would like to discuss this matter in three parts. First, I will speak about the problem of intensity coding, including the relationship of the size of an electrophysiologically measurable signal to sound intensity i, and that of absolute intensity to the difference limen for intensity Δi. Both are obviously interrelated in a rather complicated manner. Second, I would like to say a few words about the broad topic of frequency coding, again concerning absolute frequency as the difference limen for frequency, in other words, the just noticeable difference for this parameter of the auditory stimulus Δf, which, as we know, is highly specialized in mammals and in man.

As a third subtopic, the question will be discussed how a single unit within the auditory system is able, by just one parameter of excitation, namely pulse or signal density, to signal the peripheral intensity of a stimulus at least as the periodicity of a sinusoidal tone, if not its frequency. There will be restrictions in a relatively small bandwidth of 800 Hz which might be of importance to the problem of phase locking of peripheral responses to the auditory stimuli. Finally the problem of the number of necessary channels for speech understanding will be discussed with respect to electrical stimulation of cochlear implants.

As early as 1929, v. Békésy showed, under the key word "Tonhöhenverschiebung bei Ermüdung des Ohres", that the number of just noticeable differences for a given interval of intensity increases when by adaptation processes the absolute intensity threshold of the ear is increased. Based upon this observation of v. Békésy (1929), Ranke (in Keidel, 1961) then concluded, in the fifties, that the main meaning of adaptation is not, as earlier believed, the change in absolute threshold of a sensory system, but rather an improvement in the difference limen of intensity. As has been demonstrated experimentally in the last 10 years in our laboratory, this means that the intensity function of any sensory system is related to the number of just noticeable differences in intensity or, in other words, the steepness of the intensity function is an inverse correlate of the dynamic range of a sensory modality: best specialized by evolution for the visual and auditory system and worst for pain and inadequate electrical stimulation as used for directly stimulating the auditory nerve in cochlear implants. The steeper the intensity function, the less the dynamic range, as less different intensity steps can be separated by the system. Consequently, a system functions better, the smaller the steepness of its intensity function for a given sensory modality. Thus, both the visual and the auditory system have a very highly specialized dynamic range, · with up to 485 jnd in the case of vision (König and Brodhun, 1888); the figure for the ear is not much less. The vibratory sense in the context of our topic is somewhere between that of the visual and auditory system and that of pain and electrical stimulation. This is of importance for conveying language, spoken words and syllables through the skin to the human brain.

One of the first to record from single fibers of the auditory nerve, comparing auditory stimulation with electrical current stimuli, was Kiang (1965). He was able to show, together with Moxon, as early as 1972, that the dynamic range of a single fiber to auditory stimulation is in the order of 35 dB, as shown in Fig. 2, while the dynamic range for electric stimulation with a sinusoidal current of the same frequency is much smaller, the intensity function being extremely steep, ranging over only a difference of a few decibels in stimulus intensity from threshold to maximum firing rate.

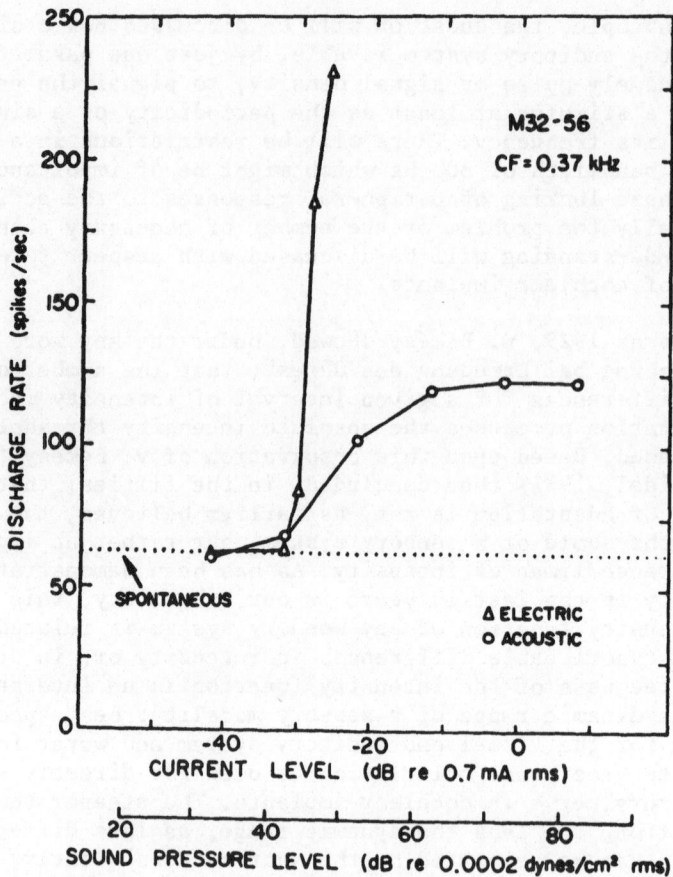

Fig. 2.   Discharge rate in a single auditory nerve fiber as a
         function of intensity for acoustic and electric stimula-
         tion (according to Kiang and Moxon, 1972).

As early as 1957, a group at MIT (Keidel et al., 1958) was
able to record the intensity functions of the auditory system in
the cat by stimulating with sinusoidal tones of different frequencies, by recording the compound action potential of the auditory
nerve with bipolar electrodes and simultaneously picking up the
evoked responses from the cortical areas AI and AII. About the same
dynamic range of some 35-45 dB for a given ensemble of neurons at
the periphery and at the cortical level could be found including
the fact that this dynamic range was improved by the adaptation.
This is shown by the differences between the dynamic curves for the
non-adapted, "dynamic", and steady-state curves for the adapted
state of the system.

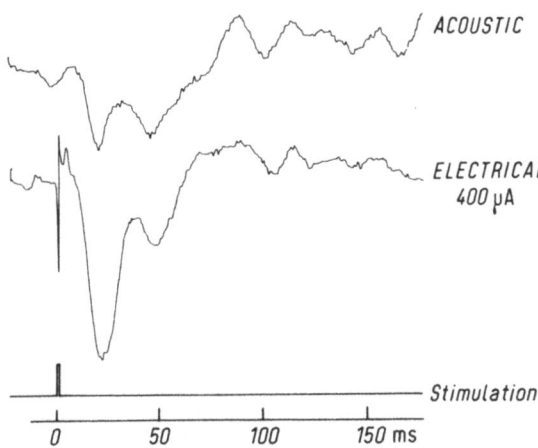

ACOUSTICAL AND ELECTRICAL STIMULATION OF THE AUDITORY NERVE

Fig. 3.   Averaged corticograms from auditory projection area AI in
          the cat, following acoustic stimulation of the auditory
          nerve.

Last year, one group in our laboratory, namely Finkenzeller
et al., 1979), compared those results with other obtained by stimulating the cat´s cochlear nerve, just as Kiang (1965) did by means
of sinusoidal electrical current and recording from the cat´s
auditory projection areas AI and AII. One result is shown in Fig. 3
comparing the evoked response to acoustic and electrical stimulation within an averaged corticogram of the cat.

Although there is some difference between the first two peaks,
in general the potentials look very similar. These experiments and
results cannot be discussed here in detail. However, I would like
to point to the results at the cortical level which are very similar
to those obtained by Kiang (1965) at the periphery. Our experiments
on anesthetized cats showed (Fig. 4) that only a 2-dB increase in
intensity would lead to maximum cortical excitation in the evoked
response for the first deflection, while in the adapted state, a
flatter intensity curve would be recorded. It could thus be proved
that the type of intensity function found at the periphery, including all restrictions that are discussed in the literature, compared
to the overall function of the auditory system, is well represented
at the cortical level.

To conclude these observations, I would like to mention another
set of experiments performed in our laboratory during recent years
by Keidel et al. (1973): the auditory-evoked responses of children
of different ages have been carefully averaged and recorded (Fig.
5). The resultant intensity functions have been drawn together in
this figure for the different ages of children on whom the studies

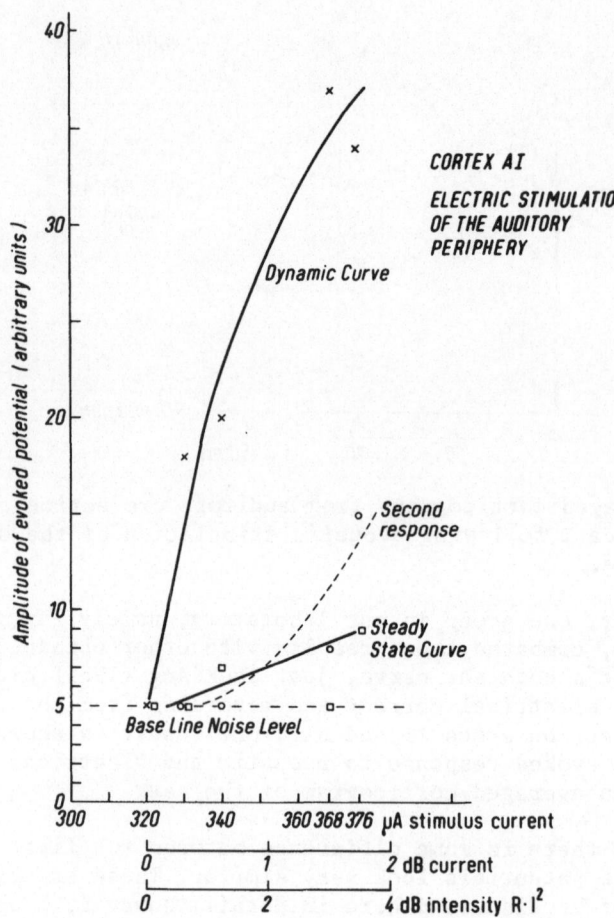

Fig. 4.   Amplitude of electrically evoked potentials at cortical
          level (AI) in the cat as a function of the applied current
          intensity in 0.25 ms pulses. After adaptation (steady-state
          curve), the cortical excitation increases much less with
          stimulus intensity in comparison to the dynamic function.

were done. Interestingly enough, Stevens‾ (1964) exponent ‾n‾of the
steepness of these intensity functions clearly depends upon the age
of the children. Thus, the steepness of the intensity function of
a relatively young child (exponent n in the order of 0.6) drops
during development to an exponent of 0.2 which is found in the
adult. This means that the dynamic range (indicated by the intensity
function) is improved by a factor of nearly 3 within a few years of
development in childhood.

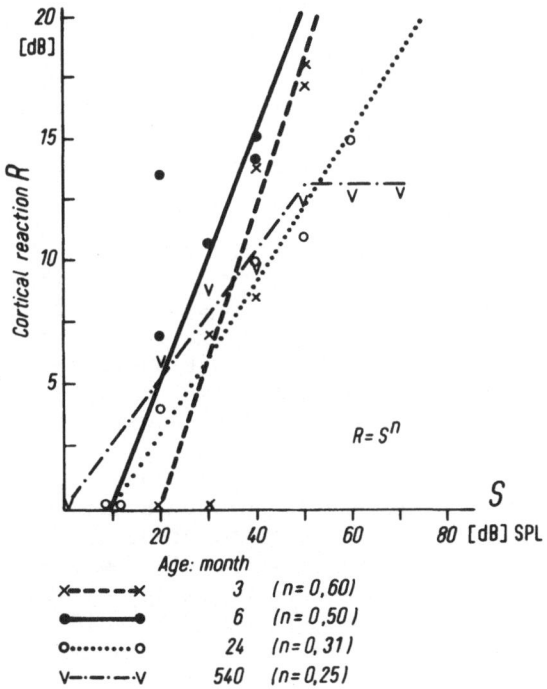

Fig. 5. Power functions of cortical reaction R versus stimulus intensity S in infants of different age (according to Stange, 1972).

Summarizing this series of experiments and interpretations, it is tempting to speculate that the functional improvement of the sensory system of any given modality can be reflected in the broadening, the ˝decompression˝, of the dynamic range during maturation (see also Stange, 1972).

Since the electrical stimulation of the auditory system shows a much steeper intensity function both at the peripheral and the central level, a part of the difficulty of conveying speech information by electrical stimulation is due to this great steepness of the intensity function with a very small dynamic range for intensity representation at the cortical level. It can easily be understood

the fine resolution of the changes in amplitude of the different
components of the complex sound used for speech communication
between two subjects might be lost by the very small dynamic range
associated with electrical stimulation. If this is so, then preprocessed acoustic information should broaden (decompress) the
statistical distribution of the input stimuli just as the hair cell
populations naturally do by their gaussian distribution of thresholds
for a given number of single fibers which are thought to be in the
order of 30 000. In our opinion this sort of ´intensity decompres-
sion´is one of the possible avenues for future research in this
area.

Our second topic considers how the frequency of an auditory
stimulus is encoded along the auditory channel. There is a tremen-
dous amount of literature available, again beginning from v. Békésy´s
experiments, to find out how frequency is represented neurophysio-
logically in the auditory nerve up to the auditory cortical level.
For the periphery, it began with the place theory, including considerations for periodicity and similar phenomena. At the central
level of the auditory system, there was Tunturi´s (1952) concept of
a tonotopic representation of sinusoidal tones. I would like to give
you just an idea of the difference between a simple auditory stimulus
as complex as a speech sound. One of my co-workers, David (1972),
built one of the numerous electronic models now available for the
function of the cochlea as a hydrodynamic system. There is a clear
shift of the place for maximum deflection from 18 Hz to 18 kHz along
the basilar membrane. There is also an obvious periodicity within
the time domain for a given frequency mainly used for the low frequencies up to 300 Hz.

It is clearly seen that both place and time, as a highly complex
periodicity, play an important role for stimulus representation
along the basilar membrane. For reasons of the complexity of this
speech one needs at least ten channels for visual speech, or for
vocoding processes, to be able to learn how to read spoken words.
However, one should bear in mind that the actual situation within
the ear along the cochlea in reality is much more complex being
based upon as many as 30 000 channels represented by a comparable
number of single fibers within the auditory nerve. It seems nearly
hopeless to construct a prosthesis which might be able to feed this
information via the 30 000 different channels toward the decoding
auditory system within the human brain.

There are, however, some possible escapes from this very difficult situation. One of the avenues for further research in this
field is the fact that, at least for vowels, there is considerable
redundancy in the form of the segments of the time course of such
a vowel which are typical for its character as a special phoneme;
other words have the task of separating a given vowel, say a from
e,i,o,u. As Kusch (1971) showed (bottom row of Fig. 6), the typical

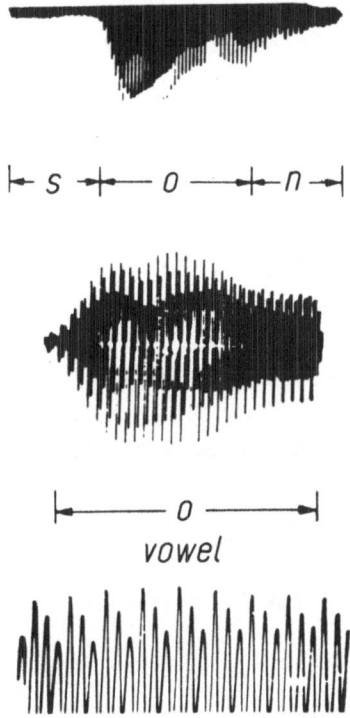

$\vdash s \dashv\!\!\vdash\!\!-\!\!o\!\!-\!\!\dashv\!\!\vdash\!\!n\!\!\dashv$

$\vdash\!\!-\!\!-\!\!-\!\!o\!\!-\!\!-\!\!-\!\!\dashv$

vowel

Fig. 6. Time course of the speech signal for the word ˘son˘. The
vowel ˘o˘ is expanded in time to show the redundancy in
speech (according to Kusch, 1971).

segment for vowel o is repeated in normal speech about 10-20 times
when human speech containing the vowel o is conveyed from the
speaker to the listener. This redundancy might be used for an
advanced construction of devices adapted to function as auditory
prostheses.

A second way out of the terribly complex situation is the fact
that, when speech communication is performed, at least for the
speaker, a relatively great variety of different sensory modalities
is used to control the tone and speech voicing. Besides the motor
innervation of the entire speech-generating organ system of man,
a tactile, mainly vibratory, control of the throat, mouth and lips
is acting; and the control and synchronization of breathing for
correct speech sound generation are a third modality. Finally, the
auditory and - for the listener - the visual systems are involved
as a further modality for producing and detecting speech sound. It
is thus tempting to look at these supplementary sensory modalities,
which are also involved in speech detection (Keidel, 1968). A con-

siderable amount of work has been done on ¯vibratese¯, lipreading, visible speech and braille speech reading.

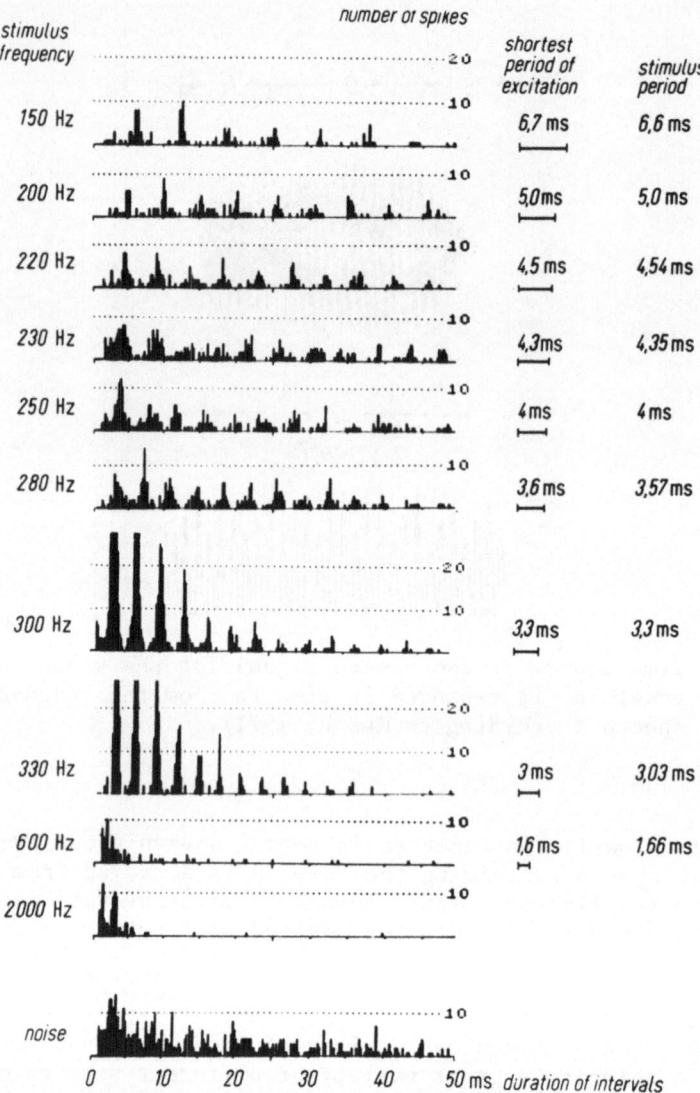

Fig. 7. Periodic distributions of interspike intervals from a single auditory nerve fiber responding to pure tones of different frequencies (according to Rose et al., 1967).

It is possible, by means of vibratory stimuli with a maximum frequency of 800 Hz and using von Békésy´s model of the cochlea, to convey speech information for monosyllables and spoken words via the skin of the forearm. Relatively little training is necessary to use such skin prostheses as speech-recognizing aids. Publications from our laboratory (Biber, 1961; Keidel, 1974) are available on this subject.

Proceeding to the "frequency aspect" of our topic in more detail, it can be demonstrated (Fig. 7) that the periodicity of the low frequencies in the auditory system is phase-locked and syn-chronized in frequency to a maximum of 1 kHz just in the same way as the skin when using vibratory stimuli. The periodicity of dif-ferent low-frequency sinusoids is clearly represented in the histo-grams, but it fails to be represented in records of responses to stimuli of frequencies between 1 and 18 kHz. This again means that this relatively primitive type of frequency coding, with a bandwidth of less than 1 kHz, is common to all types of mechanoreceptors. This small bandwidth is due to the fact that no single fiber of a myelin-ated nerve fiber can fire with a rate higher than the reciprocal time for the repolarization of the membrane which produces the nerve signal. This maximum rate is in the order of about 800 per second. Thus, this type of phase-locked 1:1 periodic frequency coding

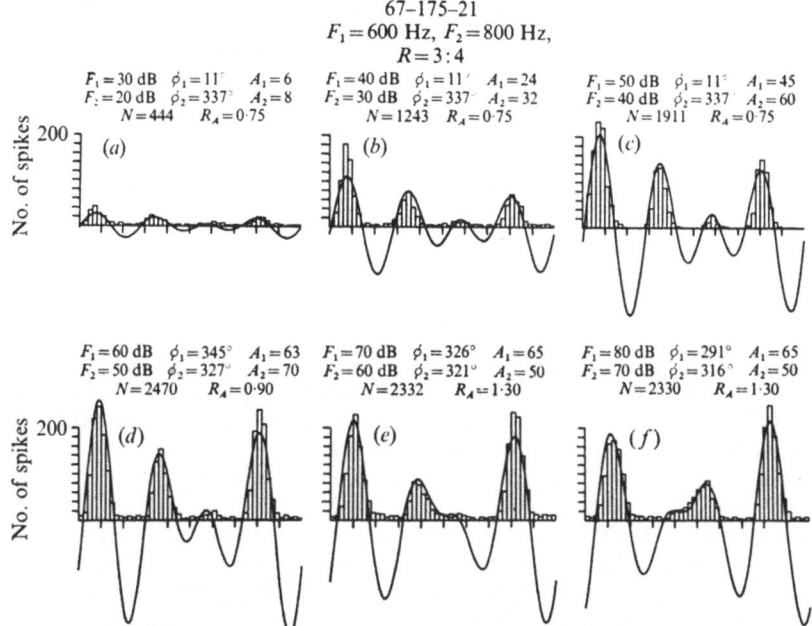

Fig. 8. Period histograms for a fiber of the auditory nerve respond-ing to a complex periodic sound when the sound pressure level is raised in 10-dB steps from a to f (according to Rose et al., 1971).

principle does not work for the remaining range (1-3 kHz) of frequencies which carry important information for speech recognition
and therefore have to be conveyed to the decoding nuclei within the
auditory system. There is, however, another problem, which has been
considered by Rose et al. (1967, 1971), namely the fact that, at
the peripheral level within single-fiber research of the auditory
nerve, the phase-locking of the single individual response, even to
a sinusoidal tone, may range around 360°, but with a clear maximum
within a smaller phase range of about 90°. In other words, the
stimulus periodicity shows a clear-cut and appreciable time jitter.
So there must be some ˉphysiological averagingˉ involved in the
auditory system, even at lower levels, able to sum up and to integrate a group of responses temporally as well as specifically to a
sharpened response. Fig. 8 shows an example of this basic principle
of auditory coding. The response type to a two-tone stimulus as a
histogram clearly shows the surprisingly exact correspondence of its
time pattern compared to that of the stimulus. This means some sort
of an improvement of the physiological encoding processes as compared to the basic functional principle of timing the triggering
for a given individual signal elicited within a single nerve fiber.
Similarly, within the visual system, the type of spatial pattern
originated by so-called contrast phenomena seems to be superior to
the pure geometric physical pattern of the image of an object
physically projected on the retina. This phenomenon has been known
for more than 100 years since it was first described by v. Helmholtz (1856).

There thus has to be some system within the auditory channel
which counteracts the time jitter of the encoding processes of the
periphery: the volley principle proposed by Wever (1949) should be
mentioned. V. Békésy first emphasized an unknown principle comple-
menting the volley principle when studying the similarities of
vibratory sensation. He labeled this basic phenomenon, or principle,
ˉdemultiplicationˉ.

A group of our laboratory (David et al., 1969) was able to
record, in single-unit studies at the colliculus level in the cat,
periodicities which are frequency-locked (not phase-locked) in the
PST histogram to a sinusoidal tonal stimulus, but in such a way
that the original periodicities of stimuli with frequencies higher
than 1 kHz are not detectable in the records. These high frequencies
have rather been demultiplicated by a factor 1:100 or more. When
constructing auditory prostheses, one has to have in mind that the
volley principle in an ear with destroyed hair cells can no longer
work. Therefore, some substitute for this special function of the
ear has to be integrated in any implant using preprocessed auditory
information.

The most effective way in which the highly specialized performance of the mammalian and the human difference limen for frequency

is realized electrophysiologically may be seen in the tuning curves
for single fibers. Although the well-known shape of the tuning
curves at the periphery has been studied very carefully (Kiang,
1965; Evans, 1972), there is still a considerable lack of knowledge
about the basic principles of the so-called second filter of the
ear (Keidel, 1969).Certainly, v. Békésy´s original idea of how the

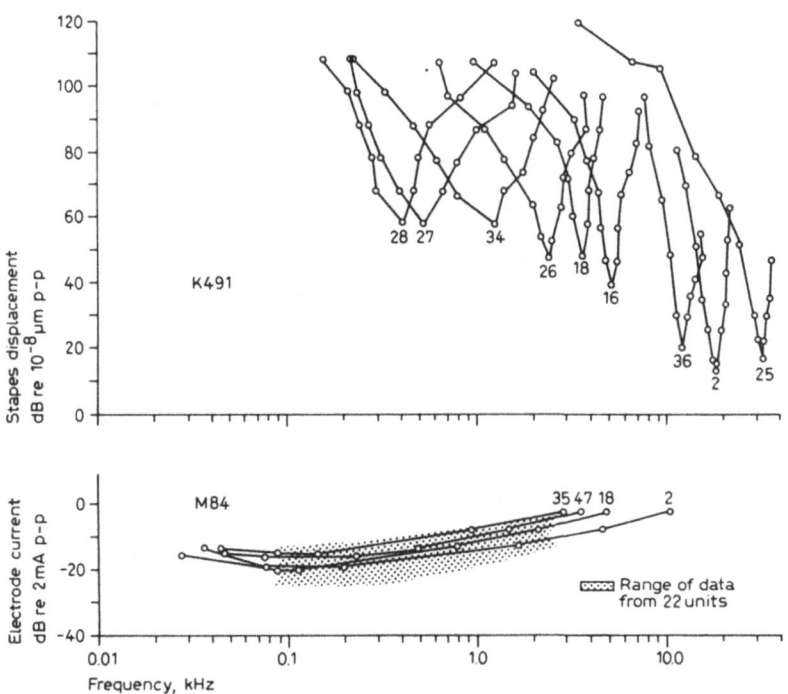

Fig. 9.   Sample tuning curves of auditory nerve fibers for tones
          (top) and sinusoidal electric currents (bottom; according
          to Kiang and Moxon, 1972).

envelope of the displacement of the cochlear partition according to
its hydrodynamics is originated does not explain the funneling
processes either electrophysiologically or mechanically between the
hydrodynamics of the cochlea and the steepness of the tuning curves
of the auditory nerve fibers. According to ˜Wien´s objection˜
(Wien, 1905), the damping factor within the perilymph is rather high
and therefore better fits the ability of the ear to follow up
relatively rapidly the speech periodicity in its pattern of displacement than the extremely small difference limen for frequency.
This in turn fits well the steepness of the tuning curves. We thus
need more basic research to understand better man´s ability for

speech communication in just this respect. There is agreement that, when electrically stimulating the auditory nerve in case of a completely destroyed cochlea, or dead hair cells, the tuning curves of the remainder (a situation which can easily be mimicked in experiments by use of appropriate ototoxic antibiotics) become flat and do no longer display the typical v-shaped type of single normal auditory nerve fibers. This has been demonstrated very clearly by Kiang and Moxon (1972; Fig. 9). The results have been confirmed in other laboratories in the last years.

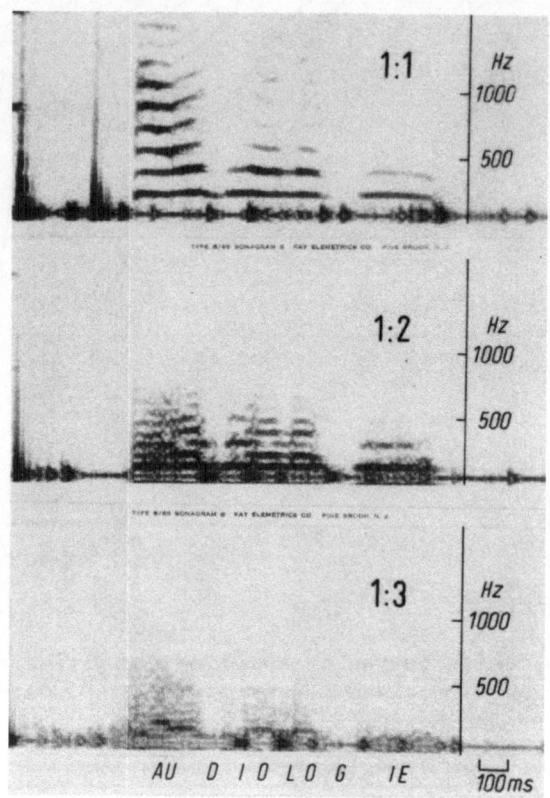

Fig. 10. Sonagram of the German word ˉAudiologieˉ, preprocessed
          with the indicated frequency transposition ratios by an
          electronic device (according to Finkenzeller, unpubl.).

Our own efforts toward the construction of an ear prosthesis have their origin in the research which has been performed in our laboratory for many years in the field of vibratese language. The idea was that with a frequency range not exceeding say 500 Hz it is possible to convey speech information via the skin by means of a mechanical system using frequency dispersion along the forearm with

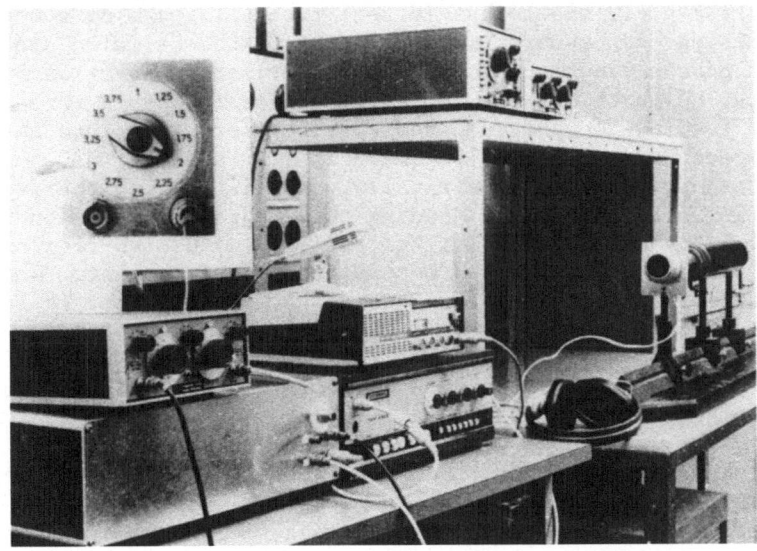

Fig. 11.   Frequency transposition apparatus. The control knob for
           the frequency transposition factor appears in the superior
           left corner. The Békésy internal ear model for vibro-
           tactile stimulation is located on the right side. In the
           middle of the figure the electronic device as it was 1
           year ago can be seen.

a Békésy model. This system works at least for single syllables and
words. It should also be possible to construct an electronic system
preprocessing speech information in such a way that the remaining
simple system of a completely deaf ear could be stimulated electrically in the same manner as is done for the skin. This included the
fact that, although the dynamic range of the electrically stimulated
skin is considerably smaller as compared to that of mechanically
stimulated forearm skin, it should be possible, by electrical
stimulation, to convey speech information via the skin. Our laboratory has already published work on this theme.

The second idea was that the redundancy in phonemes like a,e,
i,o,u should allow the frequency range (up to 3 kHz) of normal
speech to be compressed into a bandwidth of 500-800 Hz. This would
enable us to use the periodicity of these frequencies as information. This holds also for the auditory nerve, even by means of
electrical stimulation for conveying auditory information also in
this channel toward the cortical level with the goal of restoring
the ability for speech communication. The sonagram shown in Fig. 10
displays the result of frequency transposition obtained by means
of our electronic device for compression ratios of 2:1 and 3:1. The
compression factor is tunable in 12 steps from 1:1 to 3.75:1 by

means of a small computer built in our laboratory by Finkenzeller
(Fig. 11). Fig. 10 shows that in real time this type of speech
preprocessing can be performed, at least on the stimulus side, by
means of modern electronics using miniaturized computer devices of
this type. Research on this topic in our laboratory is threefold,
i.e. (i) feeding the preprocessed speech information into the
Békésy model using the tactile channel; (ii) feeding the somesthetic
channel and stimulating electrically, and (iii) using the frequency
transposition technique and feeding the preprocessed speech information either by auditory stimuli to patients with high-frequency
hearing loss or electrically to completely deaf patient. The last
series of experiments was performed on normal subjects. It is now
in progress for severely damaged ears.

For our experiments we used standardized syllables to measure
the learning curves.

As one would expect for neurophysiological reasons, in electrical stimulation experiments in man, single electrode stimuli produce
different pitch sensations dependent upon the stimulation rate
within a bandwidth of stimulus frequencies of some 500 Hz only (Fig.
12; according to Simmons et al., 1979). The reason for that behaviour is the lack of the volley-principle under those experimental
condition.

Thus, using an information preprocessing device of that type,
as we have developed it in our laboratory, would allow to encode
all ˉcompressedˉ speech frequencies by one electrode only. The
necessary difference limen for frequency (Δf), however, and in
addition a temporal and frequency pattern, sufficient for speech
perception, can be mimicked in cochlear implants by means of a
multielectrode device only. This again raises the question of the
minimal number of separate electrically stimulated channels compared
with the natural 30 000 ones. Burianˉs and Merzenichˉs laboratories
(Burian et al., 1979, 1980; Merzenich et al. 1979, Walsh et al.,
1980) have been concerned with measurements of excitation patterns
for e.g. 3 bipolar scala tympani electrodes introduced into optimal
locations for effecting discrete stimulation within the scala tympani using the place-pitch-relation underlying the "one-place-
conceptˉ inaugurated by Gildermeister and v. Békésy. The next Fig.
13 shows Merzenichˉs best results for stimuli near threshold in
man.

Kiang et al. (1979) - based upon single fiber studies - have
been able to develop an electronic model adapted for the generation
of time-frequency-patterns within an artificial ˉauditory nerveˉ.
The model allows judgements about the lower limit of the number of
channels necessary for designing auditory implants, when no information preprocessing has been utilized unlike to our system of frequency compression . In Fig. 14a and b accordingly, a shows the

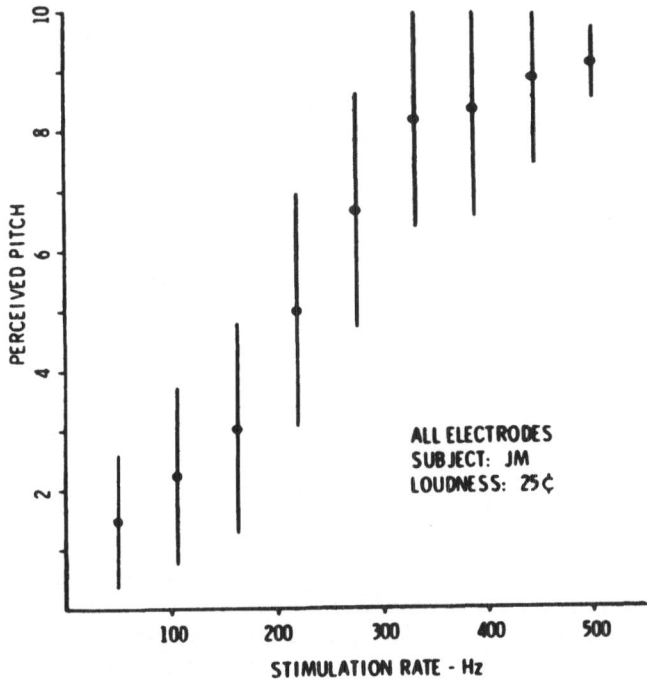

Fig. 12.   Pitch scaling for stimulus rates between 50 and 500 Hz.
            Plot points and total range bars are shown for pooled data
            from the S´s 3 electrodes. The S was given the 50 and
            500 Hz rates first, as the extremes of low and high pitch,
            then asked to rank the 50 Hz rate increments using the two
            reference rates. He was unaware of the rate intervals or
            other features of the test stimuli. (According to Simmons
            et al., 1979).

influence of the number of channels, when all fibers have the same
threshold and for each channel the response is the same as that for
the middle element, in b all fibers have the same threshold, in c -
corresponding to the natural dependence of each ´fiber´ upon the
auditory threshold curve, - when finally with the reduced number of
fibers to one the most sensitive one, responds. This model clearly
shows that the natural neurogram and sonagram to the spoken word
´SHOO-CAT´ (Fig. 14b) can be mimicked by 48 elements. The details of
the neurogram, with other words the encoded time-frequency-pattern,
drops drastically, when the number of elements is reduced to one

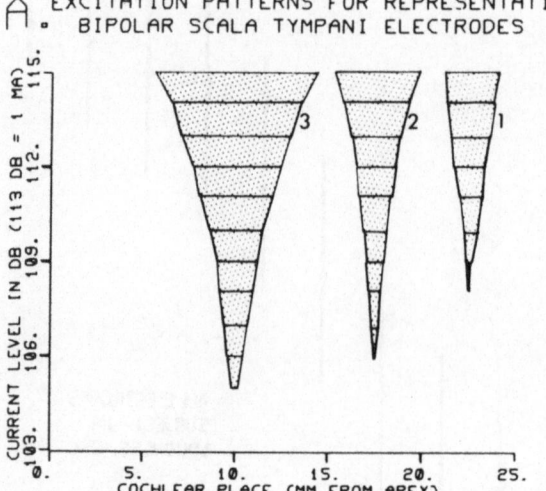

Fig. 13. Excitation patterns for 3 bipolar scala tympani electrodes
introduced into optimal locations (for effecting discrete
stimulation) within the scala tympani (see text). Excitation patterns were mapped using a brainstem evoked response mapping technique (see Merzenich et al., 1978).
Electrode pair 1 had 2 of 1.4 mm; and electrode pair 3 of
2.4 mm. (According to Merzenich et al., 1979).

channel. Some more less ⁻crude⁻ pattern, however, possibly sufficient to be decoded by the brain, to be recognized and percepted
as the spoken word ⁻SHOO-CAT⁻, could be obtained at a number
between 6 and 12 channels. The conclusion could be, that a number
of 6 to 12 bipolar electrodes might do the job.

SUMMARY

     For the design of cochlear implant the basic principles of
electric versus acoustical stimulation of the ear in the animal
and in man have been discussed. For practical and technological
application problems arise (1) from the small dynamic range for
electric stimulation. This makes necessary a ⁻decompression⁻ of
the intensity function for electric stimuli. (2) The applicable
number of bipolar electrodes is restricted and very small compared
to natural conditions. This problem has been explored by an
electronic device (Kiang et al., 1979) mimicking the behaviour of
integrated responses of up to 48 ⁻single fibers⁻ under electric
stimulation. It turns out that 6 to 12 channels might do the job.
(3) For single and multiple electrode stimulation a device has
been discussed for ⁻compression⁻ of speech frequencies into a bandwidth of 500 Hz (Keidel and Finkenzeller, unpubl.) .

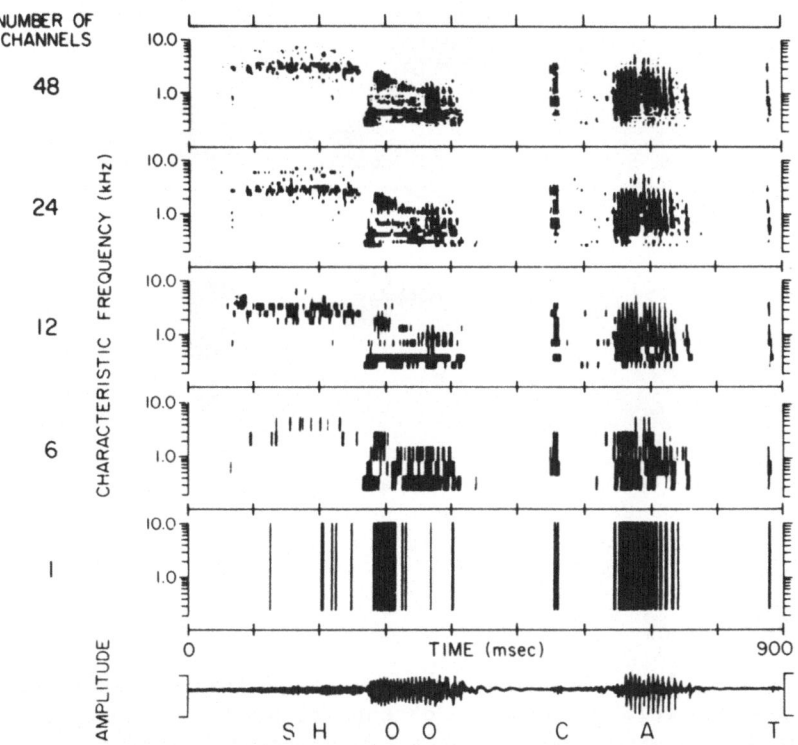

Fig. 14 a. Effect of reducing the number of channels on the model neurogram, details see text.

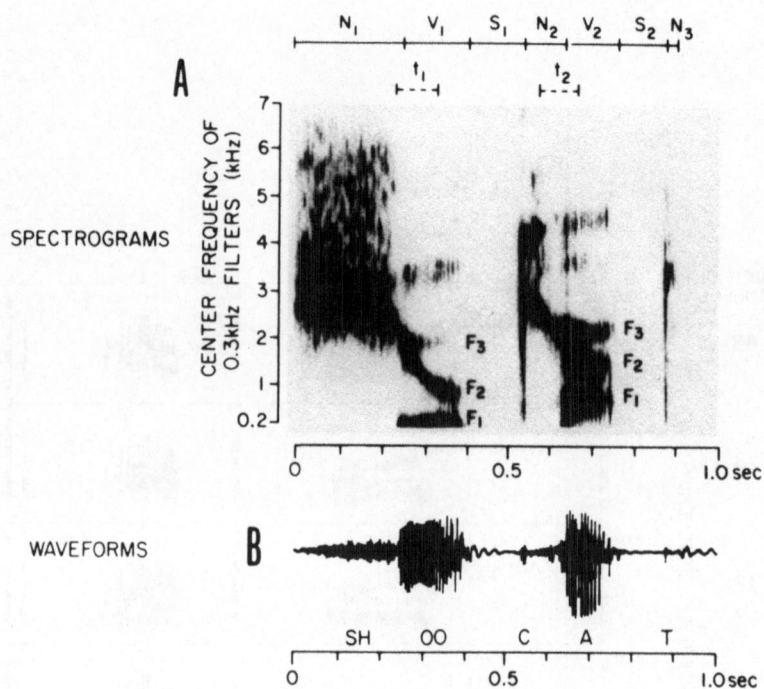

Fig. 14 b.   Spectrogram (Sonagram) and waveform of the spoken word
             ⁻SHOO-CAT⁻. Note the linear ordinate in this figure
             (according to Kiang et al., 1979).

REFERENCES

Békésy, G. v., 1929, Zur Theorie des Hörens. Über die Bestimmung
    des einem reinen Tonempfinden entsprechenden Erregungsgebietes
    der Basilarmembran vermittels Ermüdungserscheinungen, Physik.
    Z., 30: 115.
Biber, K.-W., 1961, Ein neues Verfahren zur Sprachkommunikation
    über die menschliche Haut, Diss. Erlangen.
Burian, K., Hochmair, E., Hochmair-Desoyer, I. and Lessel, M. R.,
    1979, Designing of and experience with multichannel cochlear
    implants, Acta Otolaryngol., 87: 190-195.
Burian, K., Hochmair, E., Hochmair-Desoyer, I. and Lessel, M. R.,
    1980, Electrical stimulation with multichannel electrodes in
    deaf patients, Audiology, 19: 128-136.
David, E., 1972, Elektronisches Analogmodell der Verarbeitung
    akustischer Information in Organismen, Habil. schr., Erlangen.
David, E., Finkenzeller, P., Kallert, S. and Keidel, W. D., 1969,
    Reizfrequenzkorrelierte untersetzte neuronale Entladungs-

periodizität im Colliculus inferior und im Corpus geniculatum mediale, Pflügers Arch., 309: 11-20.

Evans, E. F., 1972, The frequency response and other properties of single fibres in the guinea-pig cochlear nerve, J. Physiol., Lond., 226: 263-287.

Finkenzeller, P., Thumfart, W. and Fellner, E., 1979, Encoding processes of speech sounds in the auditory system, in: "Hearing mechanisms and speech", O. D. Creutzfeldt, H. Scheich and Ch. Schreiner, eds., Göttingen.

Helmholtz, H. v., 1856, "Handbuch der physiologischen Optik", Vieweg, Leipzig.

Keidel, W. D., Keidel, U. O., Kiang, N. Y.-S. and Frishkopf, L., 1958, Time course of adaptation of evoked responses from the cat's somesthetic and auditory system, Q. Prog. Rep. Res. Lab. Electron. MIT 48: 121.

Keidel, W. D., 1961, Rankes Adaptationstheorie, Z. Biol., 112: 411-425.

Keidel, W. D., 1968, Electrophysiology of vibratory perception, in: "Contributions to sensory physiology", Vol. 3, 1-79, W. D. Neff, ed., Academic Press, New York.

Keidel, W. D., 1969, Informationsphysiologische Aspekte des Hörens, Studium gen., 22: 49-82.

Keidel, W. D., Innitzer, J., Neuhäuser, G. and Plattig, K. H., 1973, Electroencephalographical audiometry of the new-born, J. fr. ORL, 22: 671-683.

Keidel, W. D., 1974, The cochlear model in skin stimulation, in: "Proceed. Conf. on Vibrotactile Communication", Psychonomic Society, Austin.

Kiang, N. Y.-S., 1965, "Discharge patterns of single fibers in the cat's auditory nerve", MIT Press, Cambridge.

Kiang, N. Y.-S. and Moxon, E. C., 1972, Physiological considerations in artificial stimulation of the inner ear, Ann. Otol. Rhinol. Lar., 81: 714-731.

Kiang, N. Y.-S., Eddington, D. K. and Delgutte, B., 1979, Fundamental considerations in designing auditory implants, Acta Otolaryngol., 87: 204-218.

König and Brodhun, 1889, Sber. berl. Akad. Wiss. 917 (1888); 641.

Kusch, H., 1971, Ein neues Verfahren zur Verbesserung der Sprache in Heliumatmosphäre, Acustica, 25: 42-46.

Licklider, J. G. R., 1951, Basic correlates of the auditory stimulus, in: "Handbook of experimental psychology", pp. 985-1040, S. S. Stevens, ed., Wiley, New York.

Merzenich, M. M., White, M., Vivion, M. C., Leake-Jones, P. A. and Walsh S., 1979, Some considerations of multichannel electrical stimulation of the auditory nerve in the profoundly deaf: interfacing electrode arrays with the auditory nerve array, Acta Otolaryngol., 87: 196-203.

Meyer-Eppler, W., 1950, Die Spektralanalyse der Sprache, Z. Phon., 4: 240.

Rose, J. E., Brugge, J. F., Anderson, D. J. and Hind, J. E., 1967,

Phase-locked response to low-frequency tones in single auditory nerve fibers of the squirrel monkey, J. Neurophysiol., 30: 769-793.

Rose, J. E., Hind J. E., Anderson, D. J. and Brugge, J. F., 1971, Some effects of stimulus intensity on response of auditory nerve fibers in the squirrel monkey, J. Neurophysiol., 34: 685-699.

Simmons, F. B., Mathews, R. G., Walker, M. G. and White, R. L., 1979, A functioning multichannel auditory nerve stimulator, Acta Otolaryngol., 87: 170-175.

Stange, G., 1972, Klinische Ergebnisse einer objectiven Audiometrie beim Säugling und Kleinkind, Bull. Audiophonol., Suppl.

Stevens, S. S., 1964, The psychophysics of sensory function, in: "Sensory communication", W. Rosenblith, ed., MIT Press, Cambridge.

Tunturi, A. R., 1952, A difference in the representation of auditory signals for the left and right ear in the isofrequency contours of right middle ectosylvian auditory cortex of the dog, Am. J. Physiol., 168: 712-727.

Walsh, S. M., Merzenich, M. M., Schindler, R. A. and Leake-Jones, P. A., 1980, Some practical considerations in development of multichannel scala tympani prostheses, Audiology, 19: 164-175.

Wever, E. G., 1949, "Theory of hearing", Wiley, New York.

Wien, M., 1905, "Ein Bedenken gegen die Helmholtzsche Resonanztheorie des Hörens", Festschrift für A. Wüllner, Teuber, Leipzig.

Zollner, F. and Keidel, W. D., 1963, Gehörvermittlung durch elektrische Erregung des Nervus acusticus, Arch. Ohr.-Nas.-Kehlk. Heilk., 181: 216-223.

ELECTRICAL STIMULATION OF THE HUMAN COCHLEA - PSYCHOPHYSICAL AND
SPEECH STUDIES

Y. C. Tong and G. M. Clark

Department of Otolaryngology
University of Melbourne
Parkville, 3052, Victoria, Australia

This report describes psychophysical and speech studies
conducted on two of our post-lingually deaf patients implanted with
multiple-electrode cochlear prostheses. The objectives were to study
the nature of the hearing sensations produced by the individual
electrodes, and to investigate the feasibility of the transmission
of speech information to the higher centres by means of cadences
of stimulation using one electrode at a time. Two totally deaf
patients (MC1 and MC2) participated in these studies.

In the implant operation, an array of ten electrodes were
inserted through an opening in the round window membrane for a
distance of 15mm around the scala tympani in MC1, and 19.5 mm in
MC2. The electrodes were numbered from 0 to 9 in the apical to
basal direction. These electrodes, spaced 1.5 mm apart, were driven
by a receiver-stimulator, which was positioned in a bed created in
the mastoid bone. Residual auditory nerve fibres were activated by
biphasic current pulses with each phase fixed at approximately 180
μs. Fifteen current levels ranging from 67 μA to 1 mA could be
assigned in 67 μA steps, and a maximum pulse rate of 1000 pps was
possible (Clark et al., 1979; Tong et al., 1979).

PSYCHOPHYSICAL STUDIES

A series of psychophysical studies (Tong et al., 1979) was
conducted on the two patients after the implant operation. Both
time-invariant and time-varying signals were used. Time-invariant
signals are those whose signal parameter values do not vary over
the duration of the signal. Time-varying signals, on the other
hand, are characterized by a linear variation from an initial to

a final parameter value over a specified duration. Furthermore,
signals with time-varying electrode position are called single-
electrode trajectories.

The results for time-invariant signals are summarized as
follows:
(1)   Threshold currents for 300 ms pulse trains at 150 pps ranged
from 130 μA to 600 μA for MC1, and 200 μA to 470 μA for MC2;
(2)   The loudness growth as a function of stimulus (current) level
for 300 ms pulse trains was much steeper than that for acoustic
stimulation;
(3)   For pulse rates below 200 pps, the perceived pitch for 300 ms
pulse trains increased with pulse rate. Above 200 pps, however, the
increase in pitch with pulse rate was less pronounced;
(4)   For a fixed pulse rate of stimulation, the perceived pitch
varied from low to high and the sharpness from dull to sharp for
300 ms pulse trains at single-electrodes ordered in the apical to
basal direction;
(5)   Vowel labels could be assigned to the sensations produced by
single-electrodes at a fixed pulse rate (200 ms pulse trains);
(6)   Relative difference limens of 7 to 12% were observed at pulse
rates below 250 pps for 300 ms pulse trains;
(7)   At a fixed pulse rate of stimulation, 200 ms pulse trains on
different electrodes were readily discriminated. Furthermore, the
signals at two adjacent electrodes were rarely confused within a
restrictive range of pulse rates - tested from 105 to 165 pps and
from 180 to 240 pps.

The results from time-varying signals are summarized as
follows:
(1)   Relative difference limens of 9 to 13% were observed for pulse
rate variation over 300 ms for pulse rates below 200 pps;
(2)   The discriminability of signals with time-varying pulse rate
deteriorated with decreases in the duration of the pulse rate variation. Pulse rate variations with a variable initial pulse rate
(150, 180, 210 or 240 pps) and a fixed final pulse rate (150 pps)
were tested for three durations: 25, 50 and 100 ms. The discrimination performance was the poorest at 25 ms;
(3)   100 ms single-electrode trajectories differing in the
direction (apicalward, basalward) of the electrode variation were
readily discriminated;
(4)   The discriminability of single-electrode trajectories was
independent of the duration of the trajectories. Trajectories with
a variable initial electrode position (electrodes 1, 2, 3 or 4)
and fixed final electrode position (electrode 1) were tested for
three durations: 25, 50 and 100 ms. The same performance was
observed for each duration;
(5)   Phenomenological reports from MC1 indicated that the hearing
sensations produced by the electrode trajectories were similar to
those produced by acoustic speech signals characterized by second

formant frequency transitions to which phonetic labels may be
assigned. The hearing sensations produced by time-varying pulse
rate, on the other hand, were similar to those produced by speech
signals with a variation in the fundamental (voice) frequency.

SPEECH STUDIES

    In the light of the psychophysical results, a speech process-
ing strategy, which converts the frequency of the spectral emphasis
in the mid-frequency range of speech signal to single-electrode
position, and the fundamental (voicing) frequency to pulse rates
below 250 pps, was proposed. With this strategy, the auditory system
will discriminate the electrode shifts required to convey rapidly
changing segmental speech information. The suprasegmental information contained in the fundamental frequency of a speech signal, on
the other hand, is encoded in the pulse rate which is more suited
to slowly time-varying signals. Although the discrimination perfor-
mances for electrode position and pulse rate are different, their
interaction also needs to be considered in speech processing
strategies. In this regard, the psychophysical results (see item 7
for time-invariant signals) indicated that, for speech signals
involving a single voice with a limited range of fundamental frequency (and therefore a limited range of pulse rates below 250 pps),
the information for electrode position is not likely to be confused
with that for pulse rate. Furthermore, the pitch matching experiments reported in Bilger (1977) and Eddington et al., (1978) for
implant patients with residual hearing in the unimplanted ear
indicated that, with appropriate psychophysical procedures, it may
be possible to separately identify these two types of information
for pulse rates below 250 pps. It is also possible that, in the
speech mode, the rapidly time-varying "electrode cues" for segmental
information are readily discernible from the slowly time-varying
"pulse rate cues" for suprasegmental information because of the
difference in the time scale of processing. These suggestions are
supported by the phenomenological reports of patient MC1 described
in the last section. As far as the encoding of the intensity of
the speech signal is concerned, the current level of the electrical
stimulus in conjunction with an appropriate compression scheme may
be used.

    The feasibility of this proposed speech processing strategy
in the transmission of speech information to the higher centres
was studied by means of a laboratory-based speech-processor. In
the real-time speech processor, speech parameters were estimated
and converted to electrical stimulation parameters every 10 ms,
and only one electrode was activated within a 10 ms time frame.
The speech parameters were the second formant frequency (F2) and
its amplitude (A2), the fundamental frequency (F0), and the voicing
amplitude (A0). For a given F2 estimate, the single-electrode to

be stimulated in the present time frame was selected from a predetermined F2-to-electrode conversion table based on psychophysical
results. Similarly, the current level was determined from an A2-
to-current table. Furthermore, an energy threshold detector for
AO was used to determine whether voicing was present or not. If
voicing was present, the pulse rate (250 pps) on the selected
electrode was made proportional to FO. On the other hand, if voicing
was absent a constant low pulse rate was used as it produced a
sensation described as "rough" which could be associated with that
of "noise" previously experienced by the patients when they had
hearing.

Speech comprehension tests were conducted under three conditions: electrical stimulation alone (EO), vision alone (VO), and
combined electrical stimulation and vision (EV). The first test was
a confusion study (Miller and Nicely, 1955) using twelve consonants
(/b/,/p/,/m/,/f/,/v/,/s/,/z/,/n/,/t/,/d/,/k/,/g/,) in a vowel-
consonant-vowel (VCV) context, with the vowel fixed as a /a/ as in
father. The two patients received ten randomized presentations of
the 12 VCV syllables in each condition. Scoring was based on the
number of consonants correctly identified. In the second test,
everyday sentence lists prepared at the Central Institute for the
Deaf were presented. Scoring was based on the number of key words
correctly identified. The test materials were presented at 80 dB
SPL by a female speaker.

The mean scores for the two patients in the first test were
38% (EO),37% (VO) and 65% (EV). The mean scores obtained in the
second test were 11% (EO), 24% (VO) and 83% (EV). These results
show that the scores are highest for the EV condition, indicating
the contribution of electrical stimulation of speech communication.
It should also be noted that the sentence scores in the EV condition were at a level where they correlate with a satisfactory
performance in understanding connected speech (Giolas, 1966). In
addition to these results, the consonant confusion data obtained
in the first test were also analyzed in terms of the information
transmission for the articularory features described in Miller
and Nicely (1955). This analysis showed that a significant amount
of information about voicing, nasality, duration, affrication and
place of articulation was transmitted in the EO condition (Tong
and Clark, 1980). The results of this analysis confirmed that useful
speech information can be transmitted to the higher centres of the
auditory system by means of the proposed strategy.

CONCLUSION

The results obtained in these studies indicate that it is
possible to convey useful speech information to the higher centres
of the auditory system by cadences of electrical stimulation at

different single-electrodes in the scala tympani. Furthermore, the hearing sensations produced by the electrical stimuli could be linked with those previously experienced by the patients. These results are informative because they confirmed the usefulness of a speech processing strategy employing both the "place" information relating to electrode position and the "period" information relating to pulse rate, and they formed the basis for future research. As it stands, however, the proposed speech strategy when used in conjuction with lipreading does improve significantly the patients ability in speech communication.

REFERENCES

Bilger, R. C., 1977, Evaluation of subjects presently fitted with auditory prosthesis, Ann. Otol, Supp. 38.

Clark, G. M., Pyman, B. C., and Bailey, Q. R., 1979, The surgery for multiple-electrode cochlear implantations, J. Laryngol. Otol., 93: 215-223.

Eddington, D. K., Dobelle, W. H., Brackmann, D. E., Mladejovsky, M. G., and Parkin, J. L., 1978, Auditory prostheses research with multiple channel intracochlear stimulation in man, Ann. Otol., Suppl. 51.

Giolas, T. G., 1966, Comparative inteligibility scores of sentence test and continuous discourse, J. Aud. Res., 6:31-38.

Miller, G. A., and Nicely, P. E., 1955, An analysis of perceptual confusions among some English consonants, J. Acoust. Soc. Am., 27: 338-352.

Tong, Y. C., Black, R. C., Clark, G. M., Forster, I. C., Millar, J. B., O'Loughlin, B. J., and Patrick,J. F., 1979, A preliminary report on a multiple-channel cochlear implant operation, J. Laryngol. Otol, 93: 679-695.

Tong, Y. C. and Clark,G. M., 1980, Speech comprehension with multiplechannel electrical stimulation of human auditory nerve fibres. J. Physiol. and Pharmacol. Soc. Aust., (in press).

PRELIMINARY SPEECH PERCEPTION RESULTS THROUGH A COCHLEAR PROSTHESIS

I. J. Hochmair-Desoyer, E. S. Hochmair, R. E. Fischer
and K. Burian

Techn. Univ., Gusshausstr. 27, A-1040 Vienna and
II. ENT-Dept. Univ. of Vienna, Alserstr. 12,
A-1097, Vienna, Austria

INTRODUCTION

Psychophysical as well as speech comprehension tests have been used in bilaterally deaf patients implanted with a four-channel cochlear prosthesis. The best open speech comprehension without lip-reading was obtained in a 24-year-old patient C. K. with progressive total hereditary deafness. This speech perception was achieved using a small portable stimulator activating only one of the four channels of the implant in a bipolar scala tympani electrode configuration.

STIMULATION SYSTEM

Fig. 1. shows the complete prosthesis used. It consists of an implanted part and a portable stimulator. The scala tympani electrode with eight Platinum-Iridium contacts (Hochmair-Desoyer et al., 1980a) is connected to a four-channel implanted receiver circuit. The implant is very flexible with respect to various speech processing strategies. Different signals of any desired shape can be transmitted simultaneously or with any timing relationship desired over the different channels. The portable stimulator currently used is a single-channel stimulator. It is inductively coupled to one of the implant channels via a transmission antenna coil mounted on an ear hook.

The four channels differ in the dynamic range, pitch discrimination and other psychophysical characteristics, thus differing in their suitability for speech perception. The channel best suited for speech transmission is selected through psychophysical tests.

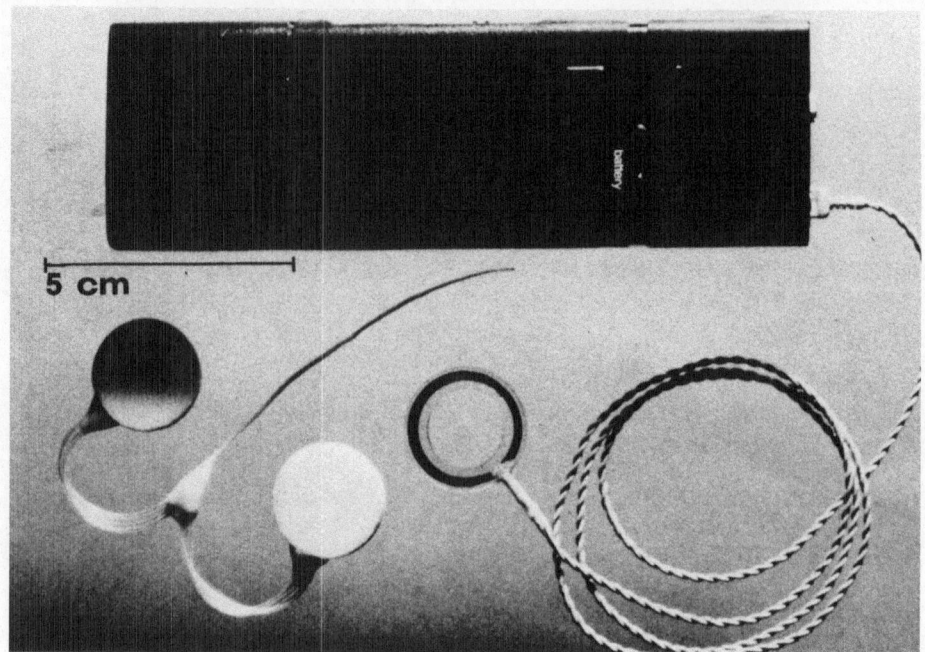

Fig. 1. Cochlear prosthesis consisting of four-channel implant and
         portable single-channel stimulator driving the antenna coil.

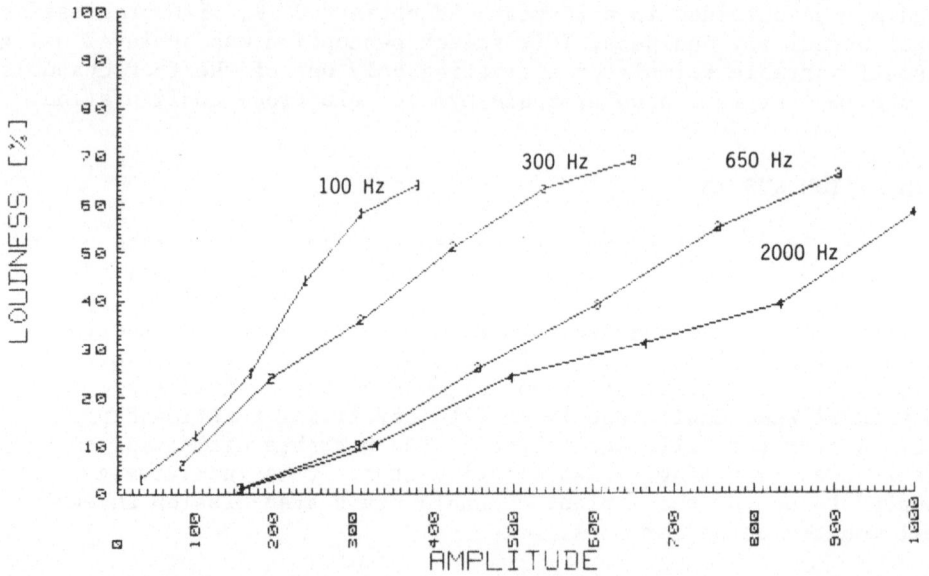

Fig. 2. Subjective loudness as a function of stimulus amplitude and
         frequency. 500 ms bursts of sine waves, 25 ms rise and fall
         time. 6 runs each. The amplitude is given in arbitrary units.

The subjective loudness as a function of stimulus intensity varies for different frequencies (Fig. 2). The external speech processor is adapted to these channel-specific characteristics with the expectation that this benefits speech perception.

The stimulator can be used for several hours without a decrease in loudness. In addition threshold decay and Bekesy threshold tests with continuous stimuli showed no adaptation.

SPEECH PERCEPTION

Preliminary speech tests with both live and taped voice have been used. The results reported here are from our patient C. K.. Complete deafness was reported during the summer of 1978, and C. K. was implanted in August 1979. x) Comparative results with our other cochlear prosthesis patients using standardized speech test material will follow in future publications.

Using the patient´s stimulator, the initial perception to speech material is that it sounded just like speech. The first speech test consists of the live-voice presentation of 100 numbers between 13 and 99. Without previous training or simultaneous feedback the only cue for recognition is the perceptual similarity between the input stimuli and auditory memory from previous years. The results for C. K. were 38% correct (40% half correct). Another postlingually deaf patient (F. W.) scored 33% correct (48% half correct) under the same conditions. These results show that there is a considerable degree of similarity between normal percepts and those electrically evoked by our stimulator.

Through training the score for numbers could be improved significantly. C. K.´s best result is 89% correct.

Another speech test consists of closed lists of 20 words, each list containing equal length words (one, two or three syllables). With two presentations of a test list, C. K. usually achieves a perfect score.

As a more demanding test, open lists of untrained everyday sentences were given. C. K.´s scores usually were between 60 and 100% (Hochmair-Desoyer et al., 1980b) for sentences spoken as one would speak to a hearing-impaired person. The list below is an example of 25 test sentences. They were presented on tape, spoken by a familiar female speaker. Each sentence was spoken twice with

---

x) An eight-min. film was presented showing the performance of C. K. in speech perception tests.

a pause of four seconds in between. The test was performed with electrical stimulation only, and thus without lipreading. The sentences presented are listed on the left. Entirely correct responses are indicated by " √ "; otherwise the patient´s response is given.

Es gibt viel zu viele Autos.	Es gibt viel ...
Warum ist die Suppe kalt?	√
Wo ist Deine Katze?	√
Heute kommt unser Onkel zu Besuch.	√
Hallo, wie geht's?	Heraus mit Dir!
Wer hat meine Tasche gestohlen?	√
Warum darf ich nicht ins Kino gehen?	√
Unsere Lage ist schwierig.	In der Lade ist ...
Zu Weihnachten gehen wir skifahren.	√
Der Wetterbericht sagt ein Gewitter voraus.	√
Wir wünschen gute Reise!	√
Der Sommer ist noch lange nicht vorbei.	√
Wo hast Du dieses Brot gekauft?	Wer hat bloss dieses Brot gekauft?
Das glaube ich nicht.	Das ... ist ...
Ich fahre mit dem Flugzeug.	
Nachts sehen Katzen besser als Menschen.	... seinen Katzen ... Menschen.
Was ist das?	√
Dieser Mann ist ein bekannter Schauspieler.	√
Die Hose hat ein Loch.	Die Vase hat ein Loch.
Die Uhr an der Wand gefällt mir.	Die ... an der Wand ist ein Tier.
Am Morgen bin ich immer müde.	√
Was kostet dieser Tisch?	√
Lass mich in Ruhe!	√
Sei nicht so laut!	Fahr mich ...
Im Westen geht die Sonne unter.	√

For open lists of words in citation form, C. K. achieved 40 - 60% correct without lipreading.

CONCLUSION

Open speech recognition has been obtained using one channel of a cochlear implant driven by a portable speech processor. The external device is approximately the size of a pack of cigarettes. While also improving her lipreading, the prosthesis allows our bilaterally deaf patient C. K. to understand 60 - 100% of everyday sentences without lipreading.

ACKNOWLEDGEMENTS

This work has been supported by the Austrian Research Council, grant No. 4151.

REFERENCES

Hochmair-Desoyer, I. J. and Hochmair, E. S., 1980a, An eight channel
    scala tympani electrode for auditory prostheses, IEEE Trans.
    Biomed. Eng., vol. BME-27:44-50.
Hochmair-Desoyer, I. J., Hochmair, E. S., Fischer, R. E. and Burian,
    K., 1980b, Cochlear prostheses in use: Recent speech comprehen-
    sion results, Arch. Oto-Rhino-Laryngol. (in press).

TACTILE AID FOR THE DEAF: SEARCH OF A CODE ALLOWING SOMESTHETIC

PROCESSING OF ACOUSTIC MESSAGES

Y. de Ribaupierre, P. Heierli, M. Holden, M. Rossi,
M. Demoulin and F. de Ribaupierre

Institute of Physiology
University of Lausanne
Bugnon 7, 1011 Lausanne, Switzerland

This preliminary study is mainly concerned with two problems:

I.   The capacity of the tactile channel
II. The evaluation of the quality of a tactile code.

I. Is the capacity of the tactile channel large enough to carry the information flow of an everyday conversation?

The flow of semantic information of an everyday conversation is not easy to measure. But in some well-defined situations the minimal information needed to perform a given task is simple to evaluate. For example the identification of a given object among a set of n objects requires at least an average information $I = lg_2 n$. The task of matching pairs of small objects belonging to two identical sets can be done using different sensory channels as information source:

a) If both sets are in view of the subject, only the visual channel is used for the matching task.

b) When one set is hidden in a bag, the subject uses the tactile channel to get the information and the visual channel to perform the matching process.

c) In a third mode, the experimenter can use one of the sets to describe the objects the subject has to search for. In this case the verbal channel is the input channel.

It can be shown that as long as the number of objects n is less than 60, the average identification time per object is proportional to $\lg_2 n$. So the capacity C (bits/s) of the different channels is: $C = \lg_2 n/T$ where T is the average time for matching correctly 50 pairs of objects from two sets of 100.

To have an idea of the effect of training on the scores, each subject performed all 3 procedures 5 times. The ratio R of the 5th score to the 2nd one gives a crude measure of the training effect. The mean of 4 subjects scores obtained at the 5th trial is the highest for the visual performance, $(C = 7.26 \pm 1.3$ bit/s$)$ and the task seems quite natural, as shown by the small effect of training $R = 1.1$. The tactile performance $(C = 1.46 \pm 0.14)$ is slightly better than the verbal one, and is still quite natural: $R = 1.2$. The verbal description score is the lowest: $C = 1.15 \pm 0.18$ and $R = 2$ indicates that the verbal description of objects is not as natural as expected. If properly coded, the semantic information of an every-day conversation can be carried by the tactile channel. But in the above task, the tactile channel is used in a sensorimotor loop, and the information derived from joints and proprioreceptors are certainly important. To investigate the contribution of these factors, a texture identification task has been used. 12 subjects had to identify 36 textures in two different conditions:

Active: the subject was allowed to palpate small disks on which different kinds of fabric or sand paper were glued.

Passive: the experimenter moved the disks under the subject's finger.

In these conditions, the information derived from non-cutaneous receptors is of no help, all the objects having the same shape. Moreover the sensorimotor loop is open in the passive case. From the scores obtained in these 2 conditions, one can say:

A) For the active situation $(C = 0.91 \pm 0.1)$ the improvement with repetition $(R = 1.74)$ allows one to predict that the tactile channel, if trained in an active mode, could reach the verbal capacity.

B) In the passive mode, both capacity and improvement with exercise are low $(C = 0.56 \pm 0.07, R = 1.38)$. It is not obvious that the tactile channel could ever reach the verbal capacity; it means that passive presentation of texture is not an efficient code for tactile aids.

II. What kind of experimental procedure is most suitable for testing the potential efficiency of a given acoustic to tactile transformation?

A simple 3 channel device with 3 vibrating pins has been designed:

The speech signal, previously recorded on a magnetic tape, is separated by means of adjustable filters in three channels: low, medium and high frequencies (see Fig. 1). The treatment of each channel is similar: the initial bandwidth is transfered by a frequency division into a lower bandwidth in the range of tactile sensitivity. This low frequency signal is modulated by the envelope of the initial filtered signal and then activates the vibrating pin. Such a system has about ten adjustable parameters.

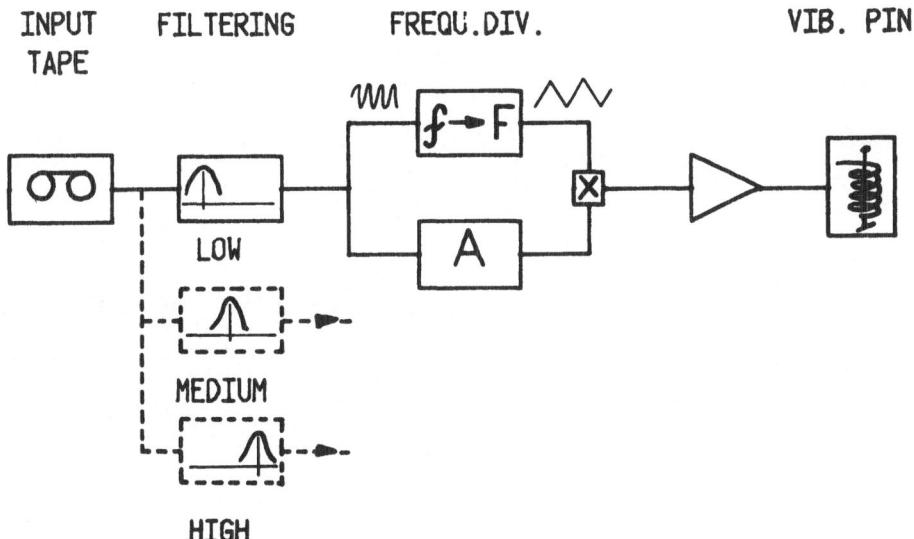

INPUT TAPE   FILTERING   FREQU.DIV.   VIB. PIN

LOW

MEDIUM

HIGH

Fig. 1.   Block-diagram of the 3 channel device.

It would be misleading to judge the quality of an aid by observing the results of a psychophysical test done by subjects with good hearing. But, when adjusting some of the parameters of the system, it would be useful to have sensitive psychophysical tests that take little time to be run and analyzed, and give better scores when the accessible information is increased. The value of different tests has been assessed by comparing the scores obtained when using one or three pins of our device. It is obvious that the information presented by three pins is greater than by one. The three following types of tests were performed.

- A discrimination test: pairs of words are presented to the subject who has to answer whether the words are the same or not.

- An identification test: first, the subject is familiarized with six words by reading them on a sheet of paper as they are tactually

presented to him. During the test, these words are presented in a
random manner, and the subject has to identify them.

- A recognition test: first, three or six words are learned in the
same manner as in the identification test; then sixty words are
presented to the subject, among which only half belongs to the
learned list. The subject has to name the words he recognizes.

   The results shown in Fig. 2 are the error averages (with ± 1
standard deviations) of four subjects with good hearing. For each
test 60 words are presented.

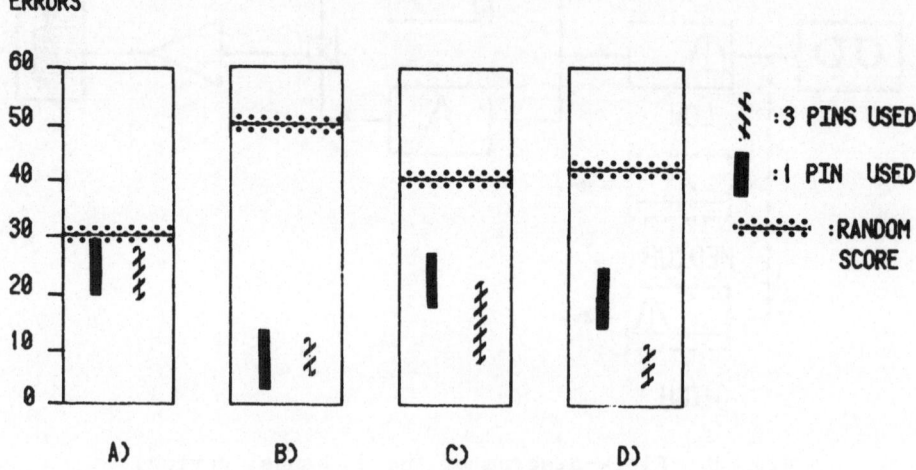

Fig. 2.   A) Discrimination test B) Identification test
          C) Recognition test; 3 words learned D) Recognition test;
          6 words learned. Scores are expressed by the number of
          wrongly-discriminated or identified words. The difference
          between 1 and 3 pins condition is not significant in C
          (p < 4 %) but is significant in D (p < 0.1 %)

   The interpretations of the results are not obvious. Only the
recognition test shows some significant difference between the one
and three pin situation and consequently, reflects the increase of
information between these two situations. Therefore, tests that do
not even show this difference, cannot be used to optimize coding
parameters.

LIST OF PARTICIPANTS

AITKIN, LINDSAY M.
    Department of Physiology, Monash University, Clayton, Victoria
    3168, Australia.
BIBIKOV, NIKOLAY, C.
    Acoustical Institute, Svernik 4, Moscow 117036, USSR.
BIEDERMANN, MANFRED
    Physiologisches Institut der FSU Jena, Teichgraben 8, 69 Jena,
    GDR.
BONKE DIETER
    Institut für Zoologie der THD, Schnittspahnstr. 3, 6100 Darm-
    stadt, FRG.
BUCHWALD, JENIFFER S.
    Brain Research Institute, UCLA, Los Angeles, California 900 24,
    USA.
BURDA, HYNEK
    Institute of Experimental Medicine, Czechoslovak Academy of
    Sciences, U nemocnice 2, 128 08 Prague 2, CSSR
CASSEDAY, JOHN H.
    Division of Otolaryngology, P.O.Box 3943, Duke University,
    Durham, N. C. 27710, USA.
ČADA, KAREL
    Department of Otolaryngology, Medical Faculty, Pekařská 53,
    Brno, CSSR.
COLBURN, STEVE
    Dept. of Electrical Engng., MIT, Room 36-759, Cambridge,
    Massachusetts 02139, USA.
COLEMAN, JAMES R.
    University of South Carolina, School of Medicine, Department
    of Physiology, Columbia, South Carolina 29208, USA.
DALLOS, PETER
    Auditory Research Laboratory, Speech Annex Building, North-
    western University, Evanston, Illinois 60 201, USA.
DAVID, EDUARD
    Institut für Physiologie der Universität Erlangen-Nürnberg,
    Universitätsstr. 17, Erlangen, FRG.
DRUGA, ROSTISLAV
    Department of Anatomy, Medical Faculty, Charles University,

U nemocnice 5, 120 00   Prague 2, ČSSR

EGGERMONT, JOSEPH J.
   Dept. of Medical Physics and Biophysics, University of Nijmegen,
   Netherland.

EVANS, EDWARD F.
   Department of Communication and Neuroscience, University of
   Keele, Keele, Staffordshire S T 55 BG, Great Britain.

FISCHER, RICHARD
   Institut für Allgemeine Elektrotechnik, Technische Univers.,
   Wien, Gusshausstr. 27, A - 1040, Austria.

FLOHR, HANS
   Department of Neurobiology, University of Bremen, Leobenstr.
   NW-2, 2800 Bremen, FRG.

FRITZE, WALTER
   2. HNO - Klinik, Universität Wien, Garnison Gasse 13, A-1090
   Wien, Austria.

FURUKAWA, TARO, T. F.
   Department of Physiology, Tokyo Medical and Dental University,
   Medical School, Yushima, Bunkyo-ku 1-5-45, Tokyo 113, Japan.

GLENDENNING, KAREN, K.
   Department of Physiology, Florida State University, Tallahassee,
   Florida 32 306, USA.

GRUBEROVÁ, JAROSLAVA
   Research Institute of Occupational Medicine, Limbova 16,
   Bratislava, CSSR.

HABERLAND, ERNST-JÜRGEN
   Universitäts - HNO - Klinik, Leninallee 12, 402 Halle/Saale,
   GDR.

HARTMANN, RAINER
   Im Wietloh 15, Schwerte 4, FRG.

HASSMANNOVÁ, JARMILA
   Department of Physiology, Medical Faculty of Hygiene, Charles
   University, Prague, Czechoslovakia.

HEIERLI, PIERRE
   Institute of Physiology, Bugnon Street 7, Lausanne 1011,
   Switzerland.

HIND, JOSEPH, E.
   Department of Neurophysiology, 283 Medical Sciences Building,
   University of Wisconsin, Medical School, Madison, Wisconsin
   53 706, USA.

HOCHMAIR-DESOYER, INGEBORG
   Institut für Allgemeine Elektrotechnik, Technische Univers.,
   Wien, Gusshausstr. 27, A-1040, Wien, Austria.

HRUBÝ, JAROSLAV
   Institute of Radiotechnique and Electronics, Czechoslovak
   Academy of Sciences, Lumumbova str. 1, 182 51 Prague, CSSR.

IVARSSON, C.
   Institute of Physiology, Bugnon Street 7, CH-1011 Lausanne,
   Switzerland.

JANDA, VRATISLAV

Department of Otolaryngology, Medical Faculty, Charles
University, U nemocnice 2, 128 08 Prague 2, CSSR.

JAVEL, ERIC
Human Communication Laboratories, Boys Town Institute, 555
North 30th Street, Omaha, Nebraska 68 131, USA.

JUNIER, NICOLA
Institute of Physiology, Rämistr. 69, CH-8001 Zürich, Switzerland.

KATSUKI, YASUJI
National Center for Biological Sciences, 38 Nishigonaka,
Myodaiji, Okazaki 444, Japan.

KEIDEL, WOLF, D.
Institut für Physiologie und Biokybernetik, Universitätstr. 17
8520 Erlangen, FRG.

KELLY, JACK, B.
Department of Psychology, Carleton University, Ottawa, K1S 5B6,
Canada.

KLINKE, RAINER
Klinikum der J. W. Goethe Universität, Zentrum der Physiologie,
Theodor-Stern-Kai 7, D-6000 Frankfurt 70, FRG.

KOČKA, JIŘÍ
Department of Otolaryngology, Medical Faculty, Charles
University, U nemocnice 2, 128 08 Prague 2, CSSR.

KORTMANN, HELGA
Department of Neurobiology, University of Bremen, Leobenstr.,
NW-2, 2800 Bremen, FRG.

KROPOTOV, YURIJ, O.
Institute of Experimental Medicine, AMN SSSR, Kirovskij
prospekt 69/71, Leningrad P-22, 197 022, USSR.

KUSÁK, Vlastimil
Institute of Experimental Medicine, Czechoslovak Academy of
Sciences, U nemocnice 2, 128 08 Prague 2, CSSR.

LANGNER, GERALD
Institut für Zoologie der THD, Schittspahnstr. 3, Darmstadt
6100, FRG.

LEIJON, ARNE
Department of Audiology, Sahlgrenska Sjukhuset, S-113 45
Gothenburg, Sweden.

MAIER, VERENA
Institut für Zoologie der THD, Schnittspahnstr. 3, 6100
Darmstadt, FRG.

MANLEY, JUDITH, A.
Max-Planck-Institute for Psychiatry, Kraepelinstrasse 2,
8000 Munich 40, FRG.

MASTERTON, BRUCE, R.
Department of Psychology, Florida State University, Tallahassee,
Florida 32 306, USA.

MATSUURA, SHIUSHI
Department of Physiology, Osaka City University, Medical School,
Asahimachi 1-4-54, Osaka City, Osaka 545, Japan.

MELICHAR, IVO
    Institute of Experimental Medicine, Czechoslovak Academy of
    Sciences, U nemocnice 2, 128 08 Prague 2, CSSR.
MERZENICH, MICHAEL, M.
    Coleman Memorial Laboratory, Departments of Physiology,
    University of California, San Francisco, California 94 143, USA.
MLADONICKÝ, PAVEL
    Department of Neurology, Komenský University, Medical Faculty,
    Limbova 5, Bratislava, CSSR.
MØLLER, AAGE, R.
    Division of Physiological Acoustics, Eye and Ear Hospital of
    Pittsburgh, 230 Lothrop Street, Pittsburg, Pennsylvania 152 13,
    USA.
MØLLER, MARGARETA, B.
    Division of Audiology and Speech Pathology, 230 Lothrop Street,
    Pittsburgh, Pennsylvania 152 13, USA.
MOLNÁR, MARK
    Institute of Psychology, Hungarian Academy of Sciences, Szondy
    utca 83-85, H-1394 Budapest, Hungary.
MÜLLER-PREUSS, PETER
    Max-Planck-Institute for Psychiatry, Kraepelinstrasse 2, 8000
    Munich 40, FRG.
MYSLIVEČEK, JAROMÍR
    Institute of Hygiene and Epidemiology, Šrobárova 48, Praha 10,
    CSSR.
NOVÁK, MILOŇ
    Department of Otolaryngology, Medical Faculty, Charles
    University, U nemocnice 2, 128 08 Prague 2, CSSR.
OSEN, KIRSTEN, K.
    Universitetet i Oslo, Anatomisk Institut, Karl Johans Gate 47,
    Oslo 1, Norway.
OSTWALD, J.
    Fachbereich Biologie-Zoologie, Posfach 1929, D 3550, Marburg,
    FRG.
POPELÁŘ, JIŘÍ
    Institute of Experimental Medicine, Czechoslovak Academy of
    Sciences, U nemocnice 2, 128 08 Prague 2, CSSR.
de RIBAUPIERRE, FRANCOIS
    Institute of Physiology, Bugnon Street 7, Lausanne 1011,
    Switzerland.
de RIBAUPIERRE, YVES
    Institute of Physiology, Bugnon Street 7, Lausanne 1011,
    Switzerland.
RINGDAHL, ANDERS, G.
    Tallboängen 66, 436 00 Askim, Sweden.
RITSMA, ROELOF, R. J.
    Institute of Audiology, University Hospital, 9752 RB Groningen,
    Netherland.
ROSS, MURIEL, D.
    Department of Anatomy and Kresge Hearing Research Institute,

The University of Michigan, Ann Arbor, Michigan 481 09, USA.
ROUILLER, ERIC
Institute of Physiology, Bugnon Street 7, Lausanne 1011,
Switzerland.
RUSSELL, IAN, J.
The School of Biological Sciences, The University of Sussex,
Falmer, Brighton, Sussex, BN 19 QG, Great Britain.
RUTTEM, W. L. C.
De Geitenkamp 66, 2421 TN Nieuwkoop, Netherland.
SAITO, NOZOMU
Department of Physiology, Dokkyo University School of Medicine,
Mibu, Tochigi 32 102, Japan.
SCHEIBE, FRED
HNO - Klinik des Bereichs Medizin (Charité) der Humboldt-
Universität, Schumannstr. 20/21, 104 Berlin, GDR.
SCHEICH, HENNING
Institut für Zoologie, Technische Hochschule, 6100 Darmstadt,
FRG.
SCHREINER, CHRISTOPH
Max-Planck-Institut für biophysikalische Chemie, Postfach 968,
D-3400 Göttingen, FRG.
SEDLÁČEK, KAREL
Department of Phoniatry, Medical Faculty, Charles University,
Žitná 24, Prague 2, CSSR.
SHIIDA, TAKAHASHI
Dept. of Psychiatr, Dokkyo University School of Medicine, Mibu,
Tochigi-Ken, 321 02, Japan.
SMOLDERS, JEAN, W. T.
Klinikum der J. W. Goethe Universität, Zentrum der Physiologie,
Theodor-Stern-Kai 7, D-6000 Frankfurt a.M. 70, FRG.
SMIEŠKOVÁ, ALICE
Department of Neurology, Komenský University, Medical Faculty,
Limbova 5, Bratislava.
von SPECHT, HELMUT
Hals-Nasen-Ohrenklinik, Medizinische Akademie Magdeburg,
Leipziger Strasse 44, 301 Magdeburg, GDR.
SUGA, NOBUO
Department of Biology, Washington University, Statford Ave
7455, University City, Missouri 631 30, USA.
SUJAKU, YASUMASA
Department of Neurophysiology, 283 Medical Sciences Bldg.,
University of Wisconsin, Medical School, Madison, Wisconsin
53 706, USA.
SYKA, JOSEF
Institute of Experimental Medicine, Czechoslovak Academy of
Sciences, U nemocnice 2, 128 08 Prague 2, CSSR.
TANIGUCHI, IKUO
Department of Physiology, Dokkyo University School of Medicine,
Mibu, Shimotsuga, Tochigi 32102, Japan.

TICHÝ, STANISLAV
    Department of Otolaryngology, Medical Faculty, Charles
    University, U nemocnice 2, 128 08 Prague 2, CSSR.
TONG, Yit-CHOW
    The Royal Victorian Eye and Ear Hospital, East Melbourne,
    Victoria 3002, Australia.
TOROS-MOREL, ANNE
    Institute of Physiology, Bugnon Street 7, CH-1011 Lausanne,
    Switzerland.
ÚLEHLOVÁ, LIBUŠE
    Institute of Experimental Medicine, Czechoslovak Academy of
    Sciences, U nemocnice 2, 128 08 Prague 2, CSSR.
VALVODA, MILOŠ
    Department of Otolaryngology, Medical Faculty, Charles
    University, U nemocnice 2, 128 08 Prague 2, CSSR.
VOIGT, HERBERT, F.
    Dept. of Biomedical Engineering, The Johns Hopkins School of
    Medicine, 720 Rutland Avenue, Baltimore, Maryland 212 05, USA.
VOLDŘICH, LUBOŠ
    Institute of Experimental Medicine, Czechoslovak Academy of
    Sciences, U nemocnice 2, 128 08 Prague 2, CSSR.
WAGNER, HERMANN
    Hals-Nasen-Ohrenklinik Bereich Medizin (Charité) der Humboldt
    Universität, Schumannstr. 20-21, 104 Berlin, GDR.
WAGNER, JIŘÍ
    Research Institute of the Sound and Picture Technique,
    Lidická 6, Prague 5, CSSR.
WOOLSEY, C. N.
    Department of Neurophysiology, 283 Medical Sciences Building,
    University of Wisconsin, Medical School, Madison, Wisconsin
    537 06, USA.
YOUNG, ERIC, D.
    Dept. of Biomedical Engineering, The Johns Hopkins School of
    Medicine, 720 Rutland Avenue, Baltimore, Maryland 212 05, USA.

Participants of the symposium "Neuronal Mechanisms of Hearing", Prague, July 20-23, 1980

INDEX

Ablation
  of auditory cortex, 277, 281,
    293-294
  of cochlea 277, 280
  of cochlear nucleus 277-281
Absolute pitch, 43
Acoustic deprivation
  morphological changes 359-361
  in rat 359-361
Acoustic striae, 121, 129
Acoustico facial anastomoses, 32
Acoustico lateralis system, 4, 8
Action potential
  of auditory nerve, 227-229
Adaptation
  at hair-cell -afferent synapse,
    37-41
Afferent discharges
  in the cochlear nerve, 37
Amplitopic representation, 204
Amplitude histogram, 188-189
Amplitude modulated tones, 81,
    98-101
  responses in frog, 347-351
Anteroventral cochlear nucleus,
    121, 129-130
  ablation of, 277-281
Artificial perilymph, 51
Auditory cortex
  in bat, 200-203
  in cat, 221-224
  effect of lesion, 281
  evoked potentials, 221-224
  and input to the IC, 142-147,
    150
  input from the MGB
  in rabbit, 257-259

Auditory cortex (continued)
  in rat, 139, 141-147, 150
  responses to vocalizations,
    308, 312-313
Auditory deprivation
  deoxyglucose uptake, 371
Auditory development, 363-366
Auditory neostriatum
  in guinea fowl, 323, 329
Auditory nerve
  in the caiman, 44
  recovery function of, 193-195
  and vowel spectra, 113
Auditory nerve fibers
  in the rat, 88
  responses to complex tonex,
    105-110
Auditory nervous system
  and coding, 87-102
Auditory neurons
  in bat, 200, 204
Auditory pathway
  and multiunit activity, 187-190
Auditory system
  in frog, 373, 379
Autonomic influences
  on inner ear, 34
Autoradiograph, 331, 337, 366, 372
Autoradiographic studies
  of the organ of Corti, 31-32, 34
AVCN, see anteroventral cochlear
    nucleus

Background noise, 81-82, 113
Band-pass filtered noise, 96
Band-stop noise, 71-72, 80
Barbiturate anesthesia

Barbiturate anesthesia (continued)
  effect on MGB neurons, 177
Basilar membrane
  in the caiman, 44
  and displacement, 8-11, 14
  and velocity, 8-11, 14
Basilar papilla
  of the alligator lizard, 43
  of the caiman, 44
Bat
  auditory cortex, 197-217
Beidler´s taste equation, 23-24
Best frequencies (see also Charac-
      teristic frequencies)
  in auditory cortex, 202, 209
  of MGB units, 165-172, 178
  in neostriatum neurons, 324, 330
  of DCN neurons, 127-128
  in torus semicircularis neurons,
      343
Binaural beat stimulus, 233
Binaural information, 277
Binaural interaction
  in auditory cortex, 204
  and IC neurons, 148-151, 156
      233-238
  in the MGB, 240
  model of, 283-287
Biomembranes
  and amino acids, 20
  and phospholipids, 20, 28
Biosonar information, 197-217
Blood plasma, 57
Body temperature
  and characteristic frequency, 43
Brachium of the IC
  effect of lesion, 269-270
Brainstem evoked response
  and electrical stimulation, 406
Brainstem potentials, 225-227
  in rabbit, 257-259
  in rat, 381
Broad-band noise, 78, 80, 88, 94,
      113-115, 127, 129, 245

Caiman, 43
Caiman hair cells
  inervation pattern, 46
Cartwheel neurons
  in DCN, 123

Cat
  auditory nerve in, 113
  cochlear nucleus, 119, 127
  inferior colliculus, 148-150,
      155
  medial geniculate body, 163-179,
      183-190
Center frequency
  of the transfer function, 92
Central nucleus
  of the IC, 137-138, 140, 144,
      146-147, 151, 155-156
  projections to MGB, 169,
      172-174,179
Centrum medianum, 304
Cerebellum, 122-124, 138, 140-141
  and sound source movement,
      296-297
Cerebral cortex
  in bat, 197-217
Cerebrospinal fluid, 57
Characteristic frequencies (see
      also Best frequencies)
  of IC neurons, 139-140, 145, 150
  of auditory nerve fibres,
      113-115
  and caiman afferent auditory
      fibres, 44
  of cochlear fibres, 70-73,
      75-76, 78-79, 81-82, 94
  of MGB neurons, 192
Chemoreception
  and lateral-line organ, 21
Chloralose anesthesia, 164, 177
Cochlea
  in bat, 202
  in guinea pig, 3-14, 51, 57
Cochlear destruction
  and inferior colliculus, 377-380
Cochlear efferents
  anatomy, 31-35
Cochlear fibres
  in the cat, 72, 75
Cochlear filter, 78
Cochlear mechanisms
  and comparative aspects, 17-30
Cochlear microphonics, 3, 7-13,
      155, 159
  and ion exchange, 29
  isointensity curves, 64

Cochlear microphonics (continued)
  longitudinal distribution,
     63-65
  and metabolism, 51-55
  in rabbit, 257-259
Cochlear nerve, 69
Cochlear nerve fibres
  dynamic range of, 69-82
Cochlear nucleus, 69, 96
  ablation of, 277-281
  and effect of deprivation, 360
  and input to the IC, 141-144,
     159
  of the rat, 100-101
  structure of, 119
Cochlear prostheses
  multiple electrode, 411, 417
Cochleotopic organization
  of the DCN, 120, 122
Coincidence network, 283-284
Colliculus inferior
  and multiunit activity, 188
Comb-filtered noise, 72
Complex sounds
  coding of, 87-102
  and peripheral auditory system,
     101
  and responses of auditory nerve
     fibers, 105-110
Condensation click, 44-45, 47
Conditioning signal, 192-195
Consonant, 414
  discrimination in birds, 317
Conspecifics, 212
Constant-frequency (CF) signal,
     198, 208-209
Corn cell
  in DCN, 123
Corollary discharge, 314
Cortical damage
  and sound source perception,
     295
Cross-correlation function, 91,
     93, 95
  interaural, 283
Cross-correlogram, 89-91
  of DCN neurons, 130-132
Cut-off frequency
  and haircell response, 5, 11-13

DCN, see Dorsal cochlear nucleus
Degeneration
  in the IC, 146-147
Delay sensitive units
  in the MGB, 245
Deoxyglucose
  injection
    in frog, 369-373
    in guinea fowl, 329
    in mice, 377
    in mynah, 336
    in rat, 365
    tracing, 263-264
  uptake
    and cochlear destruction, 378
Deprivation
  and auditory development, 359-
     361
  and inferior colliculus, 139,
     146
Destruction
  of inner ear, 370
Development
  of the auditory system, 363-366
  in rat, 381-385
Differential technique
  of CM recording, 63
Discharge rate
  of auditory nerve fibers, 108
Dithio threitol (DTT), 18-19, 26,
     28
Divalent cations
  and lateral-line organ, 18-19,
     21-22, 24, 26-27, 29
Doppler-shift echo, 198-199,
     202-204, 207-212
Dorsal acoustic stria (DAS),
     129-130, 155
Dorsal cochlear nucleus (DCN), 72
     80, 155-156, 158-160
  responses of cells, 127-128
  structure of, 119-124, 127,
     129-130
Dorsal division
  of the MGB, 164, 167, 172-174,
     177
Dorsomedial part
  of the IC central nucleus, 138,
     146

Dynamic range
  and age, 394
  for electric stimulation, 392
  problem, 69–82
  and stimulus duration, 80

Ear-plug, 364
Early auditory stimulation,
    355–358
Echo, 198-200, 203-204, 206,
    210, 214
Effects of stimulation
  in MGB, 191-195
Electrical stimulation
  of the auditory nerve, 383,
    392-393
  of brain
    and vocalization, 312
    and vocalization in birds,
      335
  of the dorsal cochlear nucleus,
    155-156
  of the inferior colliculus,
    151
  of the human cochlea, 411-415,
    417-420
Electrically-elicited vocaliza-
    tions, 336
Electrocochleography, 225, 227
Electrolytic lesions
  in the auditory cortex, 146-
    147
Electroshock therapy
  and hemisphere specialization,
    294
Endocochlear potential, 63-65
  anoxic, 65
Evoked potentials
  and age, 394
  in cat, 221-224
  effect of vigilance, 225-226
  to electrical stimulation, 393
  in man, 225-229
  in rat, 356
Evoked response
  development in rat, 382
  in rat, 381-385
Excitatory postsynaptic poten-
    tials in the primary afferent
        fiber terminals, 37-41

External ear
  transfer function of, 114
External nucleus
  of the IC, 137-138, 146, 158,
    172, 175

Facial nucleus, 32
Field L
  in guinea fowl, 323-326, 329-333
Fluorometric technique, 58
Folded histogram, 349-350
Formants, 217, 317-320
  of the vowel, 74, 113, 115, 117
Forward-masking paradigm, 191,
    195
Fourier transformation, 44, 75,
    90-91, 106, 114
Frequency-modulated (FM) sound,
    198-199, 203, 206, 214
Frequency selectivity
  of auditory nerve fiber, 90
  of the ear, 87-88, 92, 94
Frequency threshold curves
  of cochlear fibres, 70, 75, 77,
    79, 90-91
  of neurons in the cochlear nuc-
    leus, 96
Frequency transfer function, 90
Frequency-tuning curve, 203, 208
Frog, 341-345, 347-351
Fundamental frequency, 317-320,
    330, 413
Fused auditory image (FI), 289-298
  velocity of, 292
Fusiform cells in DCN, 129-130

Giant cells in DCN, 129-130
Glucose in perilymph, 57-60
Goldfish, 37
Golgi
  cells
    of the DCN, 122
  technique, 163
  type II interneurons, 185
Graham and Karnovsky method, 139
Granule cells
  of the DCN, 120-122, 129
Grassfrog (Rana temporaria),
    341-345, 369

Guinea fowl, 323-326, 329-333
Guinea pig, 57, 63
  and inferior colliculus, 138,
    148-149
  and medial geniculate body,
    191-195

Hair cells
  distribution in cochlea, 63-65
  membrane and impedance, 11
  and transmitter release, 37,40
Hard base, 19-20
Harmonic frequency, 113, 114
Harmonics, 75-78, 198, 205, 207,
  210, 212, 317, 330
Hearing disorders, 225
Hearing prostheses
  physiological background,
    389-408
Hippocampus, 221
Histochemical study
  of cochlear efferents, 32
Horseradish peroxidase
  and acoustical deprivation, 146
  injection
    to inferior colliculus, 139,
      141-146, 150, 158-159
    to MGB, 165, 169-173, 175
    to striatum in mynah, 336
Horseshoe bat (Rhinolophus
    ferrumequinum), 210
^{3}H-proline
  injection
    into the spiral ganglion, 32,
    34-35
HRP, see Horseradish peroxidase
Human brain
  neuronal populations of, 303-
    306

IC, see Inferior colliculus
Implanted electrodes, 411, 417
  in patients, 303
Inferior colliculus
  and binaural interaction, 234
  and cochlear destruction, 377-
    380
  development during ontogeny,
    365
  effect of lesion, 269-270

Inferior colliculus (continued)
  functional organization, 137-151
  input to, 137-151, 155-160
  input from DCN, 122
  in mice, 377-380
  responses of neurons, 309
  responses to vocalization, 312-
    313
  and vocalizations, 307-310
Information-bearing parameter,
  215-216
Information flow, 389-390
Inhibitory neurons
  in DCN, 127
Inner ear
  of the caiman, 43
  efferents, 31-35
Inner hair cells, 3-14
  and intracellular recording,
    4-6, 9-10, 12-13
Intensity range
  of auditory nerve fibers, 88
Interaction
  within a single unit pair,
    184-186
Interaural intensity difference,
  285-287, 289, 291
Interaural time delay, 283, 289,
  292
Interval histogram, 189, 251-255,
  285-286
Intracellular potential, 3-7, 9-10
Iontophoresis
  and HRP administration, 159
Isofrequency contours
  in the auditory cortex, 143, 145
  in the MGB, 164, 167
Isofrequency lines
  in neostriatum, 324-325, 330
Isofrequency layers
  in the inferior colliculus,
    139-141, 149

Jamming, 210-212
Joint poststimulus histogram,
  183-185
Just noticable differences
  in intensity (jnd), 391

Kanamycin, 9
  effect in rat, 381-385
Koelle technique, 32

Lacrimal nucleus, 31
Lactate in perilymph, 57
Laminar arrangement of the IC,
      137
Lateral lemniscus
  section of, 267-269
Lateral-line organ, 17-30
  and responses to ions, 18-30
Lateral superior olivary nucleus
      (LSO), 284-285
Lateral suppression, 71, 73, 81
Lesion
  in the auditory pathway, 263-274
  and sound localization, 263-274
Ligature
  of the auditory meatus, 146
  of the external ear, 359, 361
Lipreading, 415, 420
Localization of sound, 277-281
Logistic function, 106
Lowry's micromethod, 58
Lucifer yellow, 4

Magnocellular division of the MGB,
      163, 184
Masking noise, 113-119
Medial geniculate body
  in cat, 239-242, 245-248, 251-
      256
  coding properties, 239-242
  collicular input to, 169-172,
      175
  effect of lesion, 269-270
  electrical stimulation of, 221
  functional organization of,
      183-186
  in guinea pig, 191-195
  and multiunit activity, 188
  principal division of, 163-179
  responses to vocalization,
      312-313
  subdivision of, 239-242
  temporal information, 251
Membrane conductance, 29
Menière disease, 226-227

Metabolic inhibition
  and CM, 53
MGB, see Medial geniculate body
Mice, 377
Micro flame photometry, 58
Midbrain auditory region
  in frog, 347-351
Missing fundamental, 75
Model, 105
  of auditory nerve, 404
  of binaural interaction, 233-
      238, 283-287
Molecular layer of the DCN, 120-
      121
Monovalent cations
  and lateral-line organ, 18-19,
      21-22, 24, 26-27, 29
Mössbauer technique, 43
Mossy fibers in DCN, 121-122
Mouse
  and cochlear efferents, 33
Moving sound source, 291
Multicomponent tone stimulus,
      80-81
Multielectrode, 221
Multiunit activity, 187-190
  in patients, 304
Mustached bat (Pteronotus parnellii
      rubiginosus), 197-200,
      202, 212-213
Mynah bird, 317
  forebrain of, 317

Nauta and Gygax method, 139, 146-
      147
Necturus maculosus, 23
Neostriatum
  in guinea fowl, 324, 329
Neurons
  of the cochlear nucleus, 95-96
    responses to noise, 96
Noise
  background, 81-82, 113
  band-pass filtered, 96
  bandstop, 71-72, 80
  broad-band, 78, 80, 88, 94,
      113-115, 127, 129, 245
  comb-filtered, 72
  masking, 113-119

Noise (continued)
  pseudorandom, 88-91, 94-95
Noise exposure
  effect on CM, 63-65
Nuclei of lateral lemniscus, 172,
    175, 177
  and input to the IC, 141-143,
    150, 158, 160
Nucleus sagulum, 172-173, 175,
    177

Octopus cells, 142
Odotopic representation, 207
Olivocochlear efferents, 31
Olivocochlear fibers, 122
Ontogeny in birds
  and vocalization, 335
Organ of Corti, 3-4, 7-9
  and acetylcholin-esterase
    positive fibers, 33
  and perilymph, 57-60
Orientation sound in bat, 200
Ototoxic effect
  of kanamycin, 381-385
Outer hair cells, 3-14
  and intracellular recording,
    7-8
  and synapses, 31
Oxygen measuring chamber, 51
Oxygen tension
  in cochlea, 51-55

Pairs of neurons in the MGB,
    183-186
Parasympathetic inervation of
    the inner ear, 31-35
Pars ovoidea of the MGB, 179, 184
Perception of sound source
  movement hemisphere specializa-
    tion, 294
Pericentral nucleus of the IC,
    137-138, 158, 172, 175,
    177
Perilymph
  biochemistry, 57-60
Perilymphatic perfusate, 51
Period histogram, 44, 75, 80, 88,
    97-101, 106, 114-116,
    236-237, 251-255

Peristimulus-time histogram, 205,
    211, 236-237
Phase-locked activity, 399
Phase-locked responses
  of auditory nerve fiber, 113,
    115
  in the MGB, 252-255
Phase-locking
  of cochlear fibres, 74-77, 91,
    94
Phonation
  responses of neurons, 311
  in squirrel monkey, 311
Piezoelectric loudspeaker, 64, 139
Pinna movement, 151
Place theory of hearing, 87-88, 93
pO_2 changes in the cochlea, 51-55
Pontamine sky blue dye, 164
Posteroventral cochlear nucleus
    (PVCN), 120-121
Poststimulatory suppression, 191-
    193
Poststimulus histogram, 44-47,
    115-116, 148-149, 188,
    304-305, 313, 318
Potassium
  in perilymph, 57-60
Precedence performance, 281
Premotor cortex, 304
Primary auditory fibres
  in caiman, 43
  in cat, 43
Principal division of the MGB,
    163, 174
Procion yellow, 4
Prosthesis
  in the cochlea, 411, 417
Pseudorandom noise, 88-91, 94-95
Pyramidal cells
  in auditory cortex, 144, 150
  in DCN, 121-123
Pyruvate in perilymph, 57-60

Rabbit
  and inferior colliculus, 138,
    148-149
Range-sensitive neurons, 206-207,
    210, 215
Rarefaction click, 44-45, 47

Rat
  and auditory cortex, 145–147, 150
  and cochlear efferents, 32
  effect of acoustic deprivation, 359–361
  and inferior colliculus, 138–141, 148–150
  and ontogeny of hearing, 355–358
Rate-place code, 113–117
Receptor membrane
  and affinity of ions, 24, 29
Receptor potential, 3–14
  and the AC component, 4–7, 14
  and the DC component, 4–8, 14
  of inner hair cells, 3–14
  of outer hair cells, 7–8
Receptor sites
  and chemical reception, 26–27
Recruitment, 226
Reverse correlation technique, 78
Ribonucleic acid (RNA)
  content in cortex, 356–357
Round window
  and CM recording, 63

Saccule of the goldfish, 37
Scala media
  CM measurement in, 63–65
Scala tympani, 57–58, 63
  and cochlear implant, 411, 415, 417
Scala vestibuli, 57–58, 63
Shark, 18
Small cells in DCN, 129–130
Sodium
  channel, 28
  in perilymph, 57–60
Soft base, 19–20
Softness parameter, 24, 26
Somesthetic processing
  of acoustic messages, 423–426
Sonagram, 199, 217, 402–403, 408
Sound localization, 233, 277–281
  in bat, 214
  effect of lesions, 263–274
  and midbrain mechanisms,263–274
  in rat, 364

Sound source movement, 289–298
  experiments in animals, 293–294
  neuronal detectors, 295–296
Sound source perception, 289–298
Species-specific vocalization, 312, 330
  in guinea fowl, 323–326, 329–330, 333
Spectral selectivity
  and auditory periphery, 94
Speech perception
  and cochlear prostheses, 405, 413, 417, 419
Speech preprocessing, 404
Speech-processor, 413, 420
Spherical cells
  in the cochlear nucleus, 142, 359
Spike waveform, 342
Spiral ganglion
  and multipolar cells, 31
  and transport of proline, 34
Spontaneous activity
  of cochlear fibres, 71–73, 81, 107
  of DCN cells, 127–129
  of MGB neurons, 183, 185–186
  in neostriatum, 325, 330
Squirrel monkey (Saimiri sciureus), 307, 311
Stainless steel microelectrodes, 341
Stapes, 9, 11
Steady-state vowel, 113
Stellate cells in DCN, 122
Stereocilia
  of hair cells, 8–9, 14
Striatum in birds, 335–338
Striopallidar system, 304
Superior salivary nucleus
  in the rat, 32–33
Suppresive effect
  and model, 106–107
Suppresive ratio
  of ions, 22–23
Suppression
  of basilar membrane motion, 110
  in the MGB, 192–194
Summating potential, 4, 7

Superior colliculus, 140-141, 151, 159, 173

Superior olivary complex
  input from DCN, 123
  and input to the IC, 141-144, 150, 158-160
  and sound localization, 271, 273-274

Synapse
  between hair cell and afferent fiber, 37

Synaptic jitter, 255

Synchronization, 106-108
  in MGB units, 252,

Synchrony suppression, 75

Synthesized orientation sound, 199-200

Synthesized vowel, 113

Synthetic vowel
  response of auditory nerve fibers, 94-95

Tactile prostheses, 399, 402-403, 423-426

Tadpole, 18, 21

Tectorial membrane, 8-9, 14

Teleosts, 18

Temporal lobe lesion
  and sound localization, 266

Temporal-place code, 113, 115-117

Tetrodotoxin, 37

Time-constant
  of the hair cell membrane, 11-14

Time-locked responses
  in the MGB, 251

Tonotopic arrangement
  of the auditory system, 87
  of the auditory cortex, 202
  of the IC, 139, 143, 150
  of the MGB, 167, 178

Torus semicircularis, 371-372
  in grassfrog, 341-345

Transmission electron microscopy, 34

Transmitter release
  at the afferent synapse, 37, 40

Trapezoid body
  and sound localization, 265, 267

Trapezoidal frequency modulated tones, 97

Travelling wave
  and caiman inner ear, 43-47

Tree shrew (Tupaia glis), 277-278

Triphosphorinositide (TPI), 28-29

Tungsten electrodes, 156, 165, 187, 307

Tuning curves
  of auditory nerve fiber, 401
  and electrical stimulation, 401-402
  in the MGB, 240

Two-tone suppression
  in the MGB, 191, 194

Urethane, 51, 58, 63

Ventral division
  of the MGB, 164-166, 168, 173-174, 178-179

Ventrolateral thalamic nuclei, 304

Verbal signal learning
  in patients, 303

Vestibular afferents, 31

Vestibular efferents, 31-35

Vibrotactile stimulation, 403

Vocalization
  neuronal responses to, 307
  processing in the auditory system, 311-314
  in song birds, 335-338
  in squirrel monkey, 307

Voco-auditory centers
  in birds, 335-338

Vowel, 113-119, 414
  and redundancy, 396-397, 403
  spectra, 113-119

Vowels
  discrimination in birds, 317-320

Xenopus laevis, 27

X-ray microanalysis, 27